T0176324

LIVING COMPUTERS

Living Computers

Replicators, Information Processing, and the Evolution of Life

Alvis Brazma

European Molecular Biology Laboratory
European Bioinformatics Institute
Cambridge, UK

OXFORD
UNIVERSITY PRESS

Great Clarendon Street, Oxford, OX2 6DP,
United Kingdom

Oxford University Press is a department of the University of Oxford.
It furthers the University's objective of excellence in research, scholarship,
and education by publishing worldwide. Oxford is a registered trade mark of
Oxford University Press in the UK and in certain other countries

Published in the United States of America by Oxford University Press
198 Madison Avenue, New York, NY 10016, United States of America

British Library Cataloguing in Publication Data

Data available

Library of Congress Control Number: 2023933858

ISBN 9780192871947

DOI: 10.1093/oso/9780192871947.001.0001

Printed and bound by
CPI Group (UK) Ltd, Croydon, CR0 4YY

For my wife, Diana

Preface

All currently known forms of life rely on three types of molecules: proteins—life's main structural and functional building blocks; DNA—life's information carrier; and RNA—the messenger providing the link between these two. But what was life like when it first emerged on Earth billions of years ago? What will life be like in millions or billions of years if it still exists? Can a living being be made of, say, steel, copper, and silicon? And what is it that distinguishes living systems from those we do not regard as living?

An essential feature of life is its ability to record, communicate, and process information. Every living cell does this. Every animal brain does this. And even more fundamentally, evolution processes the information recorded in the entire collection of DNA on Earth using the algorithm of Darwinian selection of the fittest. Life and the recording of information emerged together; there is no life without information and, arguably, there is no information without life.

We do not know what the first carriers of information were when life had just emerged, but for the last few billions of years, the most important information storage medium on Earth has been DNA. Once life had emerged and started evolving, the information in DNA was accumulating, at least for a while. A few hundred thousand years ago, quite recently on the evolutionary timescale, human language emerged and a major transition began; for the first time, large amounts of information began accumulating outside DNA. Most likely, information in the world's libraries and computer clouds is now expanding faster than in the genomes of all species on Earth taken together. Even more importantly, language triggered evolutionary mechanisms different from and faster than biological evolution, namely cultural evolution. The emergence of human language was a transition as remarkable as the emergence of life itself.

Nevertheless, at least at our current state of evolution, the information in DNA is indispensable; if the DNA existing on Earth were to become too corrupted, all cultural information and life itself would swiftly disappear. But must life be like this? Or can future civilizations, possibly in thousands or millions of years colonizing planets of distant galaxies, be based on entirely different principles? Can there be another major transition in which DNA becomes less central?

This book, in which life and evolution are explored as information processing phenomena, is aimed at everybody interested in science and comfortable with elementary mathematics. It will also be of interest to students and scholars from a wide range of disciplines, from physics, computing, and biology to social sciences and philosophy. This book is for everyone who has ever wondered: how do living systems work? When do assemblies of non-living molecules become living organisms? Where does life's complexity come from? And where will life go after the Solar system ceases to exist? The fascinating idea of viewing living organisms as computers and the evolution of life as information processing will be enriching for everyone interested in our daily and long-term existence.

Acknowledgements

This book could not have been written without the support of the European Molecular Biology Laboratory (EMBL), and in particular, that of the European Bioinformatics Institute (EBI), where I have spent more than half my scientific life, gradually evolving from a mathematician into a biologist, though never really completing the transition. With its multidisciplinary focus, collaborative atmosphere, and striving for scientific excellence, EMBL and EBI provided me with the creative environment where the ideas ending up in this book could develop. But the book would also not have been possible without what I learnt about mathematics in my earlier scientific carrier, mostly at the University of Latvia, where I was fortunate enough to have been taught by some of the pioneers of theoretical computer science, and foremost, without my PhD supervisor and mentor Professor Jānis Bārzdiņš. I can trace the idea of writing this book to the early 1990s, when the Soviet Union was disintegrating and I got my first opportunity to travel outside the Soviet bloc and to work at the New Mexico State University, thanks to the kindness and generosity of Juris and Lauma Reinfelds.

I am enormously indebted to my friends and colleagues who read drafts of this book from cover to cover at various stages, and in particular to Paulis Ķikusts, Gerard Kleywegt, Robert Petryszak, Kārlis Podnieks, Uģis Sarkans, and Roderic Guigo. Without their invaluable advice and help, this book could not have been written. My wife Diana Brazma and my brother Gunars Brazma were also amongst those who read the entire manuscript and gave me important feedback. Others, in particular Nick Goldman, Mar Gonzàlez-Porta, Mark Green, Frank Holstege, Wolfgang Huber, Julio Collado-Vides, and Ainis Ulmanis, read parts of the manuscript, suggesting many important improvements. My thanks also go to anonymous reviewers who commented on the book's proposal and sample chapters. There are a larger number of colleagues and friends with whom I discussed the ideas behind this book; amongst them: Wilhelm Ansorge, Alexander Aulehla, Jurg Bahler, Pedro Beltrao, Ewan Birney, Graham Cameron, Marco Galardini, Cornelius Gross, Edith Heard, Des Higgins, Iain Mattaj, Cedric Notredame, Janet Thornton, Nassos Typas, and Rick Welch. Some of them were not aware that the questions I was asking were about an emerging book. Enormous thanks go to Andrew King, the librarian of the Michael Ashburner's Library of the Wellcome Genome Campus, who helped me chase up many obscure references central to this book. Several of my colleagues went out of their way to provide me with illustrations exactly the way I wanted them, in particular Osman Salih and Julian Hennies at EMBL, and Louis Scheffer at Janelia Research Campus. At the proposal stage I was helped by Caroline Palmer who corrected my English and Oana Stroe who helped me to write the Preface. I also want to thank my commissioning editor at OUP, Ian Sherman, for all his great encouragement and help, to my project editor, Katie Lakina, for dealing with questions as the manuscript was developing, as well as Rajeswari Azayecoche and Janet Walker, who did an admirably professional job in editing the manuscript. Any errors in this book are entirely my responsibility. Over time, my work at EMBL was funded mainly by the EMBL member states but also by many grants from the European Commission, the Wellcome Trust, the National Institute of Health, UK Research and Innovation, and others.

Contents

Introduction

The History of every major Galactic Civilization tends to pass through three distinct and recognizable phases, those of Survival, Inquiry and Sophistication, otherwise known as the How, Why, and Where phases. For instance, the first phase is characterized by the question 'How can we eat?', the second by the question 'Why do we eat?' and the third by the question, 'Where shall we have lunch?'

(Douglas Adams, *The Hitchhiker's Guide to The Galaxy*)

How complex does the simplest living being have to be? How many different parts it needs? Can the complexity of life be measured? As proteins are the most diverse components of all known life forms, one possible way to measure the complexity of an organism is by counting how many different protein molecules it has. Bacteria, arguably the simplest known free-living organisms, have hundreds to thousands of them. Humans have tens of thousands of different proteins, but so do other animals. Are humans unique among animals? If yes, what makes them special? On the one hand, in the number of different molecules, humans are of about the same complexity as many other animals. On the other hand, unlike other animals, humans can read, write, and play chess. But so can computers. Are humans closer to other animals or to computers?

These are not new questions. More than 2000 years ago, the Greek philosopher Aristotle thought that the difference between the living and non-living was in the presence or absence of a 'soul'. For Aristotle, there were different levels of a soul: the nutritive soul was common to all living things, whereas higher levels of soul accounted for the locomotive and perceptive capacities of animals, including humans. At the highest level there was the rational soul, unique to humans. Two-thousand years later in the seventeenth century, with the invention of the microscope, Antoni van Leeuwenhoek discovered another type of life: microbes. For a while, it was thought that microbes could spontaneously emerge from non-living matter, blurring the distinction between living and non-living. This changed when, in the nineteenth century, the French scientist Louis Pasteur demonstrated that microbes could not appear *de novo* but could only grow from the ones already in existence. The distinction between living and non-living appeared to be clear-cut again. But what exactly is the dividing line?

The developments of physics and chemistry in the nineteenth and twentieth centuries provided scientists with new ways of investigating this question, but they also imposed new demands on the acceptable answers. The answers would have to be based on atomic

Living Computers. Alvis Brazma, Oxford University Press. © Alvis Brazma (2023). DOI: 10.1093/oso/9780192871947.003.0001

theory and be consistent with the laws of physics. Towards this goal, in 1944, one of the founders of quantum mechanics, Erwin Schrödinger, wrote a book called *What is Life?*, inspiring a generation of scientists and facilitating the development of a new scientific discipline—*molecular biology*. Schrödinger described a hypothetical 'code script' that carries information passed on by living organisms from generation to generation.

Around the same time that molecular biology was emerging, the first electronic computers were invented, and a different new scientific discipline—*computer science*—was also developing. With it, the scientific concepts of *information* and *information processing* were solidifying. Computers were different from other machines in that they were performing intellectual rather than physical work. The question 'can machines think?' soon became a topic for discussion in science as well as in science fiction. To answer this question, Alan Turing, one of the founders of computer science, proposed a test: if a human talking to 'somebody' behind a curtain was not able to tell whether that 'somebody' was human or not, then by definition, that 'somebody' could think.[1] But does being able to think have to mean being alive? The converse does not seem to be true: not everything living can think.

In the 1950s the molecular structures of increasingly complex biological molecules were deciphered. The discovery of the double-helical structure of DNA in 1953 was a defining moment paving the way to understanding the molecular basis of biological information: how it is coded; how a living cell, the basic unit of all known life, uses it to function; and how this information is passed on from generation to generation. A conceptual link between biology and information/computer science began to emerge.

The initial input from computer science to biology was arguably rather limited, nevertheless computer science entered the realm of biology a few decades later, in the 1980s. This was when DNA sequencing—reading the sequences of the molecular 'letters' of DNA (reading Schrödinger's 'code script') and transcribing them onto paper or, more usefully, into a computer memory—was invented. This ability to read life's information molecule resulted in the emergence of a new scientific discipline—*bioinformatics*. Soon scientists found that analysing DNA as a sequence of letters, rather than a chemical, can reveal many secrets of life. Apparently, for life's processes, the information in DNA is at least as important as the specific chemistry. But to what extent does the particular chemistry matter at all?

All currently known life relies on the same carbon-based chemistry, which seems to suggest that chemistry matters. But is this the only way how life can exist, or was it a historic accident that life emerged on Earth based on this particular chemistry? And even if this specific chemistry was the only possible way how life could have emerged, does life have to remain based on the same chemistry in the future? Can life evolve to inhabit very different physical substances and use very different, non-carbon-based chemistry? Perhaps even no chemistry at all?

In this book I will argue that first and foremost, living means processing information. I will argue that the emergence of life and the first recording of information are intrinsically linked and perhaps are one and the same phenomenon. I will show that biological evolution can be viewed as computing where the information residing in the entire collection of DNA on Earth is processed by the *genetic algorithm* of Darwinian survival of the fittest. At least initially, as life on Earth was evolving and adapting to changing environment, information stored in DNA was growing. In a sense, life was learning from the environment, recording what it had learnt in its DNA.

For billions of years, all durable information present on Earth was almost exclusively recorded in DNA or other polymer molecules. But a few hundred thousand years ago, this changed. Human language emerged, providing an alternative means for recording and transmitting large amounts of information from individual to individual and from generation to generation. Information begun increasingly accumulating in substances different than DNA. Now, there is probably more information stored in the world's libraries and computer clouds than in the genomic DNA of all the species on Earth. Even more importantly, language enabled information processing mechanisms different from and much faster than the ones existing before. Cultural evolution, enabled by language, does not have to rely on the survival of the fittest; the information acquired over one's lifetime can be passed on to others and to future generations.

Nevertheless, at the present stage of evolution, the information in DNA and its Darwinian processing are indispensable: cultural information cannot be maintained or evolve without the information in DNA underpinning it. But can this change in the future? Could future civilizations, which in thousands or millions of years possibly be colonizing planets of distant galaxies, be based on entirely different principles, and made of entirely different molecules? Can life exist without DNA or similar molecules? Can civilization exist without life? If the main defining feature of life is information processing, then perhaps the particular substance in which this information is stored—DNA, RNA, paper, silicon chips, or anything else—becomes secondary. Or will life always need DNA and the Darwinian algorithm working in the background? What will the life eventually escaping the dying solar system be like?

The appreciation of the role of information and information processing in biology can be traced back to the mentioned Erwin Schrödinger's book *What is Life?* and to publications of one of the co-discoverers of the structure of DNA, Francis Crick. The observation of the central role of information in biology is at the basis of the so-called *gene's-eye view* introduced by the American and British biologists George C. Williams and Richard Dawkins, stating that *genes*—pieces of heritable information—are the main object of Darwinian selection of the fittest.[2] But a special place in the development of the idea of information as a central concept in biology belongs to the British biologist John Maynard Smith. In his seminal book *The Major Transitions in Evolution*,[3] written with the Hungarian biologist Eörs Szathmáry and published in 1995, the authors show that the key events in biological evolution can be seen as transitions in how life processes information.

With genome sequencing now a commodity and bioinformatics becoming a mainstream science, the central role of the concept of information in biology is becoming increasingly accepted. In his recent book, *The Demon in the Machine*,[4] the physicist Paul Davies explores the links between the concepts of entropy, information, and life, arguing that information processing is at the very centre of life. Even more recently, the Nobel

Laureate Paul Nurse writes about *life as chemistry* and *life as information* in his popular science book that has the same title as Schrödinger's *What is Life?*.[5]

In this book, *Living Computers*, I discuss how living systems work, how life and information emerged already intrinsically linked, how life and information processing evolved through RNA and DNA to neural networks and the animal brain, and continued through human language to kick-start a new kind of evolution. Throughout the book, I try to be clear where I am describing the consensus with which most scientists would agree, note where I am touching controversies, and state where I am just speculating (and to make sure that my claims can be verified, I provide many references to peer-reviewed literature). Most of the book concerns the first, consensus positions, even though I often try to show the established knowledge in new light, and I sometimes question an odd orthodox assumption. Not everything in the book is a part of the established scientific consensus. The thesis that life and the recording of information emerged jointly seems obvious to me, nevertheless there is no reference to scientific literature that I can find where this had been stated explicitly. I argue that large amounts of information can be recorded and processed only by digital-combinatorial means, which may also sound controversial to some. This then further leads to the conclusion that before human language emerged, the only means to record large amounts of information were in polymer molecules such as DNA. Probably, the most speculative is my thesis that the new means of information recording and processing resulting from the emergence of human language can take evolution in entirely new directions. But about the future of life, clearly, we can only speculate.

I

How to Clone Oneself?

Her Majesty pointed to a clock and said, 'See to it that it produces offspring'.

(Anonymous account of an alleged discussion
between René Descartes and Queen Christina of Sweden, 1649–1650)

Is it possible to make a machine that clones itself? A machine which when set in motion, goes on and assembles its own replica? In other words, a machine that reproduces? The ability to reproduce is often seen as one of the principal dividing lines between living and non-living. Allegedly, when in the seventeenth century, the French philosopher and mathematician René Descartes in a conversation with Queen Christina of Sweden suggested that an animal body could be regarded as a machine, the Queen replied that then a clock should be able to produce offspring. Obviously, clocks do not do this, but is it possible to build an automaton or a robot that can clone itself? In other words, an automaton that can *self-reproduce*.

Regardless of whether this conversation really took place, or how accurate the narrative is, it is quite likely that the phenomenon of *self-reproduction* or *self-replication* has been occupying the minds of thinkers and philosophers for centuries. However, the first truly scientific investigation of this question apparently began only with the onset of the computer age. In June 1948, the Hungarian-born American mathematician John von Neumann (Figure 1.1) gave three lectures to a small group of friends at the Institute for Advanced Study at Princeton discussing how to build a mechanical self-reproducing machine.[1]

By that time, von Neumann was already an established scientist, having made major contributions to quantum mechanics, statistical physics, game theory, and the foundations of mathematics, amongst other disciplines. In early 1940s, von Neumann joined the project at the University of Pennsylvania's Moore School of Electrical Engineering, to design the Electronic Discrete Variable Arithmetic Computer—EDVAC.[2] Von Neumann's *First Draft of a Report on the EDVAC*, published in 1945, is now widely regarded as the first description of a stored-program computer—a computer which is given its operating instructions by software, as all computers now are.[3]

In this report, von Neumann abstracted from the specific electronic parts, such as electron vacuum tubes or electric relays, used in computers at the time, and instead talked about idealized *logical switches*. He used an analogy between a computer and brain, comparing his logical switches to neurons. He thought that the specific physical

Living Computers. Alvis Brazma, Oxford University Press. © Alvis Brazma (2023). DOI: 10.1093/oso/9780192871947.003.0002

Figure 1.1 *John von Neumann standing in front of computer.*
Photographer, Alan Richards. From the Shelby White and Leon Levy Archives Center, Institute for Advanced Study, Princeton (NJ).

implementation of the switches could soon change, and he was right: it was not long before the transistor was invented and replaced the vacuum tube.[4] For von Neumann, the logical basis of computing was more fundamental than the specific physical implementations. In 1957, less than ten years after giving the mentioned lectures on self-reproduction, at the age of 53, von Neumann died from cancer.

Around the same time when the first electronic computers were being built, in the 1940s and 1950s, geneticists and biochemists were closing in on understanding the principles of the molecular basis of life. Although von Neumann did not play a major part in this effort, he was discussing his ideas with the leading biochemists and geneticists, and when working on the problems of computing, he always kept returning to the computer–brain analogy. It is quite likely that von Neumann's interest in self-reproduction was motivated by what he saw as the analogy between machines and living organisms, and by his desire to understand the 'logical' basis of life.

For von Neumann, it was clear that a self-reproducing device was possible; after all, living organisms reproduce. For him the question was how to make such a device and how complex the simplest possible self-reproducing system would have to be. Von Neumann noticed a paradox. For a device A to be able to build another device B, device A would have to contain, in some sense, a description of B. If so, it appeared that A would have to be more complex than B. Thus, a device could only build devices simpler than itself. Von Neumann called this phenomenon the 'degenerative trend'. But this contradicted the observations that living organisms not only manage to produce descendants as complex as themselves but, through evolution, also increasingly complex ones. How could a device build something as complex or even more complex as itself?

Looking for a solution, von Neumann came up with the concept of the *Universal Constructor*—an abstraction of a general-purpose robot which, given appropriate instructions and the necessary materials, could build whatever it has been instructed to build. The Universal Constructor could be given instructions to build itself, though to complete the reproduction cycle the new device would also have to be equipped with a new copy of the respective instructions. Von Neumann argued that the Universal Constructor would inevitably have to be quite a complex device, and consequently, a self-reproducing machine—a robot that clones itself—could not be simple. He also speculated how random errors in such self-reproducing devices could potentially lead to evolution towards increasing complexity. Thus, von Neumann postulated that there is a complexity dividing line below which only the degenerative trend could occur but above which evolution towards increasing complexity was possible.[5]

A mechanical replicator

In the mentioned lectures at Princeton, and a year later in the Hixon Symposium on *Cerebral Mechanisms in Behavior*,[6] von Neumann discussed the general principles at the basis of such a mechanical self-reproducing device. He called this a *kinematic* replicator. *Kinematics* is a branch of physics that studies motion of bodies and their systems. Von Neumann was interested in self-reproduction in its 'logical form' and wanted to abstract from problems such as supply of energy, the nature of molecular forces, or chemistry. Therefore, the building blocks of the replicator were allowed to be more complex than individual '*molecules, atoms or elementary particles*'. In a lecture at the University of Illinois in 1949 he said:

> *Draw up a list of unambiguously defined elementary parts. Imagine that there is a practically unlimited supply of these parts floating around in a large container. One can imagine an automaton functioning in the following manner: It also is floating around in this medium; its essential activity is to pick up parts and put them together . . . .*[7]

This process should result in a new copy of the automaton being built. Von Neumann suggested a list of eight different *elementary parts*, which he called the *stimulus producer, rigid member, coincidence organ, stimuli organ, inhibitory organ, fusing organ, cutting organ,* and *muscle*.[8] The *stimulus producer* served as a source of discrete pulses of signals to make the other parts act. The *rigid member* was the physical building block from which the frame of the automaton was constructed. The next three 'organs'—the *coincidence, stimuli,* and *inhibitory organs* were what we would now call the *logical gates* AND, OR, and NOT—the basic building blocks of a computer.[9] Von Neumann used these elements to design the control module—the brain of the replicator, which processed the signals generated by the stimulus organ to guide the construction. The last three organs were there to follow the instructions of this control module and to perform the actual physical construction. The *fusing organ*, when stimulated by a signal, soldered two parts together; a *cutting organ*, when stimulated, unsoldered a connection; and finally, the *muscle* was used to move the parts. The automaton built from these elements would be 'floating in a container' alongside spare parts and following the instructions recorded on a tape, telling it which parts to pick up and how to assemble them.

On a closer look, there are three main ideas at the basis of von Neumann's self-reproducing automaton. The first comes from a simple observation that if one has to copy a complex three-dimensional object, for instance to make a replica of a historic steam-engine, it is usually easier first to make a blueprint of this object and then build by the blueprint, rather than to try to copy the three-dimensional object directly. This, indeed, is how things are usually manufactured. It may also help to supplement the blueprint with a sequence of instructions describing the steps of the construction process. For instance, flat-pack furniture usually comes with a blueprint and an instruction manual describing the order in which the pieces should be put together. If these instructions are to be executed by a robot, then every step will need to be encoded in a precisely defined way. Nowadays, such instructions would normally be stored on some electronic information carrier, such as a *flash* memory-card, or streamed wirelessly, but in the early days of computing, a punched tape, such as the one shown in Figure 1.2, would be used. Conceptually, such a tape can be regarded as a one-dimensional sequence of symbols encoding the instructions.

The second idea represents another simple observation that it is easier to copy a one- or two-dimensional object, such as a punch-tape or this printed page, than a three-dimensional object. For instance, to copy the punch-tape in Figure 1.2, one could put it on top of an unused (unperforated) tape and punch holes where they already are in the original. It should not be too difficult to build a purely mechanical device automating this process.

Von Neumann described a device that would mechanically copy a chain encoding instructions as a sequence of two different types of links, such as shown in Figure 1.3A. Conceptually, such a chain can be viewed as a sequence of 0s and 1s, and as we will see later, any piece of information can be encoded as such a binary string. In fact, it may be even easier to copy a chain if it is designed as two 'complementary' chains, as shown in Figure 1.3B, each unambiguously defining the complement. We simply need to decouple the complementary chains and use each as a template to make a new complement, as shown in Figure 1.3C. Once the process is finished, we have two copies of the original chain. A reader with some knowledge of molecular biology may notice a similarity between this method and how a DNA molecule is copied in a living cell.

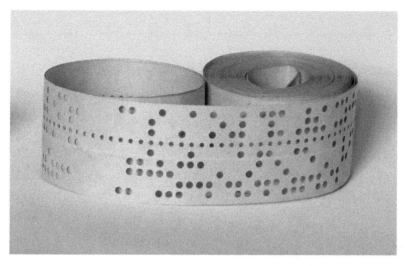

Figure 1.2 *Punched (perforated) tapes were widely used in the 1950s and the 1960s in electronic computers for data input. Although such a tape is essentially a two-dimensional object, abstractly we can view it as a one-dimensional sequence of symbols.*
(Adapted from Wikimedia, author Ted Coles.)

The third idea, the most fundamental and non-trivial of the three, was that of the already mentioned Universal Constructor. Von Neumann assumed that this machine would read a sequence of instructions given to it encoded on a tape or as a chain of elements as shown in Figure 1.3, and by following these instructions, it would build any specified three-dimensional object. This object could be another machine. A robot able to read instructions given to it on a perforated tape had already been invented in the eighteenth century by the French weaver, Basile Bouchon of Lyon. Bouchon built a loom that followed instructions encoded on such a tape, translating them into weaving patterns. Of course, that was not a general-purpose robot but a loom. One can say that von Neumann's Universal Constructor was an extension of Bouchon's loom towards Turing's Universal Machine, discussed later in this chapter.

Let us see how a combination of these three ideas can be used to make a self-reproducing device. I will roughly follow von Neumann's description, which is somewhat abstract, but it will later help us to see the analogy between his self-reproducing automaton and the main principles how a living cell replicates.

Suppose X is an arbitrary object made of von Neumann's eight elementary parts. This object, just like any object, can be manufactured by the Universal Constructor, therefore there exists a sequence of instructions following which the Universal Constructor produces X. Let us denote this sequence of instructions by code(X). Conceptually code(X) is a sequence of 0s and 1s, physically this is encoded ono a perforated tape or any way which this Universal Constructor can read. What matters is that on the input of code(X), the Universal Constructor outputs object X.

Figure 1.3 A) *A chain of two different types of links encoding a binary string.* B) *Two 'complementary' chains of such links forming a double chain.* C) *A possible mechanism of information copying that exploits complementarity: the complementary chains are separated, and a new compliment is assembled onto each of them from 'free' links 'floating' in the media. Once both chains are fully separated and the complements assembled, the neighbouring elements of the newly assembled complements only need to be 'stapled' together and a new copy of the entire chain is obtained. It is not clear if such complementarity based approach had occurred to von Neumann; he does not mention this possibility. As we will discuss in Chapter III, in DNA replication a similar mechanism is used.*

For brevity, let us denote the Universal Constructor by U. To complete the construction of a self-reproducing machine, in addition to U we need two other, simpler robotic devices, which we denote by R and C. The device R, called *code replicator*, given the tape code(X), makes a new copy of code(X). That is, R makes another copy of the tape containing the same instructions. As already noted, copying a tape is quite a simple procedure.

The third robot C controls both U and R in the following way. Given code(X), C first runs replicator R, thus making a new copy of code(X). Then C runs U on code(X), thus producing object X. Finally, C attaches the new copy of code(X) to the newly created X and releases the new object X+code(X). In von Neumann's own words, it 'cuts the new object loose'. Here, following von Neumann, we use symbol '+' to denote that the two objects are somehow 'tied' together.

Now start with the combined robot U+R+C+code(X). First, C instructs U to run on code(X). The result is:

$$U+R+C+code(X), X$$

Next, C instructs R to make another code(X); the result is:

$$U+R+C+code(X), X, code(X)$$

Finally, U 'ties' X and code(X) together; the result is:

$$U+R+C+code(X), X+code(X)$$

But now recall that X was an arbitrary object. In particular, X can be some sort of a device, for instance, we can take U+R+C for X (that is, X=U+R+C). If we do this, and give code(U+R+C) to our three-part robot as an input, we start with:

$$U+R+C+code(U+R+C)$$

and we end with:

$$U+R+C+code(U+R+C), U+R+C+code(U+R+C)$$

The robot U+R+C+code(U+R+C) has managed to build its own copy; it has reproduced itself. The 'degenerative trend' has been broken—we have a robot that built another robot as complex as itself. One may wonder how exactly it did this, but it did. We needed two separate devices: the Universal Constructor U and code replicator R. Could the Universal Constructor do both? For von Neumann's idea to work, the process R of replicating the tape had to be carried out separately from process U (a sceptic reader can try to merge both processes into one and see the difficulties). As we will see later, this is also what happens when a living cell divides; the tape there is the genomic DNA, which is a polymer 'chain' of four different elements. DNA replication is a process that is implemented in a living cell by a distinct machinery.

Note that the object U+R+C+code(U+R+C) is an object that contains its own description: code(U+R+C) describes the device U+R+C. This is what has allowed it to beat the 'degenerative trend'. But this also turns out to be at the basis of life. As Sydney Brenner, who shared the Nobel Prize in physiology and medicine with Robert Horvitz and John Sulston in 2002 'for their discoveries concerning genetic regulation of organ development and programmed cell death', wrote in 2012:

> *Arguably the best examples of Turing's and von Neumann's machines are to be found in biology. Nowhere else are there such complicated systems, in which every organism contains an internal description of itself.*[10]

What von Neumann described in his lectures was only an idea; he did not build a physical mechanical self-reproducing machine. The complexity of such a machine would have been an enormous engineering challenge in 1950s and still is today. Some parts of von Neumann's device, as he proposed it, have been built, and suggestions have been made that completing the cycle must be possible. However, as the English say: 'the proof of the pudding is in the eating'. The kinematical replicator is still not there and, as we will discuss later, possibly for a good reason—the number of mechanical steps such a device would have to carry out to complete the reproduction cycle may be prohibitive. To prove that self-reproduction is logically possible, as a mathematician, von Neumann turned to developing a rigorous mathematical model.

Self-reproduction without hand-waving

In experimental sciences, the ultimate proof that an apparatus does what it is claimed to do is an experiment. Will the machine build its replica or not? But even in experimental sciences, non-trivial reasoning may be needed to conclude that the experiment has indeed proven what is claimed. Is the replica authentic? Was it the machine that cloned itself or was it the complex environment around it that made the copy? In mathematics, the proof is achieved by defining the problem and the rules for solving it rigorously, and then finding a solution within the rules. The reasoning that we used to argue how von Neumann's kinematic replicator built its own copy did not quite fit this bill of rigour. For instance, the + sign was not defined rigorously. Obviously, this was not an arithmetic addition, it was a kind of attachment between the objects, not defined precisely. Can we define and solve the problem of self-reproduction mathematically? What is the simplest mathematical model in which the phenomenon of self-reproduction can be described rigorously—without what mathematicians call *hand-waving*?

   Those who have studied computer programming may have come across an exercise: write a code that prints its own text. Or to be precise, a code that instructs the computer to print a text identical to this code. Such a program is sometimes referred to as a *quine*.[11] In computer programming it is difficult to cheat—if the program does not do the job, no amount of hand-waving will make it work. Those familiar with computer programming can check that coding a *quine* is a non-trivial task. However, it was proven by one of the founders of computer science, the US mathematician Stephen Kleene, that it is possible to code a quine in any general enough programming language. An in-depth discussion of quines and their relevance to biology can be found in the classic book *Gödel, Escher, Bach* by Douglas Hofstadter.[12]

   We can modify a *quine* so that instead of printing the text, the program copies its own code over a computer network to another computer. This is what computer viruses do. A computer virus is a self-reproducing code. Does such a virus achieve what von Neuman set out to do? Only in a rather limited sense. A computer virus replicates because the code is executed by computer hardware using an operating system, which usually is more complex than the code of the virus itself. Those who are familiar with biology will notice that the same is true in respect to biological viruses—to replicate viruses need the machinery of a living cell, which typically is more complex than the virus. Von Neumann's kinematic replicator was supposed to 'float' in an environment simpler than itself.

To have a full analogue to *quine* in the physical world of atoms and molecules, we would need to have a program for a computer equipped with a three-dimensional printer (or a robot), which in addition to copying itself would also have to instruct the computer and printer to 'print' another computer and printer, and finally load itself into the new computer. This would be a version of von Neumann's self-reproducing device, where the program would be analogous to its tape.

As von Neumann noted in his lectures, computers output a class of objects that is rather different than themselves—they can print a text or send a code electronically to another device but they do not make physical objects. Von Neumann wanted to find a mathematical model of a device that would make an object of the same class as itself.

Inspired by his friend and colleague Polish-American mathematician Stanislaw Ulam, von Neumann turned to *cellular automata*—two dimensional structures or configurations (drawn on a plane) that develop according to well-defined rules. The goal was to find a set of rules and a particular initial structure such that after some time, two structures identical to the original would emerge on the plane.

Cellular automata and *Life*

Probably the best-known example of cellular automata is the *Game of Life*, or simply *Life*, invented by the British mathematician John Horton Conway at Cambridge University. This game was introduced to the general public in the 1970 by Martin Gardner, the legendary editor of the column *Mathematical Games* in the popular science magazine *Scientific American*.[13] At that time, computers were becoming cheaper and accessible to many scientists, and increasingly to everyone interested, which most likely contributed to the popularity of the game. Now there are many websites offering *Life* online; to see what fun *Life* can be, the reader not yet familiar with this game might find and try one of these websites out. Less well known than the game itself is that Conway invented *Life* thinking about von Neumann's problem of self-reproduction. In 1986, after leaving Cambridge, Conway became the John von Neumann Chair of Mathematics at Princeton University.

The *Game of Life*, unlike real life, is governed by simple rules, but like real life, it is very rich in possibilities. Although to see the richness of *Life* one needs a computer, it can be played on a sheet of squared paper by one player. Conceptually the sheet is of infinite size, as it can always be extended as necessary by adding another sheet. Each cell can be in one of two *states*; occupied—*alive*, or empty—*dead* (*quiescent* state). Each square is a cell with eight neighbours—one to the left, one to the right, on top, beneath, and four diagonally. The creative part of the game is to fill in the initial configuration—to choose which cells are *alive* at the start. After that, the configuration develops following strict deterministic rules in discrete steps or time-points. For the description of *Life*'s rules and how some initial configurations develop see Figure 1.4.

Amongst the simplest configurations in the figure, *glider* takes a special place: every five steps *glider* rebuilds itself, shifted by one position, thus moving diagonally on the grid. *Gosper glider gun* or *cannon* is a more complex configuration, which periodically

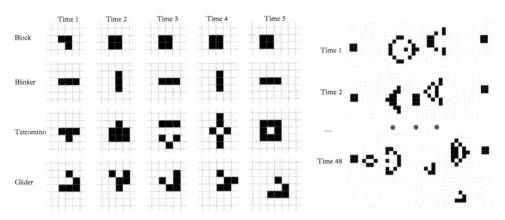

Figure 1.4 *Left. Simple configurations in Conway's Game of Life and how they develop. At every consecutive time-point, a new configuration is drawn by the following rules. A cell is* alive *(filled) in one of two cases only: 1) it was* alive *in the previous time-point, and it had two or three* alive *neighbours, or 2) it was* not alive *(was empty) in the previous time-point and had exactly three* alive *neighbours. (Thus, exactly three* alive *neighbours of an empty cell give birth to a new* alive *cell, while if an* alive *cell has less than two or more than three* alive *neighbours, it dies of 'lonesomeness' or 'overcrowding', respectively.) Following these rules, a block of four cells stays intact. A row of three consecutive cells (*blinker*) periodically turns itself into a column of three cells and back. A* glider *moves diagonally through space—it is easy to check that after four steps it will have reconstructed itself by one position to the right and downwards. The reader may try out what happens to* tetromino *after time 5.* **Right.** *Gosper glider gun (or* cannon*) periodically shoots gliders. The first glider appears at time-point 16, and then again, every consecutive 20 time-points. To appreciate the richness of such configurations in full, one needs to use a computer (for instance, Wikipedia offers an interactive website).*

shoots *gliders* in a particular direction. It may be exciting to watch on a computer screen what happens when two cannons are shooting gliders at each other; depending on their relative position, impressive fireworks can happen.

Most *Life* configurations will eventually become static (like *block* in Figure 1.4) or get into a loop of some periodicity (like *blinker*). These can be viewed as *settled* configurations as nothing new can happen. Some configurations, however, may go on developing in a nontrivial way forever; for a mathematically minded person it can be an exciting exercise to try to find some of them. In general, it is not possible to tell if a configuration will settle or not in any other way than following its development until it settles, or maybe it does not. If it settles, we can spot this, but it is impossible to know how long one should keep following a configuration that refuses to settle. This is a mathematical fact (a theorem).[14]

But here we are primarily interested in cellular automata as a model for *self-reproduction*—is there a configuration such that following the rules, at some point two configurations identical to the original would emerge?

Self-reproducing structures

To build a model for a self-reproducing device, von Neumann used a version of cellular automata different from Conway's *Life*. Cells in his automata had 29 different states[15] rather than two, as in *Life*. His construction, schematically shown in Figure 1.5, was designed to emulate a two-dimensional robot.[16] The robot had a 'construction arm', which performed the actual task of assembling a new device, like a robotic arm would be used to build a physical structure. Von Neumann described how to simulate various elements of this robot, including wires through which control signals were sent. These signals were processed in the control units built from the logical gates (also implemented

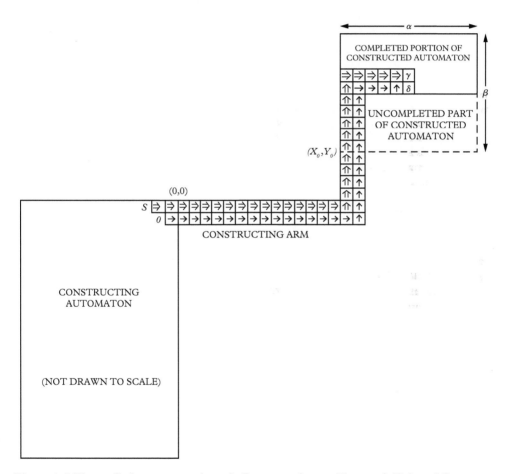

Figure 1.5 *The tessellation structure schematically representing von Neumann's Universal Constructor* (From Burks, 1970).

as cellular automata configurations), mentioned earlier in this chapter. Von Neumann described how a one-dimensional tape could be simulated using his cellular automata, and how this tape could be copied. He sketched how these components could be assembled to make the Universal Constructor, which would read instructions from the 'tape' to build a new two-dimensional object, all following the rules.

Von Neumann realized that creating a device that is already operating while it is still being built would be rather difficult in this framework; its operation would interfere with its further construction. This would be like trying to manufacture a car with the engine already running. The solution he found was first to build a static device and then, when its construction was finished, it was activated by an 'ignition' signal. This is like Frankenstein's creation—the body was first assembled and then set alive by an electric spark. To implement this 'setting alive' idea, every cell needed an unexcitable state and a live state that was induced by the ignition signal, which is one of the reasons why von Neumann needed cells with as many as 29 states.[17]

Von Neumann's two-dimensional self-reproducing robot was quite a baroque construction, which he did not finish; his collaborator Arthur W. Burkes finished it already after von Neumann's death.[18] Burkes too did not draw the construction explicitly but designed its various elements and gave a mathematical proof that the Universal Constructor and a self-reproducing configuration could be built from these. Together this proof filled more than 200 pages. Burkes later simplified the design to the extent that its description could fit on 64 pages.[19] In 1955, another of von Neumann's collaborators, John Kemeny, wrote an article in *Scientific American* estimating that a self-reproducing automaton based on von Neuman's principles could fit into a $80 \times 400 = 32,000$-cell rectangle, plus a 150,000-cell long 'tail' representing the tape.[20] Some other estimates range from 50,000 to 200,000 cells.[21]

Although various further simplifications have been reported, little has been published in peer-reviewed scientific literature since the 1970s. It seems that none of the designs has been fully implemented as a computer program allowing for observable simulations.[22] Perhaps one reason for this is that the answer to the most important fundamental question was clear from von Neumann and colleagues' early work: self-reproduction is logically possible, but the structures required to achieve this are not simple. The exact versions of the possible structures are less important as they do not model the laws of physics, and thus what they can tell us about physical self-reproducing devices is limited.

Are there simpler ways of achieving self-reproduction if we do not follow von Neumann's design? In Chapter II we will look at a simple form of self-reproduction found in nature: growing crystals. Whether a crystal growth can be truly viewed as self-reproduction is debatable, but as we will see there is some analogy. Atoms or molecules form periodic crystal lattices the structure of which is reproduced when crystal is growing. Unsurprisingly, it is not difficult to design cellular automata that grows in a way resembling a growing crystal lattice,[23] as shown in Figure 1.6.

An important feature that distinguishes a growing crystal from a living organism is in the amount of information that each of these can carry. The repetitive structures of a pure periodic crystal, as we will discuss in Chapter II, cannot carry much information.

Time 1   Time 2   Time 3   Time 4   Time 5

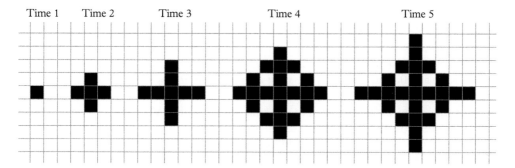

Figure 1.6 *A simple growing tessellation structure. Here we assume that a cell has four neighbours—the cell above, beneath, left, and right—and in that in each next generation a live cell is born if and only if it is a neighbour to exactly one* live *cell in the previous time-point. Following such a rule, starting from one live cell, the structure will keep growing like a snowflake. The real snowflakes have hexagonal-symmetry, which can be modelled on a honeycomb-like grid.*

In contrast, biological organisms carry large amounts of information, and moreover, like von Neumann's automata, their replication is guided by this information. John Maynard Smith and Eors Szthmary, in their seminal book *Major Transitions in Evolution*,[24] refer to *unlimited heritability*, contrasting it to *limited heritability*. A living organism can carry and pass on potentially unrestricted amounts of information to their descendants,[25] pure periodic crystals cannot. It seems that von Neumann had discovered a similar distinction in the mathematical 'world' of cellular automata a couple of years before biologists uncovered the mechanisms of inheritance. But what is *information*?

The world of information *bits* and *bytes*

Information, like many fundamental concepts, is hard to define. It can be argued that a definition that includes all important features of what we intuitively understand by this concept has not yet been found. Intuitively, most of us will agree that information is something that reduces uncertainty. Reduction in uncertainty can be quantified; 1 *bit* is the amount of information needed to answer one *yes/no* question in absence of prior knowledge. A *byte*—a unit used to measure the information storage capacity of a computer or a size of a file—equals 8 *bits*, thus providing sufficient information to answer eight *yes/no* questions.

As will be argued in Chapter VII, any piece of information can be encoded as a sequence of answers to *yes/no* questions. If we use 1 for *yes* and 0 for *no*, then a series of such answers can be encoded as a sequence of 0s and 1s, for instance, as strings like 00101101, which are often referred to as *bit-strings*. Sometimes using a larger alphabet, for instance, Latin letters or Chinese characters, or perhaps pixels of different colours, may be more convenient way to present a particular piece of information. In DNA information is encoded as a sequence of four different small molecules—four nucleotides—traditionally denoted by letters A, T, G, and C, as will be discussed in Chapter III. Nevertheless, sequences in larger alphabets can always be represented as *bit-strings*. In DNA each nucleotide contains 2 *bits* of information, as it encodes an answer to two consecutive yes/no questions. For instance, A can be represented as 00, T as 01, G as 10, and C as 11. We may also want to arrange the characters in more than one dimension, for instance as two-dimensional configurations of the *Game of Life*, or as pixels to represent an image in a digital photography. Multi-dimensional symbol arrangements too can always be represented as *bit-strings*: conceptually, everything we see on a computer screen is encoded in as a *bit-string* in the computer's memory. Some may ask—what about analogue information, for instance film photography? We will discuss this question in some more detail in Chapter VII, but the short answer is that such information too can be encoded as sequences of 0s and 1s.

Is there a distinction between a random sequence of 0s and 1s, and what we would normally call *information*? Establishing such a distinction is the core challenge in defining information in a way that reflects our intuitive understanding of this concept. A potential answer lies in acknowledging that *we can talk about information in a meaningful way only in the context of an interpreter*. To talk about information, there has to be a mechanism that converts the particular sequences of symbols representing this information into something else, for instance, into some kind of action. Von Neumann's Universal Constructor reads the information from the tape and builds an object. Those familiar with molecular biology will know that in a living cell molecular machinery reads information in DNA to make proteins. A human can read information in a cookbook to prepare dinner or a molecular biology textbook to get ready for an exam. Thus, let me postulate that *a string or an arrangement of characters is information if and only if it can be interpreted*. Though I have not seen such a postulation formulated in the literature exactly this way, it is very much in line how other authors have treated this concept. This is consistent with the approach taken by Douglas R. Hofstadter in his famous book *Gödel, Escher, Bach*, though Hofstadter does not explicitly refer to term information.[26] Obviously, as the concept of interpreter is not defined, I do not claim here to have provided a new formal definition of information. We will return to this discussion in Chapter VII.

Finally, note that although information is defined as an arrangement of symbols, in the physical world information exists only as a material pattern. The same information can be encoded by different physical means and as different patterns, for instance as an arrangement of symbols printed on a paper or an arrangement of holes in a perforated tape. The physical interpreters will have to be different, but the result of their action can be the same; for instance weaving a rug containing the same pattern.

Church–Turing thesis

Paradoxically, it is easier to define *information processing* than to define information. *Information processing is what computers do*; most scientists working in the field will agree with this statement. Examples of information processing can be as simple as the adding or multiplying of two integers, or as complex as facial recognition where two different images of a human face are compared to determine whether they are of the same person. However, what exactly do computers do? And are all computers the same?

The short answer to the last question is that they are. If we assume that our computer has unlimited memory, and if we have unlimited time to wait while the computer completes its work, then this computer can do whatever any other computer can. A different computer perhaps can do the same work faster, but it cannot compute anything that our computer with limitless memory could not. Strikingly, a rather simple mechanical device, now called the *Turing Machine* (named after British mathematician Alan Turing), can compute this too. To many readers, Turing will probably be better known for his contribution to breaking German code during the Second World War and the ENIGMA machine that helped to achieve this. However, arguably, the paper that Turing published in 1936, introducing and exploring what is now known as the Turing Machine, can be seen as the beginning of computer science.[27]

Turing invented his simple, conceptual machine when trying to understand what a mathematician does when proving a theorem. Turing realized that conceptually all what was needed was a device that had a finite number of different states (like a gearbox, which has, say, six different gears) and a tape of unrestricted length (memory). Symbols, such as 0 and 1, can be written on or read from this tape, one by one. The device can move along the tape in either direction, one symbol at a time. Finally, this device has a finite set of instructions telling it what to do in each next step, depending on the current state it is in and the particular symbol it sees on the tape. For instance, the instruction can be that if the machine is in state #5 and the symbol it sees on the tape is 0, it should switch to sate #3 and move one notch to the right. Different machines have different sets of instructions, each designed to solve a particular information processing problem. The information the machine is asked to process is given to it on the tape, and the machine writes the answer back on the tape before it stops.

The above description of the Turing Machine may appear somewhat technical, but the important take-home message is simple: *any information processing task can be executed by a simple mechanical device with memory of unrestricted capacity.* Thus, information processing can be defined as a mechanical manipulation of symbols. The terms computing, information processing, and mechanical manipulation of symbols, are all synonyms.

Turing went on to prove that there is a particular machine, the *Universal Turing Machine*, or UTM for short, in fact, quite a simple one[28] which can do anything that any other Turing Machine can. The trick is that appropriate instructions have to be written on the machine's tape alongside the information that the UMT is given to process. In other words, the set of instructions do not have to be hardcoded in the machine, the instructions can be given to the UTM alongside the information it is asked to process. This is how all modern computers are built to operate: they have general purpose hardware (a memory-limited analogue of UTM) on which specialized programs, such as text editors, computer games, or chatbots can be run. Every smartphone includes a general purpose computer onto which specific applications (apps) are installed allowing it to execute specific tasks.

Importantly, Turing also showed that not every question that one might want to ask about information can be answered by his machine (and by implication, by a computer). For instance, no software can tell a computer how to find out if an arbitrary configuration in the *Game of Life* will ever settle.[29] Obviously, if the configuration settles the computer can spot it, but if the configuration goes on developing, how long does the computer need to keep following it? Some configurations will go into a loop (like *blinker*), and this too can be spotted (in fact, computers can spot this more reliably than a human), but some will not. Information processing problems that can be solved by a computer are known as *computable problems*. Whether an arbitrary chosen configuration in *Life* will eventually settle is not computable.

For some of *Life's* configurations, a human could prove that they will not stop, even if they keep developing in a non-periodic and non-trivial way. Does this mean that the human brain does more than a computer? Entire books have been written on this subject, but the short answer is no. It is possible to write a computer program that instructs a computer to go looking for a proof that the configuration wont settle, the only problem is that we cannot be sure if the proof will eventually be found. Arguably, the same is true for a human mathematician. Regardless of whether a human brain does anything else than manipulating symbols, sufficiently fast mechanical symbol manipulation is all that is needed to beat the human world champions in chess or in the board game Go, or to recognize one face amongst millions of faces.

Shortly before Turing published his seminal 1936 paper, the US mathematician Alonzo Church introduced a rather different way of defining computability, called *lambda calculus*.[30] Although superficially there was nothing in common between the Turing Machine and lambda calculus,[31] Turing proved their equivalence: every information processing question that could be answered by a Turing Machine could also be answered using lambda calculus, and vice versa.

Turing's proof of equivalence would be no more than a technical result if not for what followed. Scientists proposed a range of different ways of defining information processing, which all turned out to be equivalent to each other and to the Turing Machine. Church generalized this observation to postulate that *any information processing problem for which 'a sentient being' would invariably come to a definite answer, the same answer would also be reached by a Turing machine or any of its equivalents*.[32] In other words, if a question has a definite answer that can be found based on the information one has, this answer can be found by manipulating symbols on a simple mechanical device. This is one of the most important discoveries of the twentieth-century science and is as important as any law of physics; as important as the Second Law of Thermodynamics, which we will discuss in Chapter II. This *law* is known as the *Church–Turing thesis*, sometimes also referred to as the *Church thesis*. As a natural law, the Church thesis cannot be proved mathematically but rather is a summary of empiric observations, just like the Second Law. The Church thesis also makes it possible to talk about an abstract Universal Computer rather than a UTM specifically, and about algorithms and algorithmically solvable problems, rather than specific computer codes.

Has anybody ever built a practical mechanical Universal Computer if it is so simple? Perhaps, not coincidentally, Church, Turing, Kleene, von Neumann, and many others were working on the problems of computing at a time when the field of digital electronics was rapidly developing. In the 1940s and 1950s, the electronic computers Colossus in Britain, and ENIAC and EDVAC in the United States were built.[33] But there was an attempt to build a mechanical programmable computer already in the nineteenth century. *Analytical Engine* was a project of the British engineer Charles Babbage in 1837. Babbage did not get very far with this project, but a working implementation of a simpler mechanical computer, *Difference Engine 2*, also designed by him, was built at the end of the twentieth century (completed in 2002) in the Science Museum in London (Figure 1.7). *Difference Engine 2* is not universal, it cannot compute everything that a Turing machine can, nevertheless, it is quite a general information processing machine. It consists of about 25,000 mechanical parts, which had to be manufactured and assembled with the precision of a fine watch and together weigh several tons. As described in the enticing book *The Cogwheel Brain* by Doron Swade,[34] the project to build this brass cogwheel machine turned out to be rather challenging even in the late twentieth century.

The UTM is conceptually simple, but the number of steps it needs to carry out for any but the most trivial computation is enormous. Even in electromechanical

Figure 1.7 *Difference Engine 2, designed by Charles Babbage in 1847–1849. The engine in the picture was built in the London Science Museum in late twentieth century, completed in June 1991.* (Figure from Wikimedia, CC-BY-SA licence by geni.)

computers, macroscopic mechanical parts are too heavy to be moved around fast enough to complete complex computations in reasonable time. Electronic computers make such computations possible by moving electrons, which are much lighter than mechanical parts.

Using electrons, however, requires parts very different to cogwheels. Von Neumann showed that just two different types of *logical switches* or *logical gates*, AND and NOT, and an unlimited supply of memory elements each able to record 1 *bit* of information, are all that is needed to build a practical Universal Computer. These logical elements can be implemented using electron vacuum tubes or transistors. Microprocessors used in contemporary computers are circuits of these elements. We will see later that these logical gates and memory elements can also be implemented using biological molecules such as proteins, and thus that a living cell—the basic unit of all known living organisms—has everything needed at its disposal for processing information.[35]

Although speeding up a computer will not solve non-computable problems, the speed matters, nevertheless. If the speed of information processing and the available memory did not matter, we would not be so keen always to get the latest laptop, tablet, or smartphone. The recent advances in artificial intelligence (AI) would not have been possible without sufficient computational power, and this also applies to biological systems. When an antelope recognizes an approaching cheetah, she has to process this information almost instantaneously, otherwise soon she may not be able to process any information. As Maynard Smith and Szathmáry noted, a key feature of biological evolution is providing living organisms with increasing computational efficiency.

Let us return to von Neumann's self-reproducing automaton and see how it is different from crystal-like growth. How can the requirement that an automaton has to be capable of unlimited inheritance be formulated rigorously? Above I postulated that information needs an interpreter—an agent that knows how to act on the information it receives. In the context of a mathematical structure, acting on information means processing it. Von Neumann postulated that his self-reproducing cellular automata configuration has to be such that it not only builds the replica but that it can also execute any information processing task. In other words, it needs to include an equivalent of a UTM.[36] This may seem to be a rather hefty requirement but, given that biological systems have everything needed to process information, it is not an excessive one. The ability to process information is an essential feature distinguishing biological systems from crystals.

The complexity of reproduction

In the mathematical 'world' of cellular automata, we can invent any rules we like; if we try to design a mechanical self-reproducing device in the real world of atoms and molecules, we are constrained by physics. Von Neumann designed the rules that enabled him to build a self-reproducing structure. Conway took a different approach; he first invented the rules that were simple and elegant and allowed for rich dynamic structures to be encoded, and then asked: is there a configuration in *Life* that would self-replicate?

Indeed, Conway's rules are so elegant that one can almost say they define some kind of alternative two-dimensional 'physics'. Do self-reproducing configurations capable of unlimited inheritance exist in this two-dimensional world of Conway's 'physics'?

Conway outlined the proof of the existence of a configuration that encoded the UTM in 1982 in the last volume of the popular book about mathematical games, *Winning Ways*, co-authored with Elwyn Berlekamp and Richard Guy.[37] His proof relied on highly abstract results from the theory of computing and some complex and ingenious constructions—carefully arranged *cannons* shooting fleets of *gliders* on a collision course with each other; these collisions produced new *cannons* shooting back, and the entire configuration behaving like the UTM. Conway also conjectured there how the UTM could be used to make a self-reproducing structure in *Life*.

Three years later, Conway's ideas were discussed by science writer William Poundstone in his popular book *The Recursive Universe*.[38] Poundstone estimated that the minimal size of a self-reproducing configuration in *Life* would be around 10^{13} cells. Assuming that each cell is a square of 1 mm^2, to fit in this configuration we would need $3 \times 10^6 \times 3 \times 10^6$ mm = 3 km × 3 km. It is therefore not surprising that nobody has attempted to draw it.

Interestingly, in 2010, a non-trivial configuration of *Life* that creates a copy of itself in another location on the plane, but at the expense of destroying the original, was found[39] (in a sense, a *glider* does this too). This, however, is more like teleportation than reproduction. The general view amongst *Life*'s enthusiasts seems to be that all the necessary elements to make a self-replicating configuration have already been designed and that implementing and simulating such a structure on a computer to see how it replicates should soon be possible. Time will tell, though we do not know how long we have to wait.

There are three important lessons that we can learn from research in self-reproducing cellular automata. First, self-reproducing structures capable of carrying and passing on (potentially) unlimited amount of information to its replicas by acting according to simple well-defined rules exist. Second, it is most likely that such a structure will be complex (whether we take the 80 × 400 = 32,000 cells plus the 'tail' of 150,000 binary cells from Kemeny, or 10^{13} two-state cells from Poundstone, or any other estimate, all these point to high complexity). Third, if we do not care about carrying information (if we are content with *limited heritability*), then a much simpler (crystal-like) reproducing structures are possible.

These lessons refer to the conceptual mathematical 'world' of information *bits* and *bytes* rather than to the physical world of atoms and molecules. Can we extrapolate our observations to the real world 'obeying' by the rules of physics? Can we assume that in the physical world there is a similar relationship between the complexities of a self-reproducing system with unlimited heritability, on the one hand, and the crystal-like growth, on the other hand? This does not sound like an unreasonable hypothesis. The cellular automata-based models seem to suggest that information carrying self-reproducing systems cannot be simple in any 'world' of well-defined rules. With some hand-waving, these conclusions can be summarized as follows: *information carrying self-reproducing systems (an essential part of life) cannot be very simple.* Biologists will say that they knew this all along.

Autonomous replicators and ecosystems

In the mathematical 'world' of cellular automata or information *bits* and *bytes*, self-replication is a well-defined problem: the replica needs to be identical to the original and the rules at work are precisely defined. But ultimately, we are interested in physical systems; the former is only a model of the latter. Replication of information *bits* inside a computer cannot go on for long without the physical environment supporting it, as can be demonstrated by a simple experiment: unplug your computer and see what happens.

In the physical world of atoms and molecules *reproduction* or *replication* is less well defined and sometimes different meanings are assigned to each of these terms. How similar does the copy have to be to the original? In the physical world no two objects more complex than a molecule are truly identical. No copy of da Vinci's 'Madonna' is identical to the original. The finest mechanical watches are said to be produced to a precision of about a thousand's part of a millimetre, which is more than 1000 atoms aligned on a string. Thus, $100 \times 100 \times 100 = 1,000,000$ atoms here or there may not matter for a fine watch to be a replica. Or would colour matter if the device were made of plastic LEGO blocks? Sometimes in the literature, replication is defined as making an exact copy, while reproduction is making an approximate one, but this effectively means that the replicators are compelled to the abstract world of information *bits* and *bytes*. Such a restriction on the use of these terms will not be intuitive in all contexts, and in fact, in the literature, this distinction is not used consistently. Therefore, here I treat these terms largely interchangeably. What does seem to matter in either self-replication or self-reproduction is that the essential features of the device are preserved in the replica.

Nevertheless, defining self-reproduction or self-replication in the physical world is not trivial. To start with, what kind of building blocks is the replicating machine allowed to use? Von Neumann allowed his elementary parts to possess a 'certain complexity' without going too much into the detail of what exactly he meant by this. Would microprocessors be acceptable as 'elementary' parts? He described the environment as a container in which the automaton and spare parts were floating. Does the replicator necessarily need to be floating in a liquid, or would a robot that moves on a surface, like an ant picking up parts scattered around, do?

What really seems to matter is that the parts are much simpler than the replicating device as a whole. But even more importantly is that the self-replication process does not benefit from an active (or structured) help from the environment, and the environment is simpler than the replicating device. Otherwise, the part *self* in self-replication or self-reproduction cannot be justified. Those familiar with biology may notice that this means that biological viruses would not pass this criterion because the living cell whose processes they use to replicate, will typically be more complex than the virus itself. To emphasize this, let me introduce the concept of an *autonomous replicator* which is *a physical system or a device that makes its own approximate copies for many replication cycles, using parts much simpler than itself, and without active structured help from the environment.* Obviously, this is not a rigorous mathematical definition, but here we are dealing with the physical world rather than a model.

Those familiar with the works of Richard Dawkins may note that this definition of autonomous replicator is different from that of Dawkins' *genetic replicator* (often abbreviated to *replicator*). First of all, *genetic replicator* belongs to the domain of information *bits* and *bytes* rather than the physical world. Dawkins' *replicator* corresponds to code(U+R+C) in von Neumann's setting, rather than the actual device U+R+C. The device itself corresponds to what Dawkins calls a *vehicle*. In the terminology used here, a vehicle is called an interpreter. Additionally, *genetic replicator* is defined to have several other properties, such as evolvability, which we will discuss later.

Do autonomous replicators exist in nature at all? As we will argue later, some microorganisms, including bacteria, can be regarded as fitting the autonomous replicator's definition given above. Most biologists will however say that in nature no organism is truly autonomous, that all organisms live within symbiotic communities. What do they mean by this?

Suppose we have two replicators A and B that are almost autonomous, but not quite, in the following sense. Each of them needs the same set of simple elementary parts, simple enough for A and B to be considered as autonomous, except that A needs one specific, quite complex, part a, while B needs a complex part b. On the other hand, in the replication process, replicator B makes the part a, while A makes the part b. Thus, neither A nor B can be considered truly autonomous, but the two of them together can form an autonomously replicating system. Such mutually beneficial co-habitation is called *symbiosis* and the combined system a *symbiotic* system of replicators. If the two replicators 'float' in the same medium and release the spare parts back to the environment, then two of them together will form an autonomous replicating system.

We can also think of a symbiotic replicator system that consists of hundreds or thousands of symbiotically linked replicators, dependent on each other. We could call it an *ecosystem* of replicators. What is the replicator here? Each individual, or perhaps we should consider the whole system as a replicator? When would we say that such a system as a whole has replicated? One possible definition could be to require that the number of each individual replicators has at least doubled. Possibly a better name for such a system is a *self-sustaining (eco)system of replicators*. Can the minimal total complexity of such a system be less than that of one 'solid' autonomous self-replicator? Intuitively, it seems that the answer should be negative, however I am not aware of any in-depth mathematical investigations of this question. As we will see later, the answer to this question is important for understanding how life began.

Finally, suppose the replicator B needs part b made by A, but A does not need anything from B. In that case, B is a *parasite* living on the products of A, but not doing anything for A. If B takes some of the elementary parts away from A, this will put A at a disadvantage in comparison to another community of As, not infested with parasites.

While A keeps replicating, B can replicate, but if A dies, B dies too. In evolution, it is possible that A and B initially are symbiotic replicators, but later one of them starts 'cheating' and becomes a parasite.

Chance and making a reliable device from unreliable parts

The cellular automata discussed above were deterministic in the sense that each configuration was determined by the previous one without any element of chance. An element of chance can be introduced, for instance, by tossing a coin. In the *Game of Life*, for instance, with every thousandth toss when we flip heads, the respective cell is left in the previous state regardless of what the other rules dictate. Perhaps such a stochastic algorithm would better model how humans play *Life* on a squared paper—we occasionally make mistakes, probably more than one mistake in a thousand moves. Introducing stochasticity in an appropriate way may bring our model closer to physical reality, typically however, we want to minimize stochasticity. The early electronic computer ENIAC contained over 17,000 vacuum tubes, which had limited lifespan and thus kept failing. Thinking about how to perform reliable computations on devices with such unreliable parts, von Neumann developed a mathematical theory of building reliable automata from unreliable components.[40]

Obviously, errors can disrupt a computation, but could adding stochasticity to the Universal Computer actually increase its power? Are there any non-computable problems, that a computer could solve if it was allowed to toss a coin? Presumably the human brain is, in part, stochastic and maybe this is what makes it more than a computer?

The answer depends on whether we want the result to be reproducible or not—does the same question always require the same answer? It has been proven mathematically that if we only consider answers that can be reproduced with any required precision, then everything that can be computed by a stochastic computer can also be computed by a deterministic one.[41] In other words, adding coin-tossing to the logical gates and memory elements do not expand what can be computed. Paradoxically though, some computations can be done (in a reproducible way) faster by a stochastic computer than by a deterministic one.[42]

Non-reproducible computations do happen in nature and no deterministic computer can carry them out. Biological evolution is not reproducible. As noted by one of the best-known palaeontologists of the modern age, Stephen Jay Gould, if we restarted life on Earth from what it was, say, a billion years ago, it is unlikely that the outcome of evolution would be the same.[43] Although the degree of the divergence is a contentious topic in evolutionary biology, as will be discussed later, evolution exploits random events for its benefit, and this makes the process irreproducible, at least to some degree.

Returning to von Neumann's replicating device, can we make it prone to stochastic errors? Ideas how this can be achieved and how this related to the living systems, are given in Box 1.1.

Box 1.1 How to make an infallible replicator

Consider von Neumann's kinematic replicator U+R+C+code(U+R+C). Recall that U is the Universal Constructor executing instructions given to it on tape code(U+R+C), that device R makes another copy of the tape, and that C is the controlling device that also puts everything together in the end. What happens if an error occurs when running one of these devices? As von Neumann argued, the outcome would depend on the nature of the error. If U or C is faulty, then the process will most probably stall. If there is an error on the tape code(U+R+C), the process R will replicate this error. In the next replication round, the device U will produce a device according to this new, erroneous description. If the new replicator still works, the error will be present in all future copies (which is one of the reasons why for Dawkins only code(U+R+C) is the replicator).

Is it possible to make the replicator robust to errors? Before addressing this question, let us first discuss a related matter. To make a replicating device useful, it would have to make something else in addition to its own copies. For instance, the replicator could make a bicycle with every new copy of itself, in which case we would have a self-multiplying bicycle production factory. In fact, such ideas have been proposed in the context of colonizing other planets. To achieve this, we need to add the right information to the code on the tape code(U+R+C). If we take code(U+R+C+D), where D codes for instructions to build a bicycle, and we feed it to the Universal Constructor, then starting with U+R+C+code(U+R+C+D) the device would make U+R+C+D+code(U+R+C+D). We will have a new bicycle in every replication cycle.

How is this relevant to making an error-resistant replicator? If a mistake happens in copying D, the bicycle will be faulty, but the replication process itself will not necessarily stop. But suppose device D is making spare parts for the replicator; for instance, D is making a particular cogwheel used in the replicator. Now the replicator is allowed to make specific types of errors when copying the code for U+R+C: if an error disrupts the production of that cogwheel, we can use the one supplied by D. Moreover, if the supply of these cogwheels is what limits the speed of replication, then it is not inconceivable that the device U+R+C+D+code(U+R+C+D) will replicate faster than U+R+C+code(U+R+C). To advance this idea further, suppose the device D consists of 'sub-devices' D_1, D_2, ..., D_n, which together make all the necessary parts used in U+R+C. This would allow for redundancy to many different production errors.

Going even further, we could design a replicator U+R+C+D+code(U+R+C+D), where D= $D_1+D_2+...+D_n$, and the components U, R, C, D_1, D_2, ..., D_n are all relatively independent of each other; for instance, 'swimming in the medium' and interacting with each other via the complementarity principle, similar to that shown in Figure 1.3c. Obviously, such a device would be stochastic as we do not know how long it would take for each new part to find its destination. To keep the parts together, we could enclose them in a membrane that keeps the parts inside, lets the building materials in, and any possible waste out. For a self-replicator to be autonomous, the membrane itself would have to be made of some of the parts D_1, ..., D_n.

The above description admittedly contains significant hand-waving, nevertheless something similar exists in nature. This is a living cell. If the subsystems D_1, ..., D_n jointly make

continued

Box 1.1 *continued*

all the parts that are needed for U+R+C+D+code(U+R+C+D), then U, R, and C become redundant—the device $D_1+D_2+\ldots+D_n+\text{code}(D_1+D_2+\ldots+D_n)$ is self-replicating. Those who know biology may note that $D_1,\ldots,D_n$ are analogous to proteins in a self-replicating living cell, while $\text{code}(D_1+D_2+\ldots+D_n)$ is the genome and each of the $\text{code}(D_1),\ldots,\text{code}(D_n)$ are the genes.

Microbes—molecular replicators

Decades after von Neumann's seminal work on self-replication, nobody has yet managed to make a working artificial physical self-replicating device. Why? The theoretical considerations discussed above suggest that such a replicator is likely to be quite complex, probably too complex to be built as a hobbyist's toy. But unless it also makes something useful in addition to its own copies, which would complicate the device further, practical applications of such a device are not obvious. Therefore, it is not too surprising that not enough resources and effort have gone into making a kinematic replicator. But perhaps there are more fundamental reasons why a mechanical self-replicating device cannot be built?

Is it possible that the number of mechanical steps a kinematic replicator would have to execute is prohibitively large? That it would take too long to complete the replication cycle and that a mechanical failure is likely to happen in the process? After all, as the reconstruction of Babbage's *Difference Engine* shows, building a general mechanical computer is difficult.[44] A mechanical general-purpose computer capable of non-trivial computations may be in practice impossible. Perhaps we can bypass the problem by including electronic microprocessors and use them in the control device in von Neumann's model? But then our Universal Constructor would have to be able to manufacture these microprocessors. Probably this would be more like a self-reproducing factory.

Although nobody has yet been able to build a mechanical replicator, self-replicating 'devices' do exist in nature, they are called *microbes*. There are different types of microbes, one of them is *bacteria*. One of the best-studied bacteria has the name *Escherichia coli*, or *E. coli* for short, which lives in the human gut.[45] *E. coli* is about an average-sized bacterium, and it resembles a microscopic rod or a slim cylinder of about 5 microns long and 1 micron in diameter. A micron (μm) is a thousandth part of a millimetre ($1\,\mu\text{m} = 10^{-3}$ mm). For comparison, the smallest dot clearly visible to the naked human eye is roughly a tenth of a millimetre (10^{-1} mm), thus these bacteria are in the order of 100 times smaller than what can be seen by the naked eye. The 'elementary parts' of bacteria are molecules that are still 100–1000 times smaller; their size can be measured in nanometres—one millionth of a millimetre ($1\text{ nm} = 10^{-3}\,\mu\text{m} = 10^{-6}$ mm), or even fractions of nanometres. It is largely this tiny size which allows bacteria to function and replicate fast enough to exist and to thrive—each of us may be carrying up to

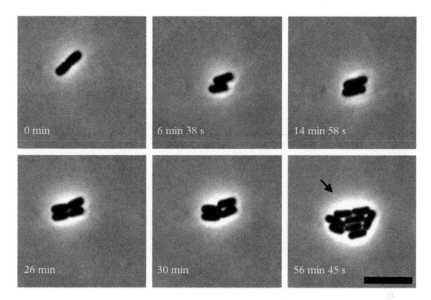

Figure 1.8 *Replicating* E. coli *bacteria. Scale bar represents 5* μm.
(Figure from Su et al., 2012, CC BY license).

a trillion of these bacteria in our bodies.[46] As we will discuss in Chapter II, life is much faster at the nano scale. Although molecules are much heavier than electrons, they are also much lighter than even the smallest macroscopic cogwheels.

Microbes replicate by growing and dividing, as shown in Figure 1.8, rather than via constructing their copies by a robotic arm. As a bacterium grows and divides, all its components are active (alive) all the time, rather than being made still and then set alive by a spark. Thus, the mechanism of microbial replication is somewhat different from that of von Neumann's kinematic replicator. Nevertheless, as we will see in Chapters III and IV, there are many deep similarities. Although microbes are hardly autonomous in the broad sense of this word, as will be discussed later, at least some of them can be viewed as autonomous in accordance with the definition given earlier.

In Chapter II we explore properties of the physical world on the molecular scale, and switch our discussion from *self-replication* to *self-organization*. Then, in Chapters III and IV, will show how a combination of von Neumann's logic of self-replication and molecular processes of self-organization provides the world with microscopic scale self-replicating systems.

II

Self-organizing Molecules

If, in some cataclysm, all of scientific knowledge were to be destroyed, and only one sentence passed on to the next generations of creatures, what statement would contain the most information in the fewest words? I believe it is the atomic hypothesis (or atomic fact, or whatever you wish to call it) that all things are made of atoms—little particles that move around in perpetual motion, attracting each other when they are a little distance apart, but repelling upon being squeezed into one another.

(*The Feynman Lectures on Physics*, Vol. 1, 1963)

Keeping things in order is hard work. Everybody who has a desk, an office, a drawer, or a garden knows this. Chaos seems to be constantly on the attack; whenever matters are left unattended, disorder emerges soon. In his seminal book *What is Life?* Erwin Schrödinger wrote about a 'natural tendency' for disorder to grow and 'living matter' resisting this.[1] When we plant a seed, after some time a green stem emerges, then possibly a flower. Things have gone contrary to this 'natural tendency'—the seed did not dissolve into the soil, instead it produced a structure more organized than what was there before.

To some readers Schrödinger will be known for *Schrödinger's cat*—a cat in a closed box, in a quantum superposition of an alive and a dead state; only when the box is opened the superposition resolves into one or the other. In the mid-1920s, Erwin Schrödinger and Werner Heisenberg independently formulated the first mathematically consistent theories of quantum mechanics that described the 'world' at the scale of atoms where phenomena of being in different states simultaneously occur. Although their formulations looked very different, they were proven to be mathematically equivalent. Heisenberg was awarded the Nobel Prize in Physics in 1932 'for the creation of quantum mechanics, the application of which has, inter alia, led to the discovery of the allotropic forms of hydrogen'. Schrödinger joined him a year later in 1933 'for the formulation of the Schrödinger equation', one of the most famous equations of physics. It was John von Neumann who later proposed the most mathematically rigorous formulation of quantum mechanics unifying the perspectives of Schrödinger and Heisenberg.

Co-discoverers of the structure of DNA Francis Crick and James Watson are among many prominent molecular biologists who have noted that Schrödinger's *What is Life?* had a profound influence on their thinking. Not everybody, however,

Living Computers. Alvis Brazma, Oxford University Press. © Alvis Brazma (2023). DOI: 10.1093/oso/9780192871947.003.0003

has been so positive; for instance, the Nobel laureate Sidney Brenner said that in his youth he had tried to read the Schrödinger's book but had not been able to understand it and later found it 'flawed' in parts.[2] As the formulation of quantum mechanics alone makes Schrödinger one of about a dozen of the most influential scientists of the twentieth century, he would not have had to worry about such criticism. Schrödinger certainly knew how to ask important questions and there is little doubt that his book had a seminal influence on the emerging field of molecular biology. But is the tendency to counter disorder unique to the living world?

We can certainly find examples in the non-living world, where what we intuitively perceive as disorder usually grows. An abandoned ghost town sooner or later turns into rubble; a snowman turns into a puddle when the weather gets warmer; a dead animal decays and disintegrates. But disorder can also grow in the living world. Chaos in a drawer is a human creation. Weeds take over a garden because they are alive and they change and grow, and a dead animal disintegrates due to microbes consuming it.

The converse can happen too: order can emerge out of disorder not only in the living world. In winter, in the right weather conditions, elegant symmetric snowflakes (Figure 2.1) form out of the water droplets, ice particles, and water molecules disorderly dispersed in the atmosphere. Or a frozen mirror-like surface of a lake on a cold winter morning surely is more orderly than the choppy waves a few days earlier. And no matter how hard we try to mix oil and water, after some time the oil will float on the top of the water in an orderly layer. But what actually is order and can we measure it?

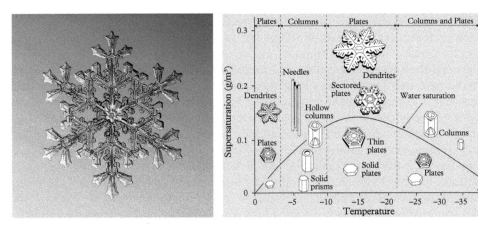

Figure 2.1 *Left. A photograph of a snowflake by Kenneth Libbrecht, Professor of Physics at Caltech. Snowflakes form by self-assembly of water molecules, ice particles, and water droplets present in the atmosphere (see Box 2.2). Right. Snow-crystal morphology diagram showing how their shapes depend on temperature and atmospheric saturation, as proposed by Nakaya (Nakaya, 1954).*
(Left: Photograph from SnowCrystals.com, curtesy of K. Libbrecht. Right: Figure from Libbrecht, 2005.)

Measuring order

Suppose we have a deck of 52 playing cards: clubs, diamonds, hearts, and spades, 13 of each suit. In how many ways can these cards be ordered? We can take each of the 52 cards as the first, each of the remaining 51 as the second, each of the remaining 50 as the third, and so on. Thus, in total there are

$$52 \times 51 \times 50 \times 49 \times \ldots \times 2 \times 1$$

different arrangements. This number is known as the factorial of 52, usually denoted by 52! or for an arbitrary positive integer n, by $n!$. If one bothers to make the calculations or looks up '52!' on the Internet, one will find that 52! = 80658175170943878571660636856403766975289505440883277824000000000000. This is a 68-digit number; it is larger than the estimated number of atoms in the entire Milky Way.

How many of these arrangements are orderly and how can we identify them? We can sort the cards first by suits: clubs, diamonds, hearts, and spades, and then each suit by ranks. Or we can first sort them by ranks and then by suits within each rank. The ranks can be ordered ascending or descending, and the four different suits can be ordered in 4! = 24 different ways. Depending on how ingenious we are in inventing orderly arrangements, we may get several hundred or perhaps thousands of different orderly arrangements. In addition, various partially ordered arrangements can be made from these where some cards are misplaced.

Do orderly arrangements have anything in common? And can the degree of order or disorder be measured? If we have a large number of items, what is common to all orderly arrangements is that they can be described in a way that is more concise than simply enumerating the items one by one. 'Order first clubs, then diamonds, then hearts, then spades, each by increasing ranks' may not be much shorter than enumerating the cards, but if we had, say, a million dollars in $1, $5, and $10 bills, then a description such as '100,000 $1, followed by 20,000 $5, followed by 80,000 $10', is obviously much shorter than enumerating all the 200,000 dollar bills all mixed up in a disorderly fashion. If the objects are arranged in complete disarray, then there is no description of the arrangement that will be much shorter than the complete list. If, on the contrary, some order is imposed on the objects, a shorter description is always possible. The observation holds for partial orderings too: we can give the description of a perfect ordering and then list the exceptions. The more orderly the arrangement, the shorter the description relative to the list.

Measuring the orderliness as the minimal description length of the arrangement was first proposed in early 1960s by the Russian mathematician Andrei Kolmogorov, one of the greatest mathematicians of the twentieth century, and independently by the US mathematician Gregory Chaitin.[3] This measure is known as Kolmogorov–Chaitin complexity, often referred to as *Kolmogorov complexity*. Kolmogorov and Chaitin proved mathematically that if the number of objects is large, then only a small proportion of the arrangements of these objects has a description shorter than their enumeration.

Is there a systematic way—an algorithm—to tell if a given arrangement has a description that will be shorter than the enumeration? If we find a shorter description, then we know it indeed exists, but how long shall we keep looking for it before we decide that it does not exist? It turns out that finding the shortest possible description is one of those problems that are not computable,[4] as discussed in Chapter I. However, the shortest description can be estimated. To roughly estimate the Kolmogorov complexity of a sequence of zeros and ones, all we need to do is to apply a good data compression algorithm to it; for instance, an algorithm that is used to compress digital photos on a computer. Nowadays most digital photography applications compress data automatically without us noticing it. Photos that contain regularities, for instance where most of the picture is of blue sky, usually compress well and thus the respective files are smaller; the pictures that are more irregular, for instance a wildflower meadow, compress poorly. Orderly pixel arrangements compress, disorderly pixel arrangements do not. Overall, the better the compression algorithm, the closer the result is to the true Kolmogorov complexity, though there may be arrangements for which standard compression methods do not work.

From order to disorder

Let us go back to our deck of playing cards. Kolmogorov's theory tells us that only a small proportion of their arrangements will have short descriptions, let assume 1 billion. Let us take a perfectly ordered deck and shuffle it. What is the chance that we will get any one of the partially ordered arrangements after the shuffling? This equals 1 billion divided by the number of all possible arrangements, which is

$$10^9/52! \; < \; 10^9/8 \times 10^{67} = 0.125 \times 10^{-58}$$

On average only one in around every $\sim 10^{59}$ shufflings will produce something that looks like a partially orderly arrangement. Assuming a reasonable shuffling speed, we can expect that on average every 10^{50} years a partially ordered arrangement shows up. The age of the universe is thought to be about 14×10^9 years. We started with perfect order, but shuffling degenerated it into a complete disorder in seconds. If all arrangements are equally likely, and given that the number of objects is large, order does not occur by chance.

In the popular science literature, when discussing the tendency of disorder to grow, the discussion often stops about here. It is declared that the measure of disorder is called *entropy* and the tendency of disorder (entropy) to grow is *the Second Law of Thermodynamics*, sometimes simply referred to as *the Second Law*. The Second Law indeed states that in an *isolated system*—in a system that does not exchange energy or matter with the external environment—the *thermodynamic entropy* cannot decrease and tends to increase until it reaches the maximum at the so-called thermodynamic equilibrium. Thermodynamics, broadly speaking, is science about heat; for instance, the concept of thermodynamic entropy can be used to calculate how much heat can be converted into mechanical work by a steam engine. But the Second Law applies only to isolated systems, while our deck of cards was not isolated—we were shuffling the cards by applying force to the deck and performing mechanical work. Nevertheless, this did not affect our conclusion that the disorder was indeed growing; it did not matter that the system was not isolated. And, indeed, in many situations disorder grows in open systems as well as in isolated ones. At the same time, as we will see in a moment, even in an isolated system, in the right conditions, order can appear out of disorder without contradicting the Second Law.

One reason why popular science literature rarely goes deeper into discussing the Second Law is that the thermodynamic entropy is a rather non-trivial and, in some ways, quite a technical concept. The 'arrangements' have to be counted in a very specific way that establishes a link between the change in a body's internal energy and its temperature.[5] Applying the Second Law to our deck of cards would be a bit of a stretch. As discussed in Box 2.1, there is a relationship between Kolmogorov complexity and thermodynamic entropy, but it is not straightforward. Some may say that the concept of entropy and the Second Law can be generalized and that my interpretation of these here is too narrow. I can accept this criticism if the generalized concepts are defined rigorously. Kolmogorov complexity is a way to do this.

Box 2.1 Kolmogorov complexity and thermodynamic entropy

Consider a drawer divided into sections or 'cells' arranged on a grid, as in drawers where we keep knives and forks, or nuts and bolts. Suppose we have a large number, say, n objects of m different types and m cells ($m \ll n$). If we arrange these objects in an orderly manner—each type in a particular cell—then we can describe this arrangement by recording which type of the object is in which cell. Thus, if we do not care about the number of objects in each cell,[8] the length of our description does not depend on n. On the other hand, if the arrangement is completely disorderly—that is, if any of the bolts or nuts can be found in any of the cells—then we cannot do much better than list all the individual objects assigning each to a particular cell, which gives us a description length proportional to n.

We can apply a similar method even if the drawer does not have separate cells; we can keep similar objects in heaps, which is how one often tries to keep a drawer orderly. Effectively we are introducing an imaginary grid. If we take objects from these heaps and put them back in the drawer without paying attention to where they go, the disorder will emerge soon—there are many more disorderly arrangements.

Until now, we have been talking about static arrangements. Although we were shuffling the cards or moving the objects in the drawer, we measured the orderliness only when the shuffling stopped. We were measuring the orderliness of a *spatial arrangement*. However, we can also measure orderliness in time. Instead of compressing a digital photograph to estimate its complexity (as discussed earlier), we can compress a movie. This is like additionally slicing the 'arrangement' along the time dimension, similarly to how we sliced the space by introducing the grid. In other words, we introduce a grid in four-dimensional space-time.

Brownian motion—small particles jiggling around chaotically in liquid, discussed in more detail later in this chapter—is an example of disorder in time and in space.

continued

Box 2.1 *continued*

In liquid, molecules also move disorderly (or chaotically)—if we could film them, it would be hard to compress the movie. The movement of people in a station after the platform has been announced will be relatively orderly. Order in time means the orderliness of a process. If a process develops according to a plan (an algorithm), we can say it is orderly.

If we were measuring the time-space disorder using Kolmogorov's approach at the resolution of individual molecules, we would be getting closer to measuring the thermodynamic entropy. However, to do this in a way that links energy and temperature, we need to introduce the imaginary grid in a rather specific way, and moreover, we need to account in the calculations for different molecular masses.

As long as the molecules all interact the same way with each other, the thermodynamic entropy works well as an intuitive measure of disorderliness. However, if different molecules interact with each other in different ways, the link between the thermodynamic entropy and our intuitive perception of disorderliness is broken, order can emerge even when the thermodynamic entropy is increasing.

From disorder to order

In our demonstration of why disorder grows, we made one important assumption—we assumed that all arrangements were equally likely. Is this always true? Suppose we have cards of just two types—red and black. If they are all otherwise equal and we shuffle them well, then using similar arguments as above we can show that both will be equally distributed in the deck. Regardless of the position in the deck, the probability that a red card is next to another red one is exactly the same as a red one is next to a black one and the same is true for permutations of more than two cards.

Now suppose the cards are not fair. Suppose that when two red cards come into contact, they tend to stick to each other but not to the black ones. More force has to be applied to separate the red cards from each other than to separate a red from a black or two black cards. When we shuffle such cards, it is quite likely that after some time the red cards will assemble in clusters. The size of the clusters will depend on the relative forces by which the cards stick together and the power we use when shuffling the cards, but in many cases a partial order will emerge: to describe such an arrangement, we only need to note the size and the order of the clusters, rather than list each card. Such a process spontaneous of emergence of (partial) order is an example of *self-organization* or *self-assembly*.[6]

In the described example, different types of objects interact with each other in different ways. Our red cards interact with each other by sticking together, but not sticking to the black ones. In principle, we could have any number of different types of objects that interact with each other by attracting or repulsing with different strength. Once we have such interactions, no longer are all arrangements equally likely, and consequently, order can emerge out of disorder spontaneously. It is not entirely obvious how to create such interacting cards, but as we will discuss in a moment, such complex interactions do indeed happen between different atoms and molecules. For instance, this is why water and oil do not mix.

Not only do interactions between the type of participating objects matter but interactions between the objects and an external force field, such as Earth's gravity or electromagnetic field, matter too. One example of such self-organization is the so-called *muesli effect* (or Brazil nut effect).[7] Different components such as fruits, nuts, and oats sometimes self-arrange predominantly in layers, very often the larger pieces are on the top. This is because during transportation there is vibration and thus shuffling of the contents of the pack. If the pack has a fixed orientation in the Earth's gravity field, different components of muesli tend to organize in layers.

A different example of self-organization observable on a macroscopic scale is the Bénard rolls or Bénard cells, shown in Figure 2.2. A honeycomb-like pattern sometimes appears on the surface of a coffee pot when it is being slowly heated from the bottom. Although this pattern looks quite static, it is in fact dynamic; if we managed to introduce coloured particles, we would see the circulation—we would observe regular order both in space and in time.

There are similarities as well as differences between our 'sticky cards' discussed earlier, the muesli effect, and the Bénard cells. Although in all examples energy needs to be pumped through the system to achieve the effect of self-organization, in the first two examples, after the energy source is switched off, the system 'freezes' in its organized state. On the contrary, in the third example, the energy constantly needs to be pumped for the pattern to be maintained. This is why the first two examples are sometimes

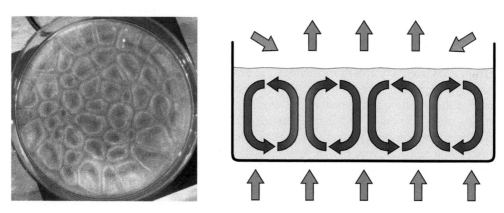

Figure 2.2 *Left. Dynamic, honeycomb-like patterns, known as Bénard rolls, in a coffee pot. **Right.** Explanation of Bénard rolls. When liquid is heated from the bottom, it become less dense, and therefore moves upwards. On the surface the liquid dissipates heat, cools down, becomes denser, and therefore moves back to the bottom. Usually the movement is chaotic, however occasionally an orderly dynamic honeycomb-like pattern emerges and can be relatively stable.*

(Left: Figure courtesy of J. Niemeyer, University of Göttingen. Right: Figure from Wikimedia by I, Eyrian, CC BY SA licence.)

referred to as *self-assembly*, while the third example is referred to as *self-organization*. Both phenomena play a role in biological systems.

The self-organizational phenomena most relevant to our discussion are those that have several different relatively stable states that can switch from one to another when there is a change in the environment. In a London train station, a few minutes before a train departs, the passengers will be organized around the notice board. As soon as the platform announcement appears on the board, the passengers reorganize in a column flowing through the ticket-checking machines towards the platform and then disappear in the train.

In the rest of this chapter, we will look at the self-organization in the nano-scale world of atoms and molecules, but before we do this, let me make use of our playing-card model one more time. Suppose we have just five cards of ranks 1, 2, 3, 4, and 5, all the same suits. They can be arranged in 5! = 125 different ways. Two of the arrangements, 1, 2, 3, 4, 5 and 5, 4, 3, 2, 1, are clearly orderly. We shuffle the cards as before. Assuming that one can shuffle a deck in 10 seconds and, given that two in every 125 arrangements are perfectly orderly, if we keep shuffling, an orderly arrangement will show up on average every 10 minutes. Such events can be described as *fluctuations*—the deviation from the expected disorderly arrangement happening by chance. Such fluctuations are likely if the number of objects is relatively small.

Molecular order of crystal lattices

Is there a limit to miniaturization? What is the size of the smallest possible part of a mechanical device, be it a toy, a fine watch, or a self-reproducing robot? How small the smallest possible cogwheel, a nut, or a bolt can be made? How many times can a piece of metal be split in half? The idea of the smallest, indivisible part of a matter, an 'atom' (from the Greek ἄτομοσ or indivisible), can be traced back to the Greek philosophers Leucippus and Democritus about 400 BC; however, the existence of atoms was proven only at the beginning of the twentieth century. One of the most important contributors to this proof was made by Albert Einstein, who in 1905 used atomic theory to explain Brownian motion. His predictions were confirmed experimentally three years later by the French physicist Jean Perrin. Einstein's theory also provided the means to estimate the mass of an atom, which turned out to be in the order of 10^{-24} gram.

Many of us will know from school that an atom can be roughly described as a tiny object consisting of a positively charged nucleus surrounded by a 'cloud' of negatively charged electrons (thus, an atom is not really indivisible). This atomic structure was first proposed and experimentally confirmed by the New Zealand-born British physicist and Nobel laureate Ernest Rutherford in 1911. There are about 80 types of stable atoms, the so-called *chemical elements*, found in nature, each defined by the number of protons, positively charged elementary particles, in its nucleus. The smallest element is hydrogen, which has only one proton and a cloud consisting of one electron. It is not easy to imagine how one particle can form a cloud, but this is where the Schrodinger's cat analogy helps—an electron is both a particle and a wave. For an atom to stay electrically neutral, the number of electrons in its 'cloud' has to match the number of protons in the nucleus, otherwise the atom becomes a positively or negatively charged ion. Less than half the 80 elements, 35 to be precise, can be found in known living organisms; moreover, only six of them—carbon, hydrogen, nitrogen, oxygen, and sulphur—contribute more than 1% to the mass of the biosphere. Although some atoms, such as helium, lead a solitary life, most atoms tend to combine in relatively stable multi-atom arrangements—*molecules*. For instance, two hydrogen atoms can combine in a pair to make a hydrogen molecule H_2, or with oxygen, to make a water molecule H_2O (Figure 2.3). Atoms in these molecules are kept together by so-called *covalent bonds*, resulting from some of the electrons being shared between the atoms.

But atoms can also assemble in structures of macroscopic size—crystal lattices. For instance, sodium and chlorine atoms can arrange in lattices forming salt crystals, shown in Figure 2.4. A macroscopic size crystal, say around one cubic millimetre in size, will contain at least some 10^{20} atoms. Some atoms can arrange in several different structures, for instance, carbon atoms arrange in three-dimensional diamond crystals, or in two-dimensional graphene planes, as shown in Figure 2.5. The distances between the atoms in such a lattice can provide an estimate of the size of the atom, which turns out to be on the order of a tenth of a nanometre (0.1 nm $= 10^{-7}$ mm). Thus, we will need to align 10 million atoms on a string to obtain about a millimetre-long segment of a line. In a diamond or in graphene, carbon atoms are bound to each other by covalent bonds, thus a diamond crystal can be viewed as one giant covalent molecule—a one-carat diamond is a molecule of about 10^{22} carbon atoms.

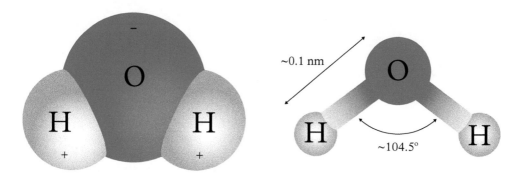

Figure 2.3 *Two different models of a water molecule: space-fill model (**left**) and ball-stick model (**right**). Atoms or molecules should not be thought as hard objects; they are more like clouds without a well-defined size. This is a consequence of nature at sub-nanometre scale, which more precisely can be described by quantum mechanics. Different models emphasize different molecular features. As noted by the statistician George Box 'all models are wrong, but some are useful'. A water molecule is an electric dipole—the negatively charged electron cloud is slightly shifted toward the oxygen atom (which tends to attract electrons) and, as a consequence, the hydrogen atom protons get partially 'exposed'. This has important implications for how water molecules interact with each other and with other molecules.* (Figure adapted from Wikipedia commons.)

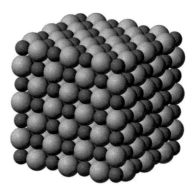

Figure 2.4 *The crystal lattice of table salt. The smaller balls represent sodium atoms, the larger ones represent chlorine. Strictly speaking, these balls represent* ions: *sodium atoms in the crystal have each lost an electron, while chlorine atoms have gained them. The attraction between the positive and negative ions holds the crystal together. The distance between ions in this crystal is around ~0.25 nm. The size of the block in the figure would be about 1.5 × 1.5 × 1.5 nm.*
(Adapted from Wikimedia commons.)

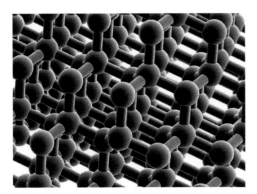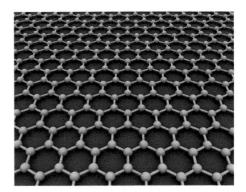

Figure 2.5 *Two different lattices formed by carbon atoms:* **left**—*diamond (three-dimensional);* **right**—*graphene (two-dimensional). The diamond crystal is somewhat more complex than table salt; in a diamond, the unit cell is a cube of eight atoms. Counting the number of atoms in a given volume or the number of atoms aligned along one of the axes can be used to estimate the size of an atom. The distance between the neighbouring carbon atoms in a diamond is ~0.154 nm, in a graphene it is ~0.142 nm. Thus, given that 1 cm = 10^7 nm, somewhere between 10^{23} and 10^{24} carbon atoms will fit in a 1 cm^3-large diamond crystal. This is close to the number of carbon atoms in 12 grams of carbon, which is known as* Avogadro's number *and is approximately 6 × 10^{23}. For comparison, if we look at the sky on a clear night, we will mostly see only the stars in our Milky Way galaxy, which is about 300 billion stars or 3 × 10^{11}. We would need all the stars in the universe to equal the number of atoms in a 1 cm^3-sized solid object.*
(Diamond model from Sciencephotos, graphene adapted from Wikimedia Commons.)

Atomic arrangements can be hierarchical—not only individual atoms can form crystals, molecules can too. An example of this is water ice, formed by water molecules arranging in an orderly lattices, as shown in Figure 2.6. In an ice crystal, water molecules are kept together by so-called *hydrogen bonds*, which form because in some orientations water molecules attract each other. Hydrogen bonds are on the order of 100 times weaker than covalent bonds. This rather different strength of different bons is why ice melts at 0 °C, while diamond at 3550 °C.

Finally, crystals can also be one-dimensional: molecules can also assemble in one-dimensional regular structures—*polymers* (though not everybody will accept that a polymer is a one-dimensional crystal). One such polymer is polyethylene, shown in Figure 2.7. Some of the most important building blocks of life, for instance proteins and DNA, are polymers.

Arguably, crystal lattices are the most orderly structures that exist in nature, not only in an intuitive sense but also according to Kolmogorov's definition. To describe a

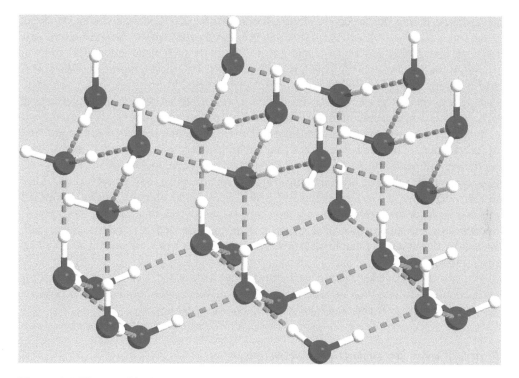

Figure 2.6 *The crystal lattice water ice. The dotted blue lines show hydrogen bonds that form between the partly negative oxygen atoms in water molecules and partly positive hydrogen atoms (Figure 2.3). While these are weak bonds, nonetheless ice is a crystal.*

(Adapted from Wikimedia, author Adam Rędzikowski, CC BY SA 3.0 licence.)

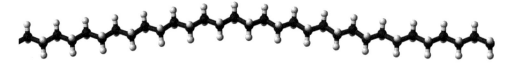

Figure 2.7 *A model of a long polyethylene polymer molecule. This polymer is a sequence of simple ethylene molecules (monomers), each consisting of a carbon atom and two hydrogen atoms (CH_2) linked in a 'chain'.*

crystal lattice, all we only need to describe one *unit-cell*, a cube of four sodium and four chlorine atoms in a salt crystal, or a cube of eight atoms in a diamond, plus the crystal dimensions. Given the number of atoms in a macroscopic size crystal, this is much shorter than enumerating all the atoms and their coordinates. But this also implies that not much information can be encoded in a pure crystal: an object cannot encode more information than the amount of information needed to describe it. If, however, we could replace an arbitrary atom in the lattice with a different atom which would nevertheless fit into the lattice, we could encode information. For instance, in graphene, if we could replace arbitrary carbon atoms (coloured grey) with a different atom (say, red), we could write any text we wanted. It is not easy to replace individual atoms in a crystal; the structure of the lattice is defined by the chemical properties of its atoms and different atoms have different chemical properties. Nevertheless, it is possible to encode information in a crystal by introducing 'impurities' in otherwise regular crystal lattices. This is how semiconductor chips are made.

One could ask what is the minimum number of atoms that is needed to encode 1 *bit* of information? This, however, is not as a straightforward a question as it may sound. First, the minimal number of atoms will depend on how permanently we want to encode this information; that is, on how long the information must stay uncorrupted. This, in turn, depends on the temperature at which we want to operate. Second, as noted in Chapter I, for an arrangement of objects to represent information, an interpreter is needed, and consequently we need to consider how the encoded information can be read. As will be discussed in Chapter III, fewer than 50 atoms per 1 *bit* can be achieved by using DNA.

But how do such orderly atomic structures form? One way to look at this is that atoms or molecules self-assemble in orderly structures driven by the tendency of a physical system to minimize its *potential energy*, as will be discussed next.

Energy and its equal partitioning

Even those who have not studied physics are likely to have some intuitive understanding of the concept of *energy*. To ride a bicycle we need energy, first to accelerate, and then to counter the friction the wheels make with the road. We obtain this energy from the food we eat. To ride uphill we need more energy, but when going downhill we get some of the 'invested' energy back. But what is energy? As with many fundamental concepts, we do not really know. However, we do know that energy is *conserved*. Different modes of energy—the *kinetic energy* of motion, *potential energy* of a force field, chemical energy, thermal energy, electrical energy, and other forms can change from one into another, but the total stays the same. As Richard Feynman's puts it:

> There is a fact, or if you wish, a law, governing all natural phenomena that are known to date. There is no known exception to this law—it is exact so far as we know. The law is called the conservation of energy. It states that there is a certain quantity, which we call energy, that does not change in the manifold changes which nature undergoes. That is a most abstract idea, because it is a mathematical principle; it says that there is a numerical quantity which does not change when something happens. It is not a description of a mechanism, or anything concrete; it is just a strange fact that we can calculate some number and when we finish watching nature go through her tricks and calculate the number again, it is the same.[9]

A pendulum shown in Figure 2.8 (left) is a textbook example demonstrating how energy is conserved in transition between the different modes. If there were no friction, once set in motion, the pendulum would keep swinging forever, converting its potential energy into kinetic energy and back with each swing. However, in the presence of friction, the pendulum gradually transfers—*dissipates*—its energy to the molecules of the surrounding environment and eventually stops. The pendulum will stop at the lowest point, at the point where the potential energy is at its minimum. In *dissipative systems potential*

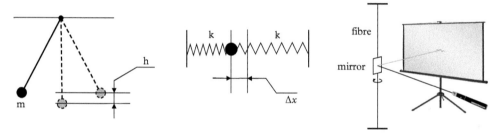

Figure 2.8 *Different forms of energy and transitions between them.* **Left.** *When a pendulum is moved to one side, it acquires potential energy* $E_p = mgh$*, where* m *is the mass of the object at the end of the pendulum,* $g \approx 9.81\ m/s^2$*, and* h *is its height above the lowest point. After the pendulum is released, it starts accelerating as it moves towards the lowest point. As it is losing its potential energy, it is acquiring kinetic energy* $E_k = mv^2/2$*, where* v *is velocity at a given point. The sum of the potential and kinetic energy will remain constant, unless some of the energy is transferred to other bodies, for instance, to the molecules of the surrounding air.* **Middle.** *A spring oscillator. When the weight held by the two springs is displaced from the centre position by* Δx*, the springs stretches/compresses, and the system acquires potential energy* $2k\Delta x$*, where* k *characterizes the stiffness of the spring. Once the displaced mass is released, it starts oscillating around the mid-point, the total energy remaining constant. Atoms in a crystal lattice can be viewed as held together by springs. At temperatures above absolute zero they are vibrating as in an oscillator.* **Right.** *A rotating oscillator. A small mirror is attached to a stretched fibre. By aiming a ray of light on the mirror and projecting it on a screen, the slightest movements of the mirror can be detected. If we wind the mirror, say, several turns clockwise, the fibre will gets stretched. After the mirror is released, it will start rotating in the opposite direction (counterclockwise), then back again (clockwise), and so on, wobbling with a decreasing amplitude. Will it ever stop completely? This depends on the temperature of the surrounding environment (see section Free energy).*

energy tends to the minimum—this is a law of physics. This tendency of dissipative systems to minimize their potential energy can be viewed as the 'driver' of *self-assembly* and *self-organization*, such as the muesli effect. In *open systems*, which exchange energy with the environment, energy can be 'pumped' into or, strictly speaking, pumped through the system, and as long as the energy is pumped, the system does not settle. An example of this is Bénard rolls.

In this example, we talked about the potential energy in the Earth's gravitational field, but we can talk about the potential energy in any force field, for instance in the field of molecular forces. Molecular forces can be compared to springs, which can either attract or repulse the interacting objects (Figure 2.8, middle). Potential *energy barriers* caused by molecular forces are what holds atoms together in molecules or in crystal lattices (see Figure 2.9).

On the molecular scale, gravity plays little role; the forces of molecular interactions are many orders of magnitude stronger. Gravity is always attractive, molecular forces can be attractive or repulsive. This is one of the reasons why on the molecular scale, the phenomena of self-organization are more pronounced and more interesting. If gravity is neutralized, for instance, if particles are placed on air-table—an upside-down hovercraft[10] shown in Figure 2.10—different self-organizational phenomena emerge.

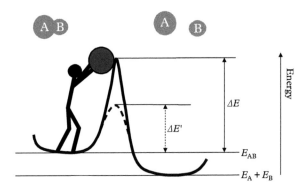

Figure 2.9 *The concept of* potential energy barrier. *Trying to split a molecule AB into two component molecules (or atoms) A and B can be viewed as trying to roll a large heavy ball over a fence or a dike: initially energy ΔE needs to be 'invested' to lift the ball, but once the ball is on the top of the barrier, it can roll over to the other side. By analogy, when parts of a molecule (for instance, atoms A and B in molecule AB) are pulled apart, initially there is resistance, but if enough force is applied, the components may break free. Splitting or joining, or more generally, recombining of atoms or molecules in different combinations, is known as a* chemical reaction. *In our fence/ball model, if the fence is high enough, the ball will tend to stay on one side for a while, even if it is being kicked around. If it is kicked around forcefully enough, sooner or later it will nevertheless jump to the other side. By analogy, at high enough temperatures the molecules will occasionally 'jump' over the energy barrier, and join, split, or recombine. Catalysts of chemical reactions, including biological catalysts—enzymes (see Chapter III), lower these barriers (as represented by broken lines: ΔE' < ΔE), allowing the chemical reactions to occur more easily.*

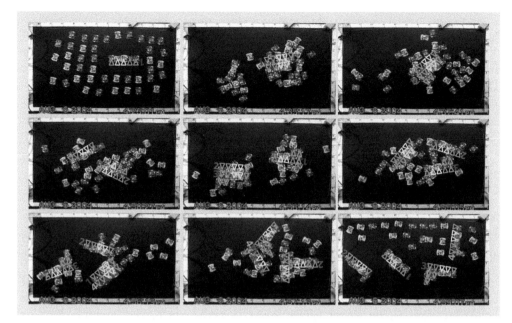

Figure 2.10 *Template-guided replication of information via self-organization on an air-table. Two types of particles (yellow and green) with magnetic interfaces are scattered on an air-table. Given a template—a sequence of such particles linked in a particular order: green, green, yellow, yellow, green (encoding a string of 5 bits), the free-floating particles self-assemble on this template to make new copies of the sequence. The figure shows repeated self-replication of a 5-bit string. Compare this to Figure 1.3c.*
(From Griffith et al., 2005).

An air-table not only counters gravity, it also keeps pumping energy through the system, constantly shuffling the particles floating on it. Atoms and molecules do not need an air-table—at temperatures above the absolute zero, above 0 degrees Kelvin, which is approximately $-273°$ Celsius ($0°K \approx -273°C$), molecules are always in motion. The higher the temperature, the faster they move (or strictly speaking, the faster they move, the higher the temperature).[11] At room temperature, the mean velocity of a molecule of the mass approximately that of a water molecule is around 500–600 metres per second. In gas, molecules move for some distance before colliding with each other; in liquid, molecules can only jiggle. In crystals, the atoms can be viewed as linked to each other in a lattice by springs and they are constantly vibrating.

Once we stop shuffling, a deck of cards freezes; molecules in temperatures above $0°K$ never stop. If we drop ink into water, the ink spreads throughout the entire volume evenly, just like the red cards introduced into constantly shuffled deck of black cards. *Mixing entropy* is the component of thermodynamic entropy that reflects the molecular 'arrangements' when more than one type of particles is present. If the molecules all interact with each other the same way, then at equilibrium they are well mixed, as shown in Figure 2.13 (left). If different molecules interact in different ways, then just like interacting playing cards, self-organizational phenomena can emerge.

In 1827 the Scottish botanist Robert Brown was investigating microscopic life when he observed that little particles of plant pollens were jiggling around in water chaotically. He figured out correctly that this was nothing to do with life, the same jiggling movement of microscopic particles could be observed inside a piece of amber that had trapped the water for millions of years. As Richard Feynman stated in his lectures:

> *we can understand it [Brownian motion] qualitatively by thinking of a great push ball on a playing field, seen from a great distance, with a lot of people underneath, all pushing the ball in various directions. We cannot see the people because we imagine that we are too far away, but we can see the ball, and we notice that it moves around rather irregularly.*[12]

The people are the molecules, the ball is the pollen. In liquid, not only do the molecules jiggle but all particles immersed in it. At *thermal equilibrium*, when all parts of the system are at the same temperature, the kinetic energy is distributed among all particles. This is the law of *equipartitioning* or equal partitioning of energy—energy is equally distributed across all *degrees of freedom*.[13] And given that every particle takes some of the degrees of freedom and that a particle's kinetic energy is proportional to its mass: the lighter the particles, the faster it jiggles, the heavier the particle, the slower it is. At room temperature, a typical protein molecule, which consists of many thousands of atoms, jiggles with the speed of about 10–20 metres per second; particles weighting around a microgram—1 millionth part of a gram—will jiggle with average speed of about one-tenth of a millimetre per second. On nanoscale, unless the temperature is close to –273 °C, everything is restless: cogwheels, leavers, or pulleys jiggle, and nanoscale mechanical robots inevitably suffer from a robotic version of Parkinson's disease. Thus, the minimum size of the smallest mechanical parts of a nanorobot is limited not only by the number of atoms needed to give the part its shape, but also by how solid and stable we need it to be at the temperature the robot is supposed to operate. And ultimately this inevitable restlessness is also why the outcome of even the most finely tuned card shuffling or dice throwing machine is unpredictable.

Free energy

As described above, in dissipative systems the potential energy tends to the minimum—the pendulum (Figure 2.8) stops at the lowest point. But how does this square up with the law of equal partitioning of energy? According to this law, if the pendulum is in thermal equilibrium with the surrounding environment, at temperatures above 0 °K, the pendulum cannot stop—it has to retain some kinetic energy. For macroscopic objects their thermal jiggling energy is negligibly small in comparison to their gravitational potential energy, therefore we could ignore the jiggling. But what if the pendulum weighs, say, a fraction of a microgram? Or, as a more practical example, take a rotating oscillator shown on the right panel in Figure 2.8. At temperatures above 0 °K, in thermal equilibrium with the environment, the oscillator will retain some energy and on the micro scale we cannot ignore this.

If the systems temperature is kept constant, the difference between the oscillator's initial and its final energy at the thermal equilibrium is its *free energy*. We could also call this the 'usable potential energy', as this is the part of the initial potential energy that can be converted into mechanical work. Some readers may know from school physics that heat engines, such as steam or internal combustion engines, can never be 100% efficient. This is because unless such an engine operates in an environment at 0 °K temperature, it cannot convert all available thermal energy into mechanical work, only the free energy. At a constant temperature, it is the free energy, rather than the potential energy, that tends to the minimum.

More generally, free energy is a concept that captures the balance between two tendencies of physical matter—the tendency to minimize its potential energy and the tendency to maximize its thermodynamic entropy. If the temperature of the system is kept constant, then these two tendencies are in conflict. On the one hand, the pendulum 'wants' to stop at the lowest point, minimizing its potential energy. On the other hand, the overall entropy of the system 'wants' to grow, as is consistent with the Second Law. Free energy minimization is the balancing act between the two.[14] Mathematically, the free energy of the system can be expressed as its total energy minus the thermodynamic entropy times the temperature: $F = E - TS$, where F is free energy, E is potential energy, T is temperature in Kelvins, and S is entropy. The higher the temperature, the more we have to subtract; if the temperature is 0 °K, we subtract nothing, and thus the free and the potential energies become the same. Note that the free energy component of the total energy is not necessarily conserved—as mechanical work is done, unless the process is reversible, some energy is dissipated and consequently the free energy component decreases. When we pump energy through the system, for instance, to 'drive' Bénard rolls, the free energy is 'pumped in', the rest of the energy is dissipated (and the total energy is 'pumped through').

If there is more than one type of particle in the system, the mixing entropy component too needs to be included in the entropy calculations, which may affect the free energy minimum point. Moreover, the atoms and molecules can chemically react to and can combine or recombine into different arrangements (Figure 2.9), which may further affect the calculations. Thus, there are two factors that determine in which direction a chemical reaction will go: the minimization of the potential (chemical) energy of molecular forces and the maximization of entropy.[15]

Creating order via the Second Law

The process of atoms or molecules self-assembling into a crystal lattice—*crystallization*—is one of the simplest examples of self-organization on the molecular scale, though some would insist on strictly calling this self-assembly. Diamond crystals form at extreme temperatures and pressures; water freezes to form ice crystals at much more moderate conditions. What happens when a bucket of water is left outside in a cold winter night? As the temperature of the surrounding air drops, either because on a clear night the molecules lose their kinetic energy by emitting photons of infrared radiation, or because colder air is moving in, the kinetic energy of the molecules in the atmosphere decreases. The equal partitioning of energy implies that the water molecules in the bucket have to slow down too. At some point, after the water temperature has dropped to 0 °C or below, a fluctuation happens—a sufficient number of slow molecules meet up and arrange themselves in a tiny orderly crystal lattice, forming the so-called crystallization nucleus. Experiments show that in pure water about 300 molecules are needed to kick-start a crystallization nucleus;[16] more often, though, the nucleation starts around particles of dust or around uneven parts of the rim (one could say, around spatial fluctuations). Extremely pure water freezes slowly and can reach temperatures well below 0 °C while still in the liquid state, known as supercooled liquid water. Once a crystallization nucleus has formed, it provides a template for increasing number of molecules to organize around it. A positive feedback loop kicks in—the larger the crystal, faster it grows. This leads to a *phase transition* and the water gradually freezes. The process of a crystal lattice growth is an example of a template-based self-assembly—the existing crystal lattice serves as the template around which other molecules organize.

By self-organizing in an orderly lattice, the molecules try to minimize the system's free energy. The minimum, however, may be a local minimum—it is possible that a different lattice may have even lower overall energy. The same type of atoms or molecules may self-organize in different types of lattices (for example, diamond and graphene) and the particular lattice that forms may depend on the particular fluctuation that kick-started the process.

Is such an emergence of order consistent with the Second Law? Our bucket of water is not an isolated system, it exchanges energy with the surrounding environment, and thus the Second Law does not apply. However, if we put a jug of water inside a large vacuum bottle filled with, say, nitrogen gas, then the Second Law does apply, the total thermal energy of the water and nitrogen remains constant. If initially nitrogen is at a temperature well below 0 °C, the water is cooling while nitrogen is warming. The thermodynamic entropy of the whole isolated system is also increasing, as is consistent with the Second Law. Whether the overall level of order of this isolated system is increasing or decreasing is debatable, but it is indisputable that islands of order—crystal lattices—are forming. Water and nitrogen molecules interact with each other differently—one can think of water molecules attracting each other more than nitrogen molecules, just like our red cards were sticking together more strongly than sticking to the black cards. In the presence of non-uniform molecular interactions, the Second Law and the emergence of order do not contradict each other. Parts of an isolated system can self-organize into increasingly orderly structures, even though the system's entropy is increasing. Such emergence of order indeed can be observed in practice; for instance, crystals forming in oversaturated solution.

An interesting example is the formation of snowflakes. Arguably snowflakes are among the most interesting structures known to form by molecular self-organization in the non-living world (see Box 2.2).

Box 2.2 Snowflakes

There is a popular saying that no two snowflakes are alike, nevertheless, all of them have a strong hexagonal symmetry (Figure 2.1). In the Middle Ages it was thought that the formation of snowflakes, like that of living beings, was guided by divine forces. One of the first scientists who tried to explain snowflakes was Johannes Kepler, best known for formulating the laws of planetary motion in the seventeenth century. He hypothesized that the hexagonal shape of snowflakes is a consequence of the hexagon being the tightest possible packing of spheres. This, however, could not explain the richness of the possible structures and Kepler had to admit a defeat.

Any scientific theory has to be preceded by gathering of data, which for snowflakes became possible in the late-nineteenth and early-twentieth centuries with the invention of photography. This was a hobby of Vermont farmer Wilson Bentley, who assembled an impressive collection of beautiful snowflake pictures. In the 1930s, the Japanese physicist Ukichiro Nakaya found a way to grow snowflakes on a thin rabbit hair. He devised the first snowflake classification—the classic Nakaya snow-crystal morphology chart,[17] shown in Figure 2.1. More recently, Caltech professor Kenneth Libbrecht has brought snowflake research into the twenty-first century. His *Field Guide to Snowflakes*[18] contains many fascinating images showing the enormous diversity of snowflakes in great detail.

Perhaps not surprisingly, it turns out that the predominant hexagonal shape of a snowflake is a consequence of the hexagonal structure of the underlying ice crystal lattice. Here I follow Libbrecht and Gravner in the description of snowflake formation.[19] All three forms of water coexist in saturated clouds below 0 °C: supercooled liquid water, vapour, and ice. Typically, minuscule dust particles that are present in the atmosphere serve as crystallization centres in small water droplets, which then freeze and initiate ice crystal growth. Further growth happens mostly by sublimation of water vapour—transition from the gas state directly to crystal state—onto the initially tiny crystal surface. Up to a size of fractions of a millimetre, such a crystal grows as a relatively flat hexagon, reflecting the lattice of an ice crystal.

So far this is a 'standard' crystal growth process; an important feature to note here is that the hexagonal edges stay straight, and the hexagon grows faster along the six edges than in the thickness. This is because any rugged parts in the edge tend to catch new molecules faster than a flatter surface. Thus, there is a negative feedback loop—the even parts of the edges grow slower than the rough ones, and thus any cavities fill in. However, a change happens once the hexagon reaches a certain size: the corners of the hexagon begin catching new water molecules faster than the edges, and thus, a positive feedback loop kicks in. Spikes, the so-called dendrites, start growing from the six corners. No convincing quantitative explanation based on the laws of physics has yet been found why such a transition happens at a particular size.

The dendrite growth is characterized by so-called branching instability, which often leads to forming of fern-like structures. But what makes all six 'ferns' look so similar? The most widely accepted qualitative explanation is based on the spatial homogeneity of the environmental parameters. As summarized by Gravner: '*Since the atmospheric conditions . . . are nearly constant across the small crystal, the six budding arms all grow out at roughly the same rate. While it grows, the crystal is blown to and from inside the clouds, so the temperature it sees varies randomly with time. . . . And because all six arms see the same conditions at the same time, they all grow about the same way*'.[20]

continued

Box 2.2 *continued*

Although we still do not have a quantitative, testable model, derived from the laws of physics, to explain the shapes of snowflakes, mathematical modelling of snowflake formation has advanced at the beginning of the twenty-first century tremendously. Partly this is helped by the advances of experimental techniques to measure various parameters and observe snowflake formation, and partly due to the advances in computer simulation techniques.[21] Notably, cellular automata, described in Chapter II, have been used to model snowflake formation and growth.[22] However, all these models still contain various speculative elements. Centuries after Keppler, we still do not have a thorough explanation from 'first principles' of why snowflakes are the way they are. What we can say, is that snowflakes form in conditions that are on the boundary between order and disorder in a changing environment, where self-organization can produce interesting phenomena.

Molecular templates and copying of information

As already mentioned, crystals may contain impurities or imperfections, potentially carrying information. Moreover, this information can be copied and thus replicated. We can think about a layer of a crystal lattice as tiles laid on a roof. A layer of tiles of just one colour cannot encode much information, but if we have tiles of two or more colours, we can arrange them to encode a message. This message can be copied by laying another layer of the same colour tiles on top of the existing one, and then 'peeling' the new layer off. Can such a similar template-based copying of information occur via self-organization?

Indeed, there is a class of crystals, known as clay crystals, where copying of impurities, and thus propagation of information, can happen. The main atom forming a clay crystal lattice is silicon. Silicon crystals are used to make computer chips, where impurities are introduced in highly controlled ways to make transistors and other features. In natural clay crystals, silicon is mixed with other atoms, often aluminium. Under specific conditions, these impurities serve as a template and as the crystal grows, they replicate. Moreover, some patterns of impurities replicate faster, some slower, thus effectively competing with each other. As we will discuss in Chapter V, such competition is one of the basic ingredients of life. It has been suggested that clay crystal growth could have played a role in the origin of life,[23] which however is not a generally accepted hypothesis (we will return to this in Chapter VI).

In his book *What is Life?*, Schrödinger referred to hypothetical *aperiodic crystals*, which were the carriers of life's information copied from generation to generation. The terms *crystal* and *aperiodic* may seem contradictory; Sydney Brenner said that this concept of aperiodic crystal was one of the reasons why he could not understand Schrödinger's book.[24] However, it has been since discovered that aperiodic crystals do exist in nature, the 2011 Nobel Prize in Chemistry was awarded to the Israeli scientist Dan Shechtman 'for the discovery of quasicrystals'.

We can also use an aperiodic polymer molecule to encode information (as noted, a polymer can be viewed as a one-dimensional crystal). For instance, take two different types of *monomers*—two different small molecules a sequence of which form the polymer. If one of the monomers represents 0, the other 1, the polymer will encode a *bit-string*. If such polymers could somehow be stack up one on top of one another, this could provide a mechanism for copying information. An example how this could be done in principle is shown in Figure 1.3. And indeed, the deoxyribonucleic acid molecules or DNA, which encode the inherited information of living organisms, enable such a mechanism. As will be discussed in Chapter III, template-based copying of information is at the very basis of life. Self-organization-driven template-guided information copying can also be implemented on a macroscopic scale, as shown in Figure 2.10.

The liquid order in water

All known forms of life need water. Although liquid water may appear a viscose featureless substance, it has quite a complex and interesting molecular structure, which is essential to the existence of life as we know it. As shown in Figure 2.3, a water molecule is an electric dipole. In particular orientations relative to each other, such dipole molecules attract and form hydrogen bonds. Below 0 °C, the molecules stay arranged in the hexagonal ice crystal lattice (Figure 2.6). At room temperature, however, molecules move around, and a bond lasts only $\sim 10^{-11}$ seconds. On average, at any given moment each water molecule is bonded to three-and-a-half neighbouring molecules. Thus, the structure of water is highly dynamic but still, importantly, there is a structure.

If 'foreign' particles, such as fatty-acid molecules are introduced into water, interesting self-organizational phenomena can occur. The reason why oil and water do not mix is that fatty-acid molecules making up oil cannot form hydrogen bonds with water. The water molecules, like our sticky red cards, stick to each other, squeezing out the fatty-acid molecules. If a water molecule happens to be next to a fatty-acid molecule, the potential hydrogen bonds on one side will remain 'unused', and as a result, the free energy of the molecules will increase. The molecules 'do not like' this. To minimize the total free energy, some of the water molecules 'sacrifice themselves' by forming a layer around such a *hydrophobic*—water-avoiding—foreign particle.[25] To minimize free energy the fatty-acid molecules clump together, and as oil is lighter than water, a layer of oil forms on top.

Even more interesting phenomena occur when the 'foreign' molecules have elongated structures with one end being *hydrophilic*—water-loving—and the other end being hydrophobic, as shown in Figure 2.11. Such bio-polar molecules, known as *amphiphiles*, can 'cooperate' with water, for instance by forming double layer arrangements with the hydrophilic ends facing water and hydrophobic ends facing each other. Moreover, the resulting bilayer membrane can form a *vesicle*, separating inside and outside environments. Thus, formation of a membrane minimizes the free energy of the system. If there are other lipid molecules present in the surrounding water, the vesicle can grow by absorbing the free lipids dispersed in the water. Distorted shapes appear, as has been demonstrated in experiments[26] shown in Figure 2.12. Such growing vesicles may be

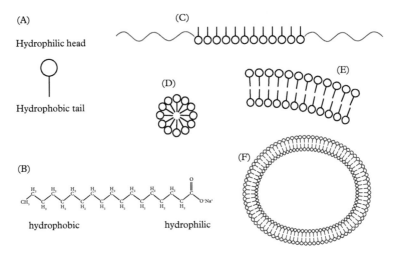

(A)

Hydrophilic head

Hydrophobic tail

(B)

hydrophobic   hydrophilic

(C)

(D)

(E)

(F)

Figure 2.11 (**A**) *Amphiphilic molecules with a hydrophilic (water-attracting) head and a hydrophobic (water-repulsing) tail.* (**B**) *One of the simplest amphiphile molecules is sodium stearate—ordinary soap. When in water, to minimize the total free energy, amphiphils self-organize so that the tails avoid water, either in a monolayer membrane floating on top of water* (**C**), *or in micelles* (**D**), *or in bilayer membranes* (**E**). *Bilayer membranes can form closed bubbles—so-called liposomes—vesicles where membrane serve as a barrier between inside and outside* (**F**).

Figure 2.12 *The shapes that result from proportional inclusion of molecules in the membrane and inside the vesicle. If both grow at the same rate, the shape of the vesicle cannot stay spherical and may snap.*
(From Lipowsky, 1991.)

prone to stochastic snapping and thus, in some sense, to self-replication. As we will see in later chapters, possibly this was an essential mechanism in early life on Earth, and even in some simple forms of the present life.

To finish the discussion of self-organization in the non-living world, let me describe one more example important to life. Every bacterial cell contains a genome, which is a long polymer molecule, which is copied when the cell replicates. Thus, at some point the two such long polymers will be present in the cell. For the cell to complete the division, the polymers need to be separated from each other. It turns out that under certain conditions, such separation can happen via free energy minimising self-organization.

A polymer molecule can be viewed as a sequence of springs attached to each other at particular 'preferred' angles. The preferred angle is defined by the molecular forces holding the particular monomers together. For simplicity let us think about this in two dimensions—as if the polymer was laid out on a flat surface. We can visualize such a polymer as a kinked line. The polymer will try to minimize its potential energy by trying to keep the 'preferred' angle between the monomers, but if it is enclosed in confined space, this may not be possible—some angles have to deviate from the preferred one. At a temperature above 0 °K the angles will keep changing dynamically.

We are particularly interested in long circular polymer molecules which, as will become clear in Chapters III and IV, are what bacterial genomes are like. What happens if two such polymer loops are placed in a small, elongated enclosure, as shown in Figure 2.13 Right? Will they predominantly stay entangled together, or will they try to separate and go to different parts of the enclosure? It turns out that, in contrast to what happens to a mixture of free small molecules, shown in Figure 2.13 Left, to minimize free energy, the polymers often separate from each other.

This may appear counterintuitive but it is not very hard to see that, at least in two-dimensional case, there are more configurations where two kinked lines are separated than where they are kept together but are not crossing each other. This turns out to be true also in three-dimensional case—the entropy of the system is higher if the polymers are separated, and as a consequence, the polymers try to separate. Computer simulations and experiments confirm this.[27] Here again we see that in presence of complex interactions, the increase in thermodynamic entropy can lead to increasing order; a system consisting of two separated polymers can be viewed as more orderly than the polymers tangled together.

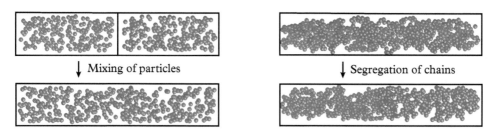

Figure 2.13 *Mixing and segregating.* **Left.** *There are many more arrangements in which two types of molecules can be distributed in space evenly than if each type is kept separate, just like red and black cards in the example described in the text. This is reflected in the concept of mixing entropy, which is a component of thermodynamic entropy and tends to increase.* **Right.** *Long polymer molecules can adopt various conformations (for instance, by bending in different ways). Unlike in the case of small molecules shown above, there are more configurations possible when the polymers are separated and located into different parts of the enclosure, than if they are 'tangled' together (see text). Therefore, when two (long) polymers are confined in small space, they tend to segregate.*
(Adapted from Jun and Wright, 2010.)

One of the main messages to take home from this chapter is that there are clear examples where molecules self-organize and self-assemble in orderly structures 'driven by' the laws of physics. Chapter III will discuss how living systems combine inherited information and self-organization, to produce a large variety of complex molecular structures and processes.

III

Informed Self-organization

The phosphate-sugar backbone of our model is completely regular, but any sequence of pairs of bases can fit into the structure. It follows that in a long molecule many different permutations are possible, and it therefore seems likely that the precise sequence of the bases is the code which carries genetical information.

(James Watson and Francis Crick, 1953)

Biology is essentially (very low energy) physics with computations.

(Sydney Brenner, 2012a)

Many different structures form and processes occur in nature driven by self-assembly and self-organization, as described in Chapter II. Some are rather transient, such as snowflakes forming in the atmosphere and dissolving on the ground, some durable, such as crystals growing in caves. But there is one class of phenomena that stands out—the structures and processes of the living world. The diversity that we can find in a rainforest, or in a coral reef, will surpass anything that we are likely to find in a similar volume of space and time in the non-living world. Where does this concentration of complex forms and patterns come from? Can this be explained by self-organization?

The answer to the last question possibly depends on the perspective we take: life on Earth has been evolving for billions of years, and perhaps evolution too can be regarded as a process of self-organization. But what if we take a shorter-term perspective, like the timescale of a human generation? Can we view the formation of patterns on the wings of a butterfly as self-organization?

The richness of patterns in the living world surpasses anything observed in the non-living world, and this is also true if we zoom in down to the scale of molecules. As we will see shortly, one of the most important classes of molecules of life—*proteins*—can be as rich and fascinating in shape and variety as the gargoyles of great cathedrals. The shapes of proteins form by self-organization, consistent with the same laws of physics as everything around us, however no class of molecule in the non-living world can match proteins in diversity and complexity. What do the molecules in the living world have that the ones in the non-living world do not?

Living Computers. Alvis Brazma, Oxford University Press. © Alvis Brazma (2023). DOI: 10.1093/oso/9780192871947.003.0004

The answer is information. Physics is the same but living systems benefit from information they inherit. Even though, as we will see in a moment, proteins assume their shapes through self-organization, they are beneficiaries of information passed on from generation to generation. The patterns on the wings of a butterfly form by self-organization of molecules defined and guided by the information inherited from the butterfly's ancestors.

The observation that living organisms inherit traits of their predecessors is probably as old as humanity; the science of inheritance—*genetics*—is much younger. Arguably, it began in the middle of the nineteenth century, when in a series of ingenious experiments the Moravian Augustinian friar and abbot Gregor Mendel demonstrated that pea plants inherit features from their previous generations following clear rules. The term *gene* was first introduced another half a century later by the Danish pharmacist Wilhelm Johannsen to describe the 'Mendelian hereditary units'.[1]

A couple of decades later still, in a separate development, three American chemists, James Sumner, John Northrop, and Wendell Stanley, showed that *enzymes*—the biological catalysts at the basis of the unique chemistry of living systems—were nothing other than molecules. This was a deadly blow to vitalism—the claim that yet unknown 'vital forces' were needed for life. However, vitalism did not fully go away for quite some time yet; for instance, the US geneticist George Williams found it necessary to argue against vitalism in his book about natural selection[2] as late as 1992. (Even today, 'neo-vitalist' arguments occasionally surface in discussions about human intelligence and whether information processing can explain it.) For their discovery, Sumner, Northrop, and Stanley shared the Nobel Prize in Chemistry in 1946.

In 1941, two American geneticists, George Beadle and Edward Tatum, demonstrated that there was a link between enzymes and genes. The hypothesis 'one gene, one enzyme' was adopted, which was later generalized to 'one gene, one protein' (still a simplification). Beadle and Tatum received the Nobel Prize in Physiology or Medicine in 1958. But what is the physical nature of a gene and how do genes encode information about the respective proteins?

These were some of the questions that Schrödinger asked in his book *What is Life?* in 1944. Schrödinger did not use the word 'information', he talked about the life's 'code-script'. The answer to the Schrödinger's questions largely came from the synthesis of two lines of investigations, an inspiring account of which can be found in the book by Ann Roller, *Discovering the Basis of Life*.[3] The first was viruses—arguably the simplest biological system that carry genetic information. In 1945, the German-American physicist Max Delbrück, along with Italian Salvador Luria and American Alfred Hershey, demonstrated that a virus stripped of everything but DNA can still be infectious. This strongly indicated that it was DNA that caried viral information. The three scientists shared the Nobel Prize in Physiology and Medicine in 1969 'for their discoveries concerning the replication mechanism and the genetic structure of viruses'.

The second major line of investigation was X-ray crystallography. At the beginning of the twentieth century, the British physicist Lawrence Bragg reconstructed shapes of crystallized molecules from the diffraction patterns that formed when X-rays were passed through their crystals. In 1915, at the age of 25, he was awarded the Nobel

Prize in Physics, which he shared with his father Sir Henry Bragg. In 1950s, the US chemist Linus Pauling used this method to uncover the structure of the first simple protein molecule—keratin. In other words, he proposed a model of what this molecule would look like, if magnified to a size visible by eye. Knowing how a part of a mechanism looks like is the first step towards understanding what the function of this part in the mechanism is. A few years later, two scientists, Max Perutz, and John Kendrew, working in the Cavendish Laboratories in Cambridge in the United Kingdom (led by Sir Lawrence Bragg), succeeded in reconstructing the structures of more complex proteins—haemoglobin and myoglobin. For this discovery all three were awarded Nobel Prizes. Later, in 1974, Kendrew became the first Director of the European Molecular Biology Laboratory (EMBL) in Heidelberg, Germany.

The breakthrough in the pursuit to find the carrier of genetic information came in 1953, when James Watson and Francis Crick, working in Cambridge alongside Perutz and Kendrew, proposed a molecular model of DNA. The double helical structure (Figure 3.7) immediately provided a clue as to how DNA could encode information and how this information could be copied and passed on from generation to generation. It took another decade, however, to understand the principles how information in DNA is interpreted by a living cell.

In the context of genetics, the term *information* first seems to have appeared in 1953 in the publication by Watson and Crick, quoted at the beginning of this chapter.[4] The science of information and its processing—computer science—was another new discipline emerging during the first half of the twentieth century[5] alongside molecular biology. The development of automated computing received a boost from the needs of the allies of the Second World War, in particular in Britain and the United States. Alan Turing was working at Bletchley Park, the top-secret home of the Second World War codebreakers in Britain, working to break the German secret code—arguably the most archetypal information processing task of them all. Meanwhile, Norbert Wiener, whose name later became synonymous with the word *cybernetics*—the science of systems control—was working at The Massachusetts Institute of Technology (MIT) in Cambridge, Massachusetts, on automating air-defence systems. An important part of air defence was the development of radar, driving advances in discrete electronics, which later played a crucial role in electronic computers.

As already noted in the Introduction to this book, initially the input of computer science to the developing molecular biology was limited to providing metaphors, which worked the other way round too—the computer–brain analogy leant metaphors to computer science. Scientists in both disciplines had joint meetings, but these did not produce much. After one such meeting in 1948, Max Delbrück wrote that he found the discussion *'too diffuse for my taste. It was vacuous in the extreme and positively inane'*.[6] Today, one can find similar comments about some meetings on 'complexity' and 'systems science', often justifiably. Nevertheless, computers and computer science became indispensable for biology by late 1970s, after Fred Sanger found a way to read the sequence of bases ('letters') in DNA; that is, to read 'genetical information'. The human genome is billions of letters long, much too long to be studied via human inspection. By now, molecular biology cannot be conceived as being possible without the use of computers. Non-trivial computer science algorithms and machine learning are playing an increasing role in understanding how life works.

As discussed in Chapter I, for a sequence of letters to become information, this sequence needs an interpreter. In a living cell, the genome is read and interpreted by processes of molecular self-organization. Driven by the tendency of a physical system to minimize its free energy, protein machines read the information stored in DNA and interpret it to make new proteins and new copies of the DNA. Proteins and DNA then self-assemble to make new copies of the very same machinery and, eventually, new copies of the entire living cell. In the rest of this chapter, we will explore how this happens.

Life's building blocks

The most diverse molecules of life, proteins, are large by molecular standards. They are often referred to as biological macromolecules; a protein molecule typically contains tens of thousands of atoms. If we compare a living cell to von Neumann's self-replicating automaton, then proteins can be viewed as the automaton's elementary parts. Proteins are the main components of the automaton's *rigid members, cutting* and *fusing organs*, as well as the *logical gates*. Proteins can be of various shapes, but an average size globular shape protein will be around 5 nanometres (5 nm = 5×10^{-6} mm) in diameter, which is two orders of magnitude larger than an atom. On the other hand, a protein is of the order of hundred times smaller than a bacterium; thus, the dimensions of a protein are about halfway (on a logarithmic scale) between an atom and a bacterial cell. There are many billions of different proteins found in nature, a single living cell contains between a thousand to a hundred thousand different proteins, each typically in multiple copies. Proteins are the second largest contributors to the mass of a living cell, after water. As a rule of thumb, we can assume that water makes up around 70% of a cell's mass, and proteins around 20%, with only 10% left for other molecules. Three different models of a protein called *phosphoglycerate kinase*, or PGK for short, are shown in Figure 3.1.

Although proteins have complex three-dimensional shapes and structures, a polymer chain is the 'backbone' of every protein. While simpler polymers, such as polyethylene (Chapter II), are chains of essentially identical small molecules (monomers), proteins are made of up to 20 different types of monomers. This is a fundamental difference. If there is only one type of a monomer to choose from, say A, then the polymer of, say, eight such monomers can be described as AAAAAAAA. Thus, the only difference between the different polymers is their length. Essentially, such a polymer is a one-dimensional periodic crystal and as such it cannot carry much information. In contrast, if we have two or more different monomers that can be linked in different orders, then the resulting polymers are like words or sentences, and like words or sentences, they can carry information. But this also means that to make a particular polymer, information is needed.

The monomers that make up a protein—*amino acids*—are small molecules made of carbon, hydrogen, oxygen, nitrogen, and sometimes sulphur atoms. Amongst the 20

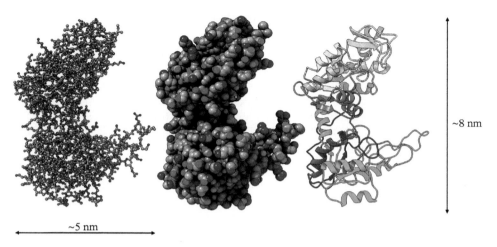

~8 nm

~5 nm

Figure 3.1 *Three different models of the structure of the protein called* phosphoglycerate kinase—*PGK enzyme (see Box 3.1). This protein has dimensions of approximately 8 × 5 × 4 nm, thus the models in the figure are on the order of 10 million times larger than the actual protein (compare it with the salt crystal cube in Figure 2.4). PGK is a polymer of 416 amino acids; if stretched out, it would be more than 150 nm (or 0.15 µm) long. In the* **left panel,** *the protein structure is shown as a ball-and-stick model, with the atoms shown as small spheres and the sticks representing the covalent bonds formed between the atoms. Nitrogen atoms are coloured blue, oxygen coloured red, carbon coloured grey. Careful inspection will also reveal a sulphur atom, coloured yellow. Orange colour indicates the phosphor atom of the ATP molecule (see text), which is bound to PGK (and is not a part of the PGK amino acid polymer). The green atom is a magnesium ion, also bound to PGK. Hydrogen atoms, which are small, are not shown. The* **middle panel** *shows a space-filling diagram, where a sphere is drawn for each atom, indicating its relative size. The colour scheme is the same as in the left panel. The* **right panel** *shows a so-called ribbon diagram, highlighting the way a protein chain folds and amino acids that may be far apart in the monomer sequence come into contact in three-dimensional space. Ribbon diagrams add representations of so-called protein secondary structure elements: spring-shaped ribbons show so-called* alpha helices *and flat arrows represent* beta strands. *This model is coloured according to the position of these elements in the sequence of the polymer chain, starting from blue (MSLSSKL . . .) through the rainbow colours green, yellow, orange, and to red at the other end (. . . LSEKK). The information that nature 'needs' to fold this polymer chain into the three-dimensional structure shown in the figure, is present in the sequence of the 416 letters given in the text. The figure is based on the Protein Data Bank (PDB) entry 3pgk.*
(Figure credit to Osman Salih, European Bioinformatics Institute of EMBL.)

different amino acids, the smallest one—glycine—consist of just 10 atoms, while the largest one—tryptophan—of 27 atoms. Thus, there is a hierarchy in nature: five different chemical elements are combined to make 20 different amino acids, which then are linked to make billions of different proteins. There is a convention to denote different amino acids by different Lattin letters. For instance, glycine is denoted by G, while tryptophan by W. To give a few more examples, M stands for methionine, S for serine, and L stands for leucine. The amino-acid sequence of the PGK protein shown in Figure 3.1, can be spelled out as:

> MSLSSKLSVQDLDLKDKRVFIRVDFNVPLDGKKITSNQRIVAALPTIKYVLEHHP
> RYVVLASHLGRPNGERNEKYSLAPVAKELQSLLGKDVTFLNDCVGPEVEAAVKA
> SAPGSVILLENLRYHIEEEGSRKVDGQKVKASKEDVQKFRHELSSLADVYINDAF
> GTAHRAHSSMVGFDLPQRAAGFLLEKELKYFGKALENPTRPFLAILGGAKVADK
> IQLIDNLLDKVDSIIIGGGMAFTFKKVLENTEIGDSIFDKAGAEIVPKLMEKAKAK
> GVEVVLPVDFIIADAFSADANTKTVTDKEGIPAGWQGLDNGPESRKLFAATVAK
> AKTIVWNGPPGVFEFEKFAAGTKALLDEVVKSSAAGNTVIIGGGDTATVAKKYG
> VTDKISHVSTGGGASLELLEGKELPGVAFLSEKK

This sequence of 416 letters[7] encodes the information needed to make the PGK protein shown in Figure 3.1. A typical protein contains somewhere between 300 and 400 amino acids, though some may be smaller and some much larger. One of the smallest known proteins microcin C7 contains just seven amino acids,[8] while one of the largest proteins titin, which is an important component of a muscle, is about 30,000 'letters' long.[9] In principle, amino acids can be joined up in a polymer in almost any order, and thus well over 20^{300} different protein molecules are theoretically possible. As the mass of the observable universe is estimated to be equal to $\sim 10^{80}$ hydrogen atoms, even a tiny fraction of the theoretically possible protein molecules would be enough to exhaust all the atoms in the universe. Obviously, the billions of different proteins found in nature form only the tiniest of fractions of the set of all theoretically possible ones, just as only a tiny minority of all possible sequences of letters form meaningful sentences in a human language.

The smallest bacteria contain a few hundred different proteins, human needs tens of thousands. Different species will usually have different proteins, but often we can match up the 'respective' proteins across species and thus talk about protein families.[10] All animals, including humans, share the same core set of roughly 10,000 different protein families. Although a one-dimensional polymer sequence is at the basis of every protein, the variety of three-dimensional shapes and structures that proteins may have is enormous; as of 2022, structures of ~180,000 different proteins have been determined and can be found in the Protein Database (PDB).[11] Many of them are quite similar to one another and can be grouped in distinct classes or folds.[12] Lately is has become increasingly clear that some parts of some proteins do not adopt a definite structure until they interact with other proteins.

How does a one-dimensional sequence of amino acids fold into a three-dimensional shape or structure? Can the same sequence fold in different structures? Is all the information about the protein's three-dimensional structure present in its one-dimensional polymer sequence?

Self-organizing sequences

In the 1960s, the US biochemist Christian Anfinsen performed a series of experiments that demonstrated that proteins could be unfolded by raising the temperature of their environment, and, importantly, after gradually lowering the temperature, proteins folded back again into their usual shapes.[13] Thus, the information about the protein's shape was not lost by unfolding the protein and apparently, the proteins' three-dimensional structure was defined by its one-dimensional sequence. In 1972, this discovery earned Anfinsen a Nobel Prize in Chemistry 'for his work on ribonuclease, especially concerning the connection between the amino-acid sequence and the biologically active conformation'.

Why does an amino-acid polymer fold into a particular structure? Just as a swinging pendulum eventually stops at the lowest point, an amino-acid polymer *self-organizes* into the conformation that minimize its free energy.[14] At high temperature that conformation is an unfolded state, but at physiological temperatures for many proteins it is a folded state. The amino acids are linked in a polymer chain by the strong chemical bonds—the same type of bonds that hold together carbon atoms in a diamond crystal. However, this polymer chain is not rigid; it can twist and fold. If a polymer folds back on itself, amino acids that are some distance apart in the one-dimensional sequence can come together in the three-dimensional space. Coming into contact, some amino acids form hydrogen bonds—the type of weak bond that holds water molecules together in an ice crystal. Each individual hydrogen bond is too weak to become permanent, but collectively they make the structure quite stable. And given that the amino acids are held together in the polymer chain by the strong bonds, it is easier to unfold a protein than to break its backbone. This is why Anfinsen's experiments could work.

But the three-dimensional structure of a protein is determined not only by the bonds between amino acids, the interactions with water molecules in the environment play as important role. Recall that 70% of a cell's mass comes from water. Different amino acids have different chemical properties, some of them are hydrophobic, some hydrophilic (Chapter II). The hydrophobic parts of the protein try to 'fold in' to avoid water. Water-avoiding 'nucleation centres' form, guiding the further folding of the protein. Thus, the process of protein folding has similarities to how amphiphile molecules self-organize in micelles to avoid water as well as to how weak bonds hold molecules together when water is freezing.

Since the free energy minimum point depends on the environment, a protein will fold 'correctly' only if the environment is appropriate. The environment inside a living cell is 'crowded' with different molecules, including many other proteins. This crowded environment may interfere with the folding of a newly made protein, which may aggregate with other proteins before it has folded as 'intended'. For instance, Alzheimer's disease is thought to be linked to erroneous protein aggregations in human brain cells. In a living cell, specific proteins (or to be precise, protein complexes, see Figure 3.3) called *chaperones*, shield newly made proteins until they fold properly. Effectively, the presence of chaperones provides a specific environment. Does this mean that *not all* information about the protein structure is contained in its sequence? Some information about the protein three-dimensional structure indeed is in the environment. The question is, how much?

To answer this question, first we need to realize that at physiological temperatures typically a protein can only assume a small number of structures distinguishable by the living cell.[15] An instructive example is provided by a class of proteins called *prions*. Prion proteins can cause a disease, for example, Bovine spongiform encephalopathy (BSE), commonly known as Mad Cow Disease, and the human versions of it—Creutzfeldt–Jakob disease (CJD). This disease is thought to be caused by a prion protein called PrPC, which can assume different distinguishable structures or states. Let us call these the *normal* and the *disease state*. In its normal state, this protein is involved in transmitting signals between neurons. However, if this protein somehow switches to its disease state, a change in the environment is triggered, starting a chain reaction making other PrPC proteins switch to the disease state too, apparently through aggregation. This is somewhat similar to a crystallization centre setting off the growth of a crystal.[16] It is thought that such an aggregation of PrPC proteins causes CJD.[17]

Although CJD is a complex disease which is not fully understood, from the point of view of information transmission, only 1 *bit* of information is needed to tell whether a disease-state PrPC is present in the environment. (As noted in Chapter I, the amount of information needed to decide between two possibilities is the definition of 1 *bit* of information.) More generally, the information that needs to be added to the sequence of amino acids to specify the particular structure is only to tell which of the few possible structures the protein should assume.[18] If a protein had four possible structures, we would need 2 *bits* to decide between them. In comparison, to describe the sequence of about a 200 amino acid-long PrPC protein, $200 \times \log_2 20 \approx 864$ information *bits* are needed (see Chapter VII). Even if not every amino acid is essential in determining the structure, nevertheless, much more information is contained in the polymer sequence than in the environment. Anfinsen's conclusion still stands, with a minor adjustment: *almost* all information that defines a protein's three-dimensional structure is in the protein's one-dimensional sequence.[19]

This is an important observation to which we will return repeatedly. As will be argued in Chapter VII, the only way to encode and transmit large amounts of information reliably is by combining a small number of distinguishable objects or states in different ways. A living cell implements this by linking together distinguishable small molecules (monomers) into polymer sequences. Small amounts of additional information are often added by other means, nevertheless most of life's information is encoded as polymer sequences. It is one of the main theses argued in the book that only the emergence of human language enabled life to transmit large amounts of information by means other than polymer molecules.

Anfinsen's experiments also demonstrated that a protein folds the same way in a test tube and in a living cell. These days, virtually any protein can be synthesized artificially in a laboratory: there is nothing special in a living cell.[20] The physics is the same.

The empiric laws of biology

Can the three-dimensional structure of a protein be computed knowing its one-dimensional sequences of amino acids? This is both a practical and a conceptual question: in practical terms, as experimentally it is much easier to determine the sequence of protein's amino acids than its three-dimensional structure, the ability to compute the structures would simplify the task of determining them. Conceptually, if (in a particular environment) all the information about the structure is in the sequence, then in principle, we should be able to reconstruct the structure computationally. Is this the case?

Given that the protein structure has to be compatible with the laws of physics, one might think that to compute the structure, we only need to consider all the possible folds and choose the one with the smallest free energy. The problem is that the number of different three-dimensional structures is enormous and grows exponentially with the length of the polymer.[21] Moreover, the free energy calculations are rather sensitive to the values of the interaction forces. Thus, to compute the protein structure, one would need to know these forces with virtually infinite precision. In practice, computing the protein structure purely from first principles of physics is possible only for very small proteins.

This however does not mean that protein structures cannot be computed by other means; for instance, using biological knowledge about how proteins relate to each other. Computing protein structures has been an unsolved problem for decades, but recently a combination of advances in machine learning and the availability of the many known protein structures, and even more sequences, has largely solved this problem. The method known as AlphFold computes the three-dimensional structures of most average-size proteins with high precision.[22]

But we cannot do this by just applying the laws of physics. Does this mean that the laws of physics are not sufficient to explain biology? No. Here is what the US theoretical physicist and Nobel Laureate Philip W. Anderson wrote about this:

> *The ability to reduce everything to simple fundamental laws does not imply the ability to start from those laws and reconstruct the universe. . . at each level of complexity entirely new properties appear, and the understanding of the new behaviors requires research which I think is as fundamental in its nature as any other.*[23]

Although the laws of physics explain the behaviour of biological systems, they alone are not sufficient to predict the behaviour of such complex systems. The rules for predicting the system's properties at the higher organization levels do not have to be the fundamental rules of physics. The discoverer of the atomic structure, Ernest Rutherford, has been quoted saying that '*all science is either physics or stamp collecting*', but this does not appear to be true. Ultimately physics is, and has to be, at the basis of the explanations of biological phenomena, but physics alone is not enough to build predictive models.

Self-assembling molecular complexes

Like most atoms combine into molecules, most proteins self-assemble into multi-molecular *protein complexes*. In *E. coli*, over two-thirds of the proteins are a part of such multi-unit complexes, with the average-size complex consisting of four proteins.

Some complexes are stable, some are transient—they form, change and dissolve in seconds or less. Models depicting complexes of different size are shown in Figures 3.2–3.4.

Protein complexes can be regarded as the smallest biologically functional units of a living cell; a single protein that has a function of its own, such as PGK, can be viewed as a special case—a one-unit complex. The biological functions of protein complexes are diverse, ranging from being building blocks of molecular scaffolds making up specific structures of a cell (analogous to von Neumann's *rigid members*), to molecular levers, wedges, or pulleys, which can use chemical energy to perform mechanical work (like von Neumann's *muscle*), and information processing units, sensing and combining different signals, or copying information from one polymer to another.

Many protein complexes are quite regular, consisting of several identical or similar units and having various symmetries,[24] though the very largest complexes typically are irregular. Some complexes include other types of molecules in addition to proteins, and thus it may be more appropriate to call them *molecular complexes*. One of the

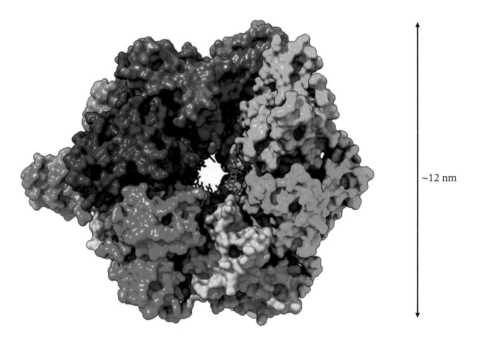

~12 nm

Figure 3.2 *A model of a replicative helicase—This is a protein complex consisting of six identical copies of the protein called DnaB, jointly forming a hexameric structure. For presentation purposes, each copy of DnaB is shown in a different colour. A helicase has a donut shape with a small hole in the middle, just large enough to enable one strand of a DNA molecule, shown in the figure in red, to pass through it. As described later in this chapter, the replicative helicase uses energy provided by small energy-rich molecules to propel itself along DNA, unwinding the DNA double helix and enabling a template-guided synthesis of new complementary strands (Figure 3.8). The figure is based on PDB database entry 4esv.* (Figure credit to Osman Salih, European Bioinformatics Institute of EMBL.)

GroES

GroEL

~20 nm

~14 nm

Figure 3.3 *A model of the GroEL-GroES chaperone is a complex of 21 proteins.* **Left.** *The complex is gherkin shaped and is composed of two distinct subcomplexes, GroEL and GroES. GroEL consists of two rings, each consisting of seven identical proteins. GroES forms one ring of seven proteins that are also identical, but different from the ones in GroEL. To show different proteins, in each sub-complex each protein is coloured differently. As mentioned in the text, in the crowded environment of a living cell, chaperones 'assist' the folding of newly assembled amino-acid chains.* **Right.** *The cross-section of the model on the right shows that this complex has a cavity—a cylindric shape opening with the diameter of about 5 nm, known as 'Anfinsen's cage'. Inside this cavity, a newly made protein is 'shielded' from the surroundings and thus can fold correctly, as aggregation with other proteins is prevented. The figure is based on PDB database entry 2c7c.*

(Figure credit to Osman Salih, European Bioinformatics Institute of EMBL.)

largest molecular complexes is a *ribosome*, shown in Figure 3.4, consisting of 54 proteins and three RNA molecules[25]. Like many large molecular complexes, a ribosome has a hierarchical structure—it consists of two subunits—the smaller and the larger subunit. Ribosomes are amongst the most essential molecular machinery components of a living cell, reading information in the genes and linking amino acids into specific protein polymers, as will be discussed later.

   Why do proteins form particular complexes? The same principles that define the three-dimensional structure of a protein are at work in forming protein complexes: minimization of free energy. Proteins that bind to each other usually have complementary shapes, fitting one into the other like hands joining in a handshake. Hundreds of hydrogen bonds form, stabilizing the complex.

   It is worth noting that molecular complexes self-assemble in a test tube the same way they do in a living cell. Smaller protein complexes assemble easily; all that is needed is a mixture of the right components in conditions similar to the physiological—water and possibly a few other small molecules or atoms. However, for larger complexes,

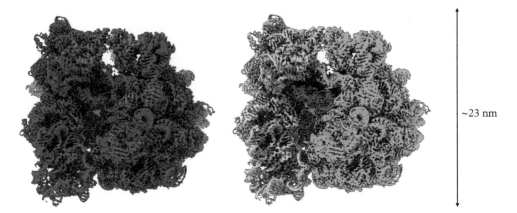

~23 nm

Figure 3.4 *Left. A bacterial ribosome is a large molecular complex consisting of ~50 protein molecules (blue) and three RNA molecules (pink); the number of proteins can vary slightly between ribosomes of different species. **Right**. A ribosome has a hierarchical structure consisting of two subunits. For the ribosome represented in the figure, the smaller subunit (yellow) consists of 21 proteins and one RNA molecule, the larger on (cyan) of 29 proteins and two RNA molecules. Ribosome is one of the most essential molecular machines in a cell, which translates the information encoded in RNA (originally in DNA) into sequences of amino acids of proteins (see Figure 3.10). The red and purple colours show two tRNA molecules, which are adapters establishing the link between nucleotide triplets and the respective amino acids, as described later in this chapter. The figure is based on EMDB database entry EMD-22586.*
(Figure credit to Osman Salih, European Bioinformatics Institute of EMBL.)

the situation is less straightforward. An instructive example is provided by a ribosome. In a milestone experiment in the late 1960s, Peter Traub and Masayasu Nomura, at the University of Wisconsin in Madison, Wisconsin demonstrated that the smaller ribosomal unit can be assembled in a test tube by adding the RNA and protein components in a particular order.[26] They concluded that all the information needed for this unit to self-assemble was encoded in the respective polymer sequences. This is an extension of Anfinsen's observation with a similar correction: here not only the information about the environment is needed in addition to the amino-acid sequence but also the information about the order in which the components need to be added. But this is a relatively small amount of information.

Assembling the larger ribosomal subunit turned out to be more difficult. Nevertheless, in 1974, Knud H. Nierhaus and Ferdinand Dohme, at the Max Planck Institute in Berlin, successfully reconstructed a biologically active larger subunit, but under rather harsh, non-physiological conditions.[27] Apparently, there is a different way that the larger subunit assembles in a living cell. Nevertheless, these experiments demonstrated that both components of one of the most essential pieces of cellular machinery can self-assemble outside a living cell.[28] In the presence of an appropriate RNA molecule, the two subunits further self-assemble into a functioning ribosome.

Another biological system that tests the limits of self-assembly is viruses. A virus typically consists of a DNA or an RNA molecule enclosed in a protein or a lipid membrane 'coat'. Viruses cause various infections, from the common cold to Ebola and AIDS, and they cause pandemics, such as the COVID-19 pandemic caused by the SARS-CoV-2 virus. Viruses cannot function by themselves but once they are in an appropriate living cell, they 'hijack' the cell's machinery by giving it 'wrong' information—just like computer viruses hijack a computer's operating system. An infected cell makes the viral proteins, the viral DNA or RNA, and then these molecules self-assemble to make new viruses. This is how viruses replicate. According to our definition, viruses are not autonomous replicators—the cellular environment in which they function is typically more complex than the virus itself. In the right complex environment, however, they can replicate very successfully.

Structurally, one of the simplest viruses is the Tobacco mosaic virus (TMV), which has a cylindrical envelope as shown in Figure 3.5. In a test tube, TMV self-assembles so easily that it has been used as an artificial structural element in manmade nanotechnology devices.[29] More complex viruses consist of tens of different types of proteins and assembling such complex viruses in a test tube may be difficult—the host cell's environment may be needed.

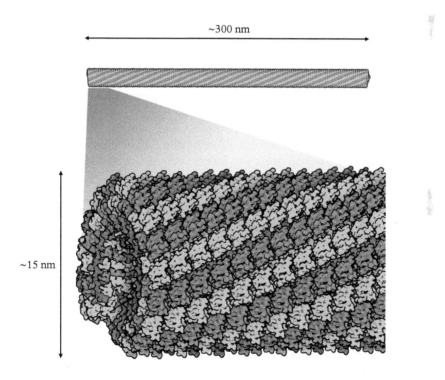

Figure 3.5 *Tobacco mosaic virus is composed of one long RNA molecule (shown in red) wrapped inside a sheath consisting of 2,130 identical copies of one protein (shown in blue). The mixture of copies of this protein and RNA easily self-assembles in the structure a model of which is shown in the figure.*
(Figure credit David Goodsell, doi:10.2210/rcsb_pdb/mom_2009_1.)

Self-organizing molecular processes

If protein complexes are to function like von Neuman's *cutting organs* and *muscles*, then they need to be able to change their shapes and exert physical force on the surrounding molecules. Indeed, changing protein shapes are at the basis of many, if not all, life's processes. We can walk because proteins in us can change their shape. The number of hydrogen bonds holding proteins and their complexes together are sufficient to make them relatively stable, however a shift in the free energy minimum point, either due to minor changes in the environment or because a minor chemical modification is inflected on the protein, may be sufficient to change the protein's conformation.[30] This is what makes life possible.

Particularly important are proteins or their complexes that have two or more relatively stable states that can switch from one to anther reversibly. A switch may occur, for instance, when a minor chemical modification is inflicted on a protein through an action of an *enzyme* (see Box 3.1). One such chemical modification, important for processes of life, is phosphorylation—attaching the so-called phosphoryl group, a small molecule consisting of one phosphor and three oxygen atoms (PO_3), to an amino acid.[31] An example of such a two-state protein is the protein called DnaA. In bacteria, when DnaA is phosphorylated, it binds to the genomic DNA in a specific location, setting off the process of genome replication—the equivalent of the replication to the tape in von Neumann's automaton. This change is reversible, DnaA can be dephosphorylated and ready to signal the next genome replication cycle.

By being able to switch between two or more different stable states reversibly, a protein can record information. A two-state protein can record 1 *bit*. If we denote the states of the protein by 0 and 1, switching it to a particular state can be seen as a recording of the particular digit. Some multistate proteins can combine incoming signals and thus they can work as the logical gates. For instance, by switching the state if and only if two specific amino acids are phosphorylated, the protein would implement the AND gate.[32] As was discussed in Chapter I, a sufficient number of such logical gates and memory elements are all that is needed to build a circuit that performs any given information processing task. Thus, proteins provide a living cell with the tools needed for information processing.

Box 3.1 Enzymes

An important class of proteins or their complexes essential to life is the already mentioned enzymes, which in analogy with von Neumann's cutting and fusing organs, implement chemical reactions splitting or joining other molecules. Unlike von Neumann's 'organs', enzymes usually are quite specialized in what they cut or fuse: there are over 600 different enzymes in *E. coli* bacteria and over 1600 different ones in humans.[33] This is like modern manufacturing, where every tool is specialized for a particular operation, though some have multiple uses. The PGK protein shown in Figure 3.1 is an enzyme that belongs to the class of enzymes know as kinases. Kinases catalyse the reversible transfer of a phosphoryl group from an ATP molecule to other molecules. PGK specifically transfers phosphoryl groups to or from a molecule called glycerate. This specificity provides a cell with means to control production processes—if a particular enzyme is not present, the respective process cannot happen. However, unlike von Neumann's cutters and fusers, each enzyme can always do both—cutting and fusing.

Chemical reactions always go in the direction that minimizes the free energy of the system, enzymes catalyse these reactions by lowering the energy barriers between the states (Figure 2.9). The direction of a chemical reaction can be manipulated by coupling it with a different, auxiliary reaction—it is the sum of the free energy balances of both reactions that tends to the minimum. However, enzymes may change the speed in which reactions happen billions of times. For instance, hydrogen peroxide, H_2O_2, which we use for wound cleaning, is a relatively stable molecule, staying intact for months or even years. The enzyme called catalase, splits these molecules into water and oxygen (more precisely, catalyse the reaction: $2 H_2O_2 \rightarrow 2H_2O + O_2$) speeding up the process about a hundred billion times. Thus, enzymes are in complete control of all the cell's manufacturing processes and by regulating how much of each enzyme is present in the cell, the cell regulates what it does.

The specific mechanisms how enzymes work, are not simple, but essentially, the so-called *catalytic centre* of the enzyme surrounds the molecule(s) on which the enzyme acts, and then a slight change in the enzyme's three-dimensional structure breaks up or joins the substrates.

We can distinguish between two types of protein state-changing events: the ones that work like semaphores to send signals (like the above mentioned DnaA), and the ones that perform mechanical work. For signalling, only the energy barrier (Figure 2.9) separating the states needs to be overcome, but to perform mechanical work, free energy needs to be spent equal to the work done. An archetypal example of a protein that can perform mechanical work, is actin, which amongst other functions, plays an important role in muscle contraction. Another example comprises molecular 'pumps' that transport molecules through a membrane.[34] In bacteria, a protein called FtsZ plays an important role in dividing a bacterium into two to complete the bacterial replication cycle. FtsZ has two states, which can be regarded as open and closed, like a door hinge with an in-built spring. When closing, FtsZ can exert force.[35] As will be discussed in Chapter IV, in a bacterial cell, FtsZ proteins self-assemble into long filaments forming a ring around the bacterium. At the right time this ring contracts, 'snapping' the bacterium in two.[36]

Where does the energy needed to perform mechanical work come from? Actin gets its energy from small, energy-rich molecules, adenosine triphosphate, or ATP for short. FtsZ gets energy from a similar widely used energy-rich molecule called GTP. Molecules like ATP and GTP can be thought of as coiled springs that can be either in a suspended (energized) or a relaxed state.[37] When switching from the suspended to the relaxed state, such a spring releases energy, and if it is inserted into a specific 'slot' of a mechanism (for instance, a specific protein, such as FtsZ), it releases energy, transferring it to this mechanism (to the protein), making the mechanism perform physical work.

In a living cell, proteins switching between states, and through this making other proteins switch, is the mechanism at the basis of many dynamic self-organizational processes. One fascinating example in bacteria is the so-called Min system,[38] which possibly provides a bacterium with the means to 'know' how long it has grown (see Chapter IV).

Figure 3.6 *Self-organization of Min proteins on an artificial membrane. In the bacterium E. coli, the Min proteins oscillate between the poles of the cell. The formation and maintenance of these patterns, which persist for hours, required ATP. On the artificial membrane the waves are significantly longer than in a living cell, possibly due to differences in diffusion constants.*
(Figure from Loose et al., 2008).

In this system, three two-state proteins called MinC, MinD, and MinE utilize energy from ATP to change each other's states, and in the process create waves that travel along the length of the bacterium from one end to the other and back, like waves in a bath (one can spot an analogy between formation of these waves and Bénard rolls, see Chapter II). Of note is the fact that the Min protein waves can be reproduced on an artificial membrane, as shown in Figure 3.6. The Min system is just one example of many different dynamic self-organizational processes in a living cell; a supply of free energy and the right set of amino-acid sequences (defined by the information in the genome) are all that is needed to drive the processes of life.

Energy and information

To make the clone, an autonomous replicator needs to be able to harvest energy from the environment and manage the flow of this energy through the system, to drive the manufacturing processes. Almost all energy available to life on Earth comes from the Sun; photosynthetic plants and some microbes convert sunlight into chemical energy of carbohydrate molecules; carbohydrates are then consumed by other organisms, transferring part of this energy to a variety of different molecules, consumed in turn, and so forth along the food chain. When moving down the chain, some free energy is dissipated and lost at every step—we can say that life consumes free energy, even though the total energy is conserved. In a living cell, fascinating molecular machinery combines small energy 'quants' released by gradually breaking up energy-rich molecules such as carbohydrates or fats to charge energy mediators such as ATP or GTP. These are indispensable mediators because they can release large energy 'quants' in single chemical steps. A living cell uses these mediators and energy they store to perform mechanical work, to drive biochemical reactions in the desired direction, and to enable various self-organizational processes.

However, having machinery to utilize free energy is not sufficient; as was discussed in Chapter I, a replicator also needs information instructing it how to make its replica. Sometimes, a trivial mistake is made in the popular science literature and even in some biophysics textbooks, equating free energy to information, which may be mis-interpreted as implying that information somehow magically flows from the Sun in the form of what is referred to as 'negentropy'. As we will discuss later in the context of evolution, indeed there is a link between Sun's energy and life's information, nevertheless equating the negative entropy term in the free energy expression (Chapter II) to information is a form of 'neo-vitalism'. Although photons arriving from the Sun carry information about the thermonuclear reactions generating them, these photons do not carry information about what proteins the living cell needs.[39]

When describing how amino-acid polymers self-organize and fold into three-dimensional proteins, we assumed that the amino-acid sequences were given *a priori*. As noted, the simplest living cell needs at least a hundred, possibly over a thousand, different specific sequences of hundreds of amino acids each; tens of thousands of such sequences are needed to make a human. Where does this information specifying the sequences come from?

Protein polymers are assembled by a ribosome reading the information encoded in the genomic DNA. When a cell replicates, each daughter cell gets a copy of the genomic DNA, like a newly made von Neumann's replicator gets its tape. Thus, the information comes from the parent cell. Obviously, the parent cell gets its information from its parent, and so on. Copying mistakes can happen occasionally and, as we will discuss in later chapters, genomes have been changing and evolving for billions of years. We will postpone the discussion of the big question of the ultimate source of the information in the genomes (and the role of the Sun's energy there) until later in the book; in the rest of this chapter, we will focus on more technical questions: first, how a molecular self-replicating unit—a living cell—gets its information from its parent and, second, how it interprets this information to function and to build its replica.

Life's information molecule

While proteins are made of twenty different types of amino acids, life's information molecule, DNA, and the related RNA, are polymers of only four different monomers. In DNA, these monomers are four *nucleic acids*, or *nucleotides*, specifically *adenine*, *thymine*, *guanine*, and *cytosine*, conventionally denoted by letters A, T, G, and C. For instance, ATTCG would denote a polymer where adenine is linked to two consecutive thymines, followed by cytosine and guanine. The links between nucleotides in DNA are asymmetric, for instance, ATTCG represents a polymer A→T→T→C→G, which is different from A←T←T←C←G, which would be spelt out as GCTTA. Nucleotides are somewhat larger molecules than amino acids, consisting of just over 30 atoms. (Like amino acids, nucleotides contain carbon, hydrogen, oxygen, and nitrogen atoms, but instead of sulphur they include phosphorus atoms.) The chemicals represented by these the letters A, T, C, and G, are sometimes referred to as *bases*, as Crick did in the letter quoted in the epigraph to this chapter. Strictly speaking, the bases are the parts of nucleotide molecules that distinguish one nucleotide from another.

There is a fundamental difference between DNA and protein molecules; RNA is somewhere in between. In contrast to the diverse structures of proteins, the three-dimensional structure of DNA is uniform and depends little on the particular nucleotide sequence. The DNA structure is the famous *double helix* shown in Figure 3.7; many will

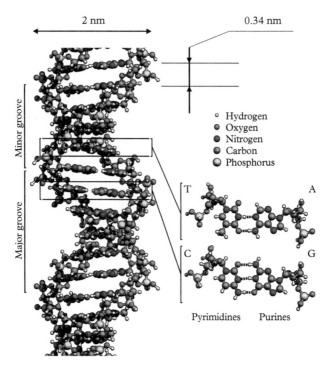

Figure 3.7 *DNA structure. Two nucleotide strands are wound around each other in a double helix. Nucleotide A forms hydrogen bonds with T, and C with G. A and G are somewhat smaller and are called* purines, *while T and C are larger and called* pyrimidines. *The dotted lines between purines and pyrimidines represent the hydrogen bonds. One full helical turn is about 10 steps, each step taking about 0.34 nm, thus a full turn is about 3.4 nm. According to Watson's account, learning about this number 3.4 nm from Rosalind Franklin's X-ray images of DNA crystals played a critical role in narrowing down the possible structures from which to choose. Franklin's crucial role in the discovery of DNA's double helix is therefore undeniable.*

(Figure adapted from Zephyris, Wikimedia, CC BY-SA 3.0 licence.)

agree that this is an elegant structure. Describing his personal account of the discovery, Watson later wrote in his book *The Double Helix*:

> *. . . a structure this pretty just had to exist.*
> (Watson, 1968)

It is noteworthy that Watson and Crick came up with their double helix largely by toying with macroscopic size models rather than by crystallography experiments or deep mathematical analysis; the crystallography experiments were done by Rosalind Franklin working in Maurice Wilkins' laboratory at King's College in London. Crick and Watson shared the Nobel Prize in Physiology and Medicine with Wilkins in 1962. Coincidentally, this was the same year when Max Perutz and John Kendrew got their Nobel Prize in Chemistry, and Linus Pauling was awarded a Nobel Peace Prize. Tragically, Franklin had died in 1958 from cancer, at age 37; Nobel Prizes are not awarded posthumously.

The double helix is formed by two interwound nucleotide chains or strands, oriented in the opposite directions, each chain fully determining the other: A is always paired with T, and C with G; thus A→T→T→C→G will be paired with T←A←A←G←C. This means that all the information is encoded in any one of the two sequences: the two sequences are *complementary*. Two complementary single-stranded DNA chains are held together by hydrogen bonds, forming a stable structure. Confined in a small space filled with water, two complementary single strands of DNA are likely to self-assemble into a stable double-stranded helix.

Recall that an average size globular shape protein has a diameter of about 5 nm; the width of a double helix is just below half of this—about 2 nm. However, the length of a DNA molecule can be just about anything. For instance, the *E. coli* genome is one long DNA molecule consisting of about 4,600,000 nucleotide pairs.[40] Multiplying this number by the distance between the helical 'steps', which is 0.34 nm (Figure 3.7), we obtain the total length of the *E. coli* genome of about 1.5 mm. This is hundreds of times longer than the length of the *E. coli* bacterium itself. If we magnified *E. coli* a million times, it would look like a 2-metres-long cistern, while its genome would resemble a 1.5 km-long telephone cable of two copper wires wound round each other, each about 1 mm in diameter. Thus, for many purposes, the genomic DNA molecule can be viewed as a one-dimensional tape. To fit it into a cell, the genome needs to be folded up and tightly packed.

DNA is a perfect medium for storing information. In a four-letter alphabet, each letter can encode 2 *bits* of information, therefore the information density in DNA can reach 6×10^8 *bits* or around 75 gigabyte (GB) per centimetre (cm). Thus, the equivalent of several high-definition movies can be stored in 1 cm of DNA. Moreover, DNA is a stable molecule and can keep this information durably and reliably. DNA extracted from 30,000-year-old mammoths has turned out to be preserved well enough to reveal how mammoths relate to contemporary elephants. Given such durable information storage potential, it has even been suggested to use DNA to preserve information generated in cultural evolution, such as books, music, or paintings.[41] Given that a nucleotide pair contains just over 60 atoms, we can estimate that in DNA 1 *bit* of information takes about 30 atoms. Even if we wanted to have some redundancy, less than 50 atoms per *bit* would suffice, as was mentioned in Chapter I.

Even more important than the DNA's information storage capacity is the relative ease with which the stored information can be read out. As discussed above, the amino-acid sequence of a protein encodes information too, however to read this information from a protein, or to copy it, is difficult. Obviously, nature uses this information to fold the protein, but it does so without reading it out as a sequence of letters. In a living cell, a ribosome writes information into a protein, however a living cell does not have the means to read this information back. To read the letters, a protein would have to be stretched out, at least locally. It is not clear how this can be achieved without raising the temperature, as in Anfinsen's experiments—a method not available to a living cell. But a living cell can both write information to and read it from a DNA molecule. This observation is at the basis of Crick's *central dogma*, which we will discuss in a moment.

While information in DNA is encoded as a sequence of nucleotides in a stable and permanent way, smaller amounts of information can be added to DNA less permanently. The letter C, cytosine, can be chemically modified or *methylated*. This provides a cell with the means to add extra information to the DNA sequence in a form of so-called *epigenetic marks*. If we compared the sequence of nucleotides in DNA to the printed letters in a book, then the methylation can be compared to pencil marks. As we will discuss below, the nucleotide sequence can be copied very reliably. The epigenetic marks can also be copied, but less reliably, just like when photocopying pages of a book, any pencil marks made will gradually fade out with repeated copying. Also, like pencil marks in a book, the methylation marks can be erased without changing the underlying text. Cells can add or erase DNA methylation marks dynamically, as they processes information.

The central dogma

In many ways RNA is similar to DNA. One difference is that instead of thymine, RNA uses a different nucleotide—uracil, denoted by U. There are some other minor chemical differences as well, together leading to significant differences in the polymer structure. RNA normally does not form a double helix but remains a single stranded polymer. On the other hand, hydrogen bonds can form between RNA nucleotides, causing RNA to form loops and to adopt quite complex three-dimensional structures (see Figures 4.10 and 4.11). We will return to this in Chapter IV; for now, what matters most is that RNA can pair up with a single strand of DNA: U pairs with A, and C with G. Thus, for instance, the RNA sequence A→U→U→C→G can pair with a single-stranded of DNA T←A←A←G←C. This complementarity is exploited by a living cell to *transcribe* information from DNA to RNA, which then can be *translated* into amino acid sequences of proteins.

In 1958, Francis Crick proposed a hypothesis, which he named *the central dogma of molecular biology*,[42] describing the possible pathways of information flow between the biological polymers RNA, DNA, and proteins, shown in Figure 3.8. Considering all possible paths, shown in the left panel, Crick hypothesized the most likely paths of the flow of information were from DNA to DNA, from DNA to RNA to proteins, and

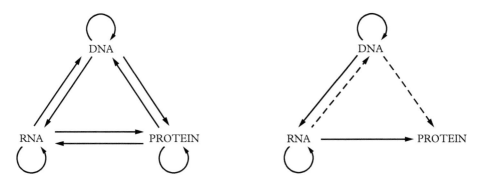

Figure 3.8 *The* central dogma of molecular biology. *The left panel shows all possible information transfers between these three polymers. The right panel shows the information transfer paths in a living cell that Crick thought to be occurring (solid lines) or possible (dotted lines).*
(Figure from Crick, 1970.)

possibly from RNA to RNA, as shown by the solid arrows in the right panel. Crick also allowed for the possibility that information could flow from RNA to DNA and from DNA to proteins directly (dotted arrows). For him, the most important part of the 'dogma' was the absence of three particular arrows—there is no flow of information from the amino acid sequence of a protein to either DNA, RNA or another protein.

Now we know that Crick was right—in every living cell, information indeed flows from DNA through RNA to proteins, and in replicating cells, also from DNA to DNA. Information can flow from RNA to DNA, as well as from RNA to RNA, which, for instance, happens in the context of RNA-based viruses. The only dotted arrow in Figure 3.8 that does not seem to exist in nature is the direct information flow from DNA to proteins. Most importantly, the sequence information never flows from proteins to RNA or to DNA. As already noted, reading the sequence information from a protein would be rather difficult, which essentially is the reason why, as we will see later, the features acquired during organism's lifetime do not normally get inherited by their descendants.

Interestingly, when reverse transcription of RNA→DNA was discovered, Crick's central dogma was criticized as '*likely to prove a considerable over-simplification*', prompting Crick to explain his rationale again,[43] stressing the importance of the negative part of the dogma:

> *This states that once 'information' has passed into protein it cannot get out again. In more detail, the transfer of information from nucleic acid to nucleic acid, or from nucleic acid to protein may be possible, but transfer from protein to protein, or from protein to nucleic acid is impossible. Information means here the precise determination of sequence, either of bases in the nucleic acid or of amino acid residues in the protein.*[44]

The famous twentieth-century philosopher Karl Popper has said that science is '*the art of systematic over-simplification—the art of discerning what we may with advantage omit*'. If Popper is right, and I largely subscribe to his views, we should not be surprised that

Crick's 'oversimplification' had such a strong impact on science: these days his central dogma is a part of every molecular biology textbook.[45] No doubt recognizing that some arrows were missing did speed up the advances of molecular biology, but even more fundamentally, the central dogma introduced the concept of information in biology.

The 'implementation' of the central dogma in living cells consists of two largely independent processes; first, copying DNA to DNA to ensure inheritance, and second, making RNA and then proteins according to the information encoded in DNA, to make the organism to grow, function, and replicate. The first is analogous to the replication of the tape in von Neumann's replicator (process R), the second is analogous to following the instructions encoded on this tape (process U).

Copying the cell's 'tape'

DNA's double-stranded structure provides a living cell with a conceptually simple way to copy information, implementing the idea shown in the Figure 1.3c. As each strand fully describes the other, each strand can serve as a template onto which its complement can be assembled, and thus the entire molecule can be duplicated preserving the original information. There is a famous quotation by Watson and Crick in the publication describing the DNA structure: '*It has not escaped our notice that the specific pairing we have postulated immediately suggests a possible copying mechanism for the genetic material*'.[46]

To copy a DNA molecule in this way, first the helix needs be unwound, and the strands separated. This is done by the doughnut-like protein complex called replicative helicase, shown in Figure 3.2, which propels itself along one of the strands using energy from ATP.[47] The moving helicase leaves behind a *replication fork* shown in Figure 3.9, first observed in 1963 by the British molecular biologist John Cairns.[48] Cairns also demonstrated that the bacterial genome is one long circular DNA molecule.

Protein complexes called DNA-*polymerases* are following the helicase, assembling the complementary strand on each of the separated original DNA strands. Two other doughnut-like structures called *sliding clamps* 'slide' over each of the newly synthesized double-stranded DNA, keeping the polymerases in place.[49] Overall, the DNA replication is carried out (strictly speaking, catalysed) by hierarchical molecular machinery called *replisome*, consisting of over a dozen dynamic sub-complexes. This enormous molecular machinery[50] moves along the DNA with an amazing speed of about a hundred helix turns per second. Tania Baker and Stephen Bell from the Massachusetts Institute of Technology described this process as follows:

> *If the DNA duplex were 1m in diameter, then the following statements would roughly describe E. coli replication. The fork would move at approximately 600 km/hr (375 mph), and the replication machinery would be about the size of a FedEx delivery truck. Replicating the E. coli genome would be a 40 min, 400 km (250 mile) trip for two such machines, which would, on average make an error only once every 170km (106 miles). The mechanical prowess of this complex is even more impressive given that it synthesizes two chains simultaneously as it moves. Although one strand is synthesized in the same direction as the fork is moving, the other chain*

(the lagging strand) is synthesized in a piecemeal fashion . . . and in the opposite direction of overall fork movement. As a result, about once a second one delivery person (i.e., polymerase active site) associated with the truck must take a detour, coming off and then rejoining its template DNA strand, to synthesize the 0.2 km (0.13 mile) fragments.[51]

This may sound like science fiction, but on the molecular scale life is much faster—recall that at room temperature an average protein has kinetic energy corresponding to a velocity of 10–20 metres per second. This is the advantage of miniaturization.

As every error made in copying the DNA will be passed on to future generations, DNA replication has to be extremely accurate. Indeed, *E. coli* genome replication makes only about one to two mistakes per every 10 billion letters copied. Maintaining high accuracy at such a speed is difficult, therefore other molecular complexes are deployed in a living cell, continuously 'proofreading' the copy and correcting errors. This is like speed-typing with a back-up of a computer-based auto-correction. One important DNA repair protein is MutT, which looks out for wrongly paired A to G and repairs them.[52] Methylation marks are used by the cell to tell apart the original and the newly synthesized

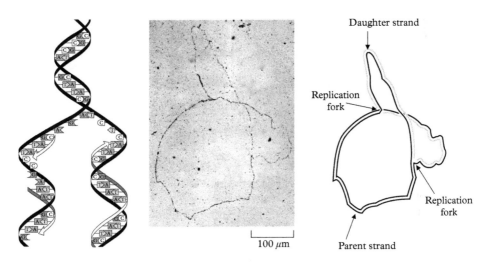

Figure 3.9 *DNA replication fork.* **Left.** *Each DNA strand serves as a template onto which its complement is synthesized. This essentially implements the principle shown in Figure 1.3C. The arrows show the direction in which the nucleotide compliment is assembled by a protein complex called DNA-polymerase. Note that one of the strands (so-called* lagging *strand) is assembled piece by piece in a direction opposite to the direction of the moving fork, because DNA-polymerase works along a DNA strand in one direction only.* **Middle.** *Electron microscopy image showing replication of an E. coli genome in a test tube, first reported by Cairns in 1963.* **Right.** *The interpretation of Cairns' image. There are two replication forks moving in opposite directions. A new complement, the so-called daughter strand, is synthesized on the parent strand—the black line. Once both forks meet, the entire circular DNA is replicated.*

(Left: Adapted from Wikimedia, drawing by Madeleine Price Ball, CC0 licence. Middle: Cold Spring Harbor Symposia on Quantitative Biology 28:44, 1963.)

DNA strands, thus helping the cell to know which nucleotides are likely to be the wrong ones. As we will see in Chapter V, when the gene coding for MutT is damaged, specific types of DNA replication errors accumulate.

Although the DNA replication machinery is sophisticated, the principle remains simple—the strands need to be separated and the complement assembled on each separated strand. The DNA replication machinery of contemporary life-forms has been evolving for billions of years to reach the present level of perfection: even in the simplest known living cell, at least 36 proteins are needed for DNA replication. Nevertheless, it is quite possible that simpler DNA replication machinery can be sufficient in cells with small genomes.

Assembling by the manual

In addition to DNA replication, in a separate process called *gene expression*, different molecular machinery reads information encoded in DNA to make proteins in a two-stage process. First, on the order of a thousand letter-long stretches of DNA called *genes* are *transcribed* into so-called *messenger* RNA molecules. In the second stage, these are *translated* into polymers of amino acids to make proteins. Thus, genes can be viewed as pages in a manual (genome) telling the cell what proteins, and eventually, what protein complexes to make.[53]

While a bacterial genome is replicated as one long molecule, there are many genes; for instance, in the approximately 4.6 million letter-long genome of *E. coli* there are about 4300 genes, on average each about 1000 nucleotides long. Biologists give names to genes and proteins coded by them, like astronomers give names to celestial objects. We already encountered quite a few such names: PGK, DnaA, FtsZ, MutT, amongst others. One way to tell a geneticist or a molecular biologist apart from their mathematician, physicist, or chemist collaborators, is by how many different genes they know by name. A mathematician working in the field can usually pick up a couple of dozen, may be a hundred names, but a proper geneticist often knows hundreds to thousands of genes and proteins, along with their function in a living cell. Usually, the gene has the same or a similar name as the protein it encodes. In scientific literature different fonts or capitalization conventions are used to tell genes and proteins apart; in this book we will distinguish genes and proteins they code by context.

Transcription of a gene from DNA to RNA is implemented, or to be precise, catalysed, by a molecular complex, called RNA polymerase, which is quite similar to the DNA polymerase. In bacteria this complex consists of five main protein units, though additional proteins are needed to initiate the transcription. It moves along DNA, copying the sequence of nucleotides in one of the strands into the complementary RNA. Every A in DNA is transcribed to U in RNA, T to A, C to G, and G to C. Different genes can be encoded on different genome strands. To make the single strand accessible to RNA polymerase, the DNA has to be unwound at the location of the respective gene. Gene transcription does not have to be as fast and as accurate as DNA replication, however RNA polymerases need to know where the genes are in the genome, and on which of the two DNA strands. For this, in addition to the information about the proteins, the genomic DNA also encodes 'road signs' pointing to genes. In *E. coli*, specific nucleotide sequences, such as TATAAT (known as the TATA-box), usually indicates the start of a gene.

The second step, the translation of a messenger RNA molecule into the sequence of amino acids of the protein, inevitably is a more complex process, as information encoded in a four-letter alphabet needs to be translated into a language using twenty letters. To encode one amino-acid letter, two to three nucleotide letters are needed; combinations of two nucleotides would give us only $4 \times 4 = 16$ possible codes, while we need 20. Three nucleotides give 64 possible combinations, which is more than necessary. Thus, in principle, a code is possible where some amino acids are encoded by two nucleotides, while others by three. Nature has chosen a simpler route—*triplet codon*—in a gene, every three consecutive nucleotides code for one amino acid in the protein. For instance, the 15-nucleotide sequence ATGTCTTTATCTTCA . . . codes for the first five amino acids MSLSS . . . of the PGK protein mentioned earlier.[54] Importantly for our discussion later, having 64 triplets available for coding 20 amino acids means that the code is redundant—the same amino acid can be coded by different triplets. For instance, triplets TCT, TCC, TCA, TCG, AGT, AGC all encode the same amino acid serine S. Thus, changing one letter in a gene does not always change the amino acid in the respective protein.

Having every three consecutive nucleotide letters in a gene coding for an amino acid makes it imperative that genes are read and translated in the right frame; a shift by one or two nucleotides would result in a completely different protein. Thus, a deletion of one or two nucleotides in the genome is likely to ruin the protein completely. Conversely, if a nucleotide letter is substituted for a different one, or if three nucleotides are deleted, in most cases no more than one amino acid in the protein will change (except when a stop-codon is created, see below). To position the translation machinery on RNA correctly, in addition to triplets coding for amino acids, specific 'road signs' mark the exact start and end of the nucleotide sequences coding for a particular protein. Specifically, all protein-coding genes start with a nucleotide sequence ATG, known as the *start-codon*, and end with one of the *stop-codons*: TAA, TGA, or TAG.

It was again Francis Crick, who as early as 1955 suggested that translating RNA molecules into proteins is likely to require adapter or linker molecules which, on the one hand would be able to recognize nucleotide codons, while on the other hand, bind the corresponding amino acids. Three years later, around the same time that Crick published this hypothesis,[55] a group of scientists, led by Paul Zamecnik and Mahlon Hoagland at Harvard University, showed that specific RNA molecules, now known as transfer-RNAs or tRNAs, play the role of such linkers.[56] The process of translation is schematically shown in Figure 3.10. A ribosome is reading the consecutive triplet codons in RNA, looking for a linker with the complementary triplet, and then uses the amino acid bound to a different part of this linker, to attach it to the growing polymer of amino acids. This is how a sequence of triplets is translated into a sequence of amino acids.

With few notable exceptions, all known species on Earth use essentially the same codon system: that is, the same triplets code for the same amino acids. The universality of

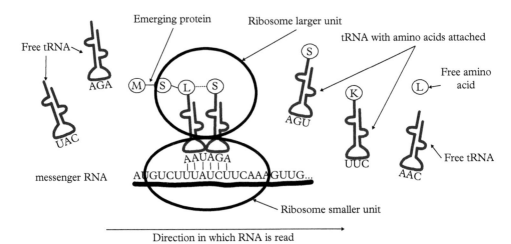

Figure 3.10 *Translating a sequence of a four-letter alphabet (RNA) into a sequence of a 20-letter alphabet (proteins). Each letter in the 20-letter alphabet (amino acid) is coded by a triplet of letters (nucleotides) in the four-letter alphabet. The translation process uses adapters—transfer-RNA (tRNA) molecules. These adapter molecules contain the complement to a particular triplet codon in one part (in the drawing, bottom) and the ability to bind the corresponding amino acid (top), thus linking the triplets of the four-letter alphabet to single letters of the 20-letter alphabet. The amino acids get attached to the respective tRNA molecules by specific enzymes—a particular enzyme for each of the 20 amino acids. The smaller ribosome subunit sits on the messenger RNA, 'grabbing' the nearby passing tRNA molecules and trying to match the current triplet in the messenger to the complementary triplet of the tRNA. If the triplets match—are complementary—the ribosome will keep the tRNA; if not, it will release it and look for another one. Once the matching tRNA is there, the larger ribosomal subunit takes the amino acid attached to the tRNA and links it to the growing amino acid chain, releasing the tRNA (on the left in the drawing). Thus, at each step, one triplet is read and one new amino acid is added, specified by the triplet in the messenger RNA. The ribosome moves 3 nucleotides to the right and the process continues until a stop codon is encountered, after which the ribosomal subunits disassociate from the mRNA and from each other, and start 'looking for' a new mRNA molecule. All these processes of 'grabbing', 'matching', 'releasing', 'linking' are self-organizational processes going in the direction to minimize the system's free energy.*

the triplet code across species is quite striking as there are no obvious physical or chemical reasons why particular amino acids would have to be coded by particular nucleotide triplets. Possibly this is a combination of a historic accident, the particular codon system was already established in the last common ancestor of all life known on Earth, and potential evolutionary advantages of some codes over others. For instance, it may be advantageous that similar codons code for amino acids with similar chemical properties.

One can wonder why a cell needs a two-stage process—why are genes not translated into proteins directly from DNA without the mediation of mRNA? Why one of the solid arrows in Crick's central dogma, the arrow from DNA to proteins (Figure 3.8), is missing? The answer is possibly that this would complicate rather than simplify the translation process. Transcription is a comparatively simple process, which separates two more complex processes of DNA replication and translation, helping to keep them from interfering with each other. Another reason may be that it is possible that early life on Earth was based on RNA, while proteins and DNA were introduced later (as will be discussed in the later chapters).

Controlling the production

A living cell does not need all proteins all the time; a gene can be switched on or off depending on whether the protein it encodes is needed or not. For instance, *E. coli* can feed on several different sugars, including glucose or lactose. Glucose is its main food. Glucose can be obtained from lactose in a chemical reaction catalysed by an enzyme called *beta-galactosidase*. Obviously, beta-galactosidase is useful to the cell only when lactose is present. And indeed, when lactose is absent, the gene coding for beta-galactosidase is switched off.

The expression of the gene coding for beta-galactosidase is controlled by a protein called *lac-repressor*. This protein binds to the DNA 'upstream' of the beta-galactosidase gene in the *E. coli* genome, preventing the transcription of this gene. The presence of lactose molecules disrupts the binding of the lac-repressor to DNA, and as the result, its place can be taken by the RNA-polymerase, thus starting the production of beta-galactosidase.[57] The described mechanism was discovered in 1961 by two French scientists, François Jacob and Jacques Monod, for which they shared the Nobel Prize in Physiology or Medicine in 1965 with André Lwoff 'for their discoveries concerning genetic control of enzyme and virus synthesis'.

Although in multicellular organisms, such as plants and animals, gene regulation is more complex, the basic idea remains the same: proteins bind to DNA and repress or activate a specific gene or groups of genes. An important class of proteins that can bind to specific locations of DNA and thus regulate transcription, are called *transcription factors*. The locations in the genome where they can bind are gene *promoters*, typically located near the respective gene, or *enhancers*, located further away. A gene can be regulated by combinations of transcription factors jointly implementing the logical gates, such as AND or OR; a repressor (as the one described above) implements the logical gate NOT. As the transcription factors binding to these promoters are themselves proteins coded by genes, regulatory feedback loops and complex information processing circuits can be encoded, as schematically shown in Figure 3.11. Such transcription regulation circuits complement the protein state-based information processing means described earlier. Thus, a living cell has a variety of tools to process information.[58]

Epigenetic information and inherited structures

Does all the information that the daughter cell inherits from its dividing mother cell reside in DNA, or does the daughter cell inherit additional information in some other

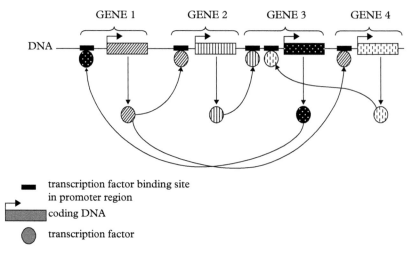

Figure 3.11 *Schematic representation of a simple, conceptual transcription factor network. All genes shown encode transcription factors that control each other's activity. A gene can be regulated by several transcription factors that can jointly implement the logical gates, such as AND (two transcription factors can bind to the promoter, the downstream gene is expressed if and only both transcription factors have bound) or OR (the downstream gene is expressed if at least one transcription factor has bound). A repressor implements the logical gate NOT—if it is present, the gene is switched off, if absent, the gene is on.*

(Figure adapted from Schlitt and Brazma, 2007).

way? Obviously, as will be discussed in more detail in Chapter IV, as the cell divides, it also passes on proteins, RNA, and other molecules to its daughter cells. How much information is passed on this way in addition to what is in the DNA? The information that resides in the sequences of the inherited RNA and proteins is redundant, as it is encoded in the genome anyway. So this does not add anything. However, additional information may be encoded in the abundances of these molecules. The number of copies that are inherited, or more realistically, the concentrations of different inherited molecules do carry information.

   Is there any information encoded in the three-dimensional structures of the inherited protein or RNA molecules? A simple example is provided by chaperones. As already noted, in a crowded environment of a cell chaperones are needed to ensure correct folding of some proteins. What if one of the proteins that is a part of a chaperone complex is itself unable to fold correctly without the particular chaperone already being present? Then to fold such a newly made chaperone protein correctly, another one already has to be there *a priori*. This would imply that it is essential that ready-made three-dimensional structures of some proteins are inherited.

   A more complicated example demonstrating the complexities of the processes of self-assembly in a living cell is provided by the bacterial flagellum.[59] The flagellum is a rotating propeller driven by a molecular engine, which helps the bacterium to move around. A flagellum consists of many thousands of proteins of about 30 different types, with copy numbers ranging from a few to a few thousand. In a test tube, most flagellar sub-complexes self-assemble relatively easily, but not the entire flagellum.

The propeller consists of flagellar filaments, which resemble nano-scale hosepipes with a narrow inner channel, about a quarter of a nanometre wide. In a living cell, these 'hosepipes' are assembled in a process guided by so-called protein-export complex, which directs newly made flagellar proteins through the inner channel of the growing 'hosepipe' towards its further end. To pass through the channel, the flagellar proteins polymer has to be in a stretched out state, that is, before the protein folds. Thus, to assemble the flagellum, the proteins have to be delivered one by one to a particular point in space and before they are folded. Can we say that all the information needed for making a flagellum resides in the amino-acid sequences of its proteins? Arguably not. Most likely specific structural elements need to be in place before the respective proteins can self-assemble to make the flagellum.

Inherited structures are a part of what we call *epigenetic inheritance*—any inherited information that is not encoded in the genome. The information that is encoded in such inherited three-dimensional structures is a part of *epigenetic information*. How much information can be inherited in an epigenetic way in addition to what is encoded in the genome? We will argue in Chapter VII that despite the complexity of some of these structures, the amount of information they carry is comparatively small.

It has been sometimes argued in the literature that the concept of information is not applicable to biological systems. If one dissects the arguments given, the recurring theme seems to be the observation that to interpret a genome, its interpreter—a living cell— itself has to be already present. But this is not specific to biology and is not a problem. As noted in Chapter I, information does not exist without an interpreter. To interpret the one-dimensional instruction-tape of von Neumann's self-replicating automata, the three-dimensional automaton has to be there *a priori*; the information on the tape will not replicate by itself. Nevertheless, few will argue that the instructions encoded in this tape cannot be regarded as information.

Genome sequencing and bioinformatics

If we had to choose one technological advance that in this century has changed biology the most, then almost certainly it is DNA sequencing. Today, if we want to study a new species, we often start with sequencing its genome. Sometimes genomes or parts of them is all we have, for instance when we sequence a mixture of microorganisms from a particular location in the ocean or soil, in an approach known as metagenomics. Advances in DNA sequencing are why we know what proteins each organism has.

Historically, scientists first learned how to sequence proteins, then RNA, and only after that DNA. Nowadays, this is reversed—to obtain the sequence of amino acids in a protein, usually the respective gene would be identified, its DNA sequenced and translated into amino acids. Both the protein and the RNA/DNA sequencing technologies were invented by the same scientist, Frederic Sanger, working at the

Laboratory of Molecular Biology in Cambridge, England. Each invention earned him a Nobel Prize in Chemistry, first in 1958 'for his work on the structure of proteins, especially that of insulin', and then in 1980, jointly with Americans Walter Gilbert and Paul Berg, 'for their contributions concerning the determination of base sequences in nucleic acids'. This made Sanger one of only three scientists awarded two Nobel Prizes in science. The other two are Marie Curie, who received Nobel Prizes in Chemistry and in Physics, and the inventor of the transistor, John Bardeen, who has two Nobel Prizes in Physics. Linus Pauling already had a Nobel Prize in Chemistry when he was later awarded the Nobel Peace Prize.

The sequencing of the first full genome, consisting of 3569 nucleotide letters belonging to an RNA-based virus named MS2, was completed in 1976.[60] This was followed by sequencing the DNA-based virus Phi-X174 in 1977, which had 5386 nucleotide letters.[61] It took almost two decades before the first complete genome of a more autonomous self-replicating organism—the bacterium *Haemophilus influenzae*—was sequenced in 1995 at the Institute of Genomic Research, led by Craig Venter in Rockville, Maryland.[62] Venter is a controversial figure in science and not particularly popular with some academics for attempting to patent the human genome. This did not happen, but Venter's attempt to compete with the publicly funded Human Genome Project and his reliance on the computational genome assembly methods arguably sped up human genome sequencing, and thus on the balance, arguably played a positive role. The genome of *H. influenzae* turned out to be 1,830,137 letters long—almost 400 times larger than Phi-X174 virus. This comparison may tell us something about the relative complexities of autonomous and non-autonomous information carrying self-replicating systems.

An automated version of the Sanger DNA sequencing technology was the workhorse of the Human Genome Project, which, at the beginning of the twenty-first century, culminated in sequencing the 3 billion letter-long human DNA.[63] It was only later, in the first decade of the new millennium, that radically new and significantly faster ways of DNA sequencing were developed. The Human Genome Project was an epic enterprise; today sequencing the entire genome of a human person is becoming a routine diagnostic procedure.

The impact of DNA sequencing would have been negligible if not for bioinformatics—using computers to analyse DNA sequences and related information. Printing the 3 billion-letter sequence of the human genome and binding it in volumes of books, as important as it is for capturing public imagination and helping us grasp the scale of the information in human genome, by itself it gives us little.[64] Now we know that the genomes of two unrelated human individuals differ in about one letter in a thousand. Imagine finding these differences by comparing volumes of printed books! But the exact letters matter; the specific letters in particular genome positions can cause a disease while knowing the specific letter may point to a cure. Moreover, without computing, genome sequencing would not only be pointless, it would also not be possible at all; DNA sequencing machines can directly read only relatively short genome fragments, which then have to be assembled computationally.

Once we have a genome sequence, it is possible to compute where the genes are, at least in some approximation.[65] The analysis of *H. influenzae* revealed 1743 genes;

in comparison the Phi-X174 virus contains 11 genes. Humans have about 20,000 protein coding genes and additional genes for functional RNA molecules, though the exact numbers are still being debated.[66] It is noteworthy that before the human genome was sequenced, it was estimated that a human would have between 50,000 and 100,000 genes. Thus, with regard to the number of genes, humans turned out to be simpler than originally thought. In fact, now we know that almost all mammalian animals have a very similar set of genes; apparently, the main difference between these species comes from how the genes are regulated.

The DNA sequencing is more than just another important technology—together with bioinformatics, the DNA sequencing has transformed biology into an information science. There is an analogy between primacy of information in the genome over its physical carrier and the importance of logics over the vacuum bulbs or transistors in the computer architecture, as was argued by von Neumann (Chapter I).

Crick's central dogma and von Neumann's replicator

The molecular processes of the central dogma make several intertwining cycles, as shown in Figure 3.12. DNA encodes information as a one-dimensional sequence of

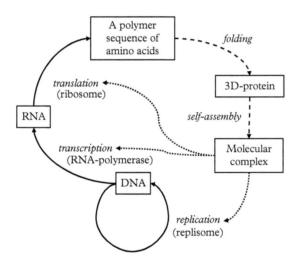

Figure 3.12 *The main principles of molecular biology in a nutshell: the central dogma extended to include the processes of self-organization (dashed lines) in addition to the flow of information (solid lines). The dotted lines depict the specific roles of molecular complexes catalysing the processes of information transfer. The processes of self-assembly and self-organization 'obey' the systems' tendency to minimize its free energy. The flow of information and interpretation of this information are intertwined—on the one hand, free energy minimizing processes of molecular self-assembly and self-organization make the machinery that reads the information in DNA, and on the other hand, this information is what ultimately defines what self-assembly and self-organizational processes occur.*

nucleotide letters, which is read and interpreted by three-dimensional molecular machinery, transcribing DNA into RNA, and then translating RNA into sequences of amino acids in processes that are 'driven' by the nature's tendency to minimize the systems free energy. To minimize the system's free energy further, these sequences of amino acids then self-organize into three-dimensional structures of proteins and assemble into molecular complexes. But the proteins and their complexes that result from this self-organization and self-assembly make the very same machinery that reads and interprets the DNA. In parallel, different molecular machinery, also encoded in the same DNA and driven by the same physical processes, makes a new copy of the DNA. Thus the cycle is completed: we started with information in DNA, this information was used to make a copy of the machinery that was interpreting DNA, and in parallel, to make the second copy of DNA carrying the very same information.

There is an evident analogy between the above-described processes in a living cell and the processes at the basis of von Neumann's self-replicating automaton, described in Chapter I. The DNA molecule is analogous to von Neumann's tape, its replication is the process R, while the gene expression is the process U, and both are integrated by C. In von Neumann's automaton, the control C is 'centralized' in one component; in a living cell this is 'implemented' via self-organization in a distributed way—there is no central control component. Noteworthy, von Neumann came up with his idea how to build an information carrying self-replicating automaton shortly before the DNA structure was discovered, and the central dogma of molecular biology was formulated. Nevertheless, the influence of von Neumann's work on the early development of molecular biology was arguably quite minor. Although he discussed his ideas with molecular biologists including Max Delbruck and Joshua Lederberg, it does not appear that von Neumann's ideas had influenced the emerging field of molecular biology in a direct way. Schrödinger's book *What is Life?* had a major impact on the future molecular biology, even though in some parts it was not entirely accurate; von Neumann's work, though visionary and, in a sense, foreseeing one of the most central principles of molecular biology, initially had little impact. One might wonder how often this happens in science; how often weak guesses turn out to be more influential than strong ideas that come before their time?

The relevance of von Neumann's work to biology, however, was acknowledged later. For instance, in 2012, a brief article published in the journal *Science*, the Nobel Laureate Sydney Brenner wrote that after the DNA structure was discovered:

> [f]or molecular biologists, the problem was how one sequence of four nucleotides encoded another sequence of 20 amino acids. It should now be evident what is needed to add to physics to account for living systems. The fundamental theory [of information processing] was formulated by Turing in his notion of universal Turing machine and developed by von Neumann in his theory of self-reproducing machines.[67]

But does the analogy between von Neumann's self-replicating automaton and a living cell go further than the basic processes of the central dogma? This is one of the topics of the next chapter.

IV

The Simplest Life

Appealing to the organic, living world does not help us greatly, because we do not understand enough about how natural organisms function. We will stick to automata which we know completely because we made them, either actual artificial automata or paper automata described completely by some finite set of axioms.

(John von Neumann, 1949)

The ability to detect single molecules in live bacterial cells enables us to probe biological events one molecule at a time and thereby gain knowledge of the activities of intracellular molecules . . . Single-molecule fluorescence tracking and super-resolution imaging are thus providing a new window into bacterial cells and facilitating the elucidation of cellular processes at an unprecedented level of sensitivity, specificity and spatial resolution.

(Andreas Gahlmann and William E. Moerner, 2014)

As can be judged from the two quotations above, times have changed in six decades. The molecular processes of a living cell and how it is growing and dividing, the so-called *cell cycle*, have been studied since the early days of molecular biology and by now a lot is known. There are two major types of living cells, *eukaryotes* and *prokaryotes*. Prokaryotes, which including bacteria, have much smaller and simpler cells. Paradoxically, we know less about the prokaryotic cell cycle than about the cell cycle in the more complex eukaryotic cells. One likely reason is that bacteria are too small for 'traditional' light microscopy. It was pointed out in 1873 by one of the founders of microscopy, the German physicist Ernst Abbe, that the resolution of an optical microscope is limited by about half the shortest wavelength of visible light, which is just under a fifth of a micron (~200 nm). As the size of a typical bacterium is on the order of a micron, bacteria can be observed via a good microscope quite well, but to see inside a bacterium, probes less 'blunt' than the photons of the visible light are needed.

The quantum wave–particle duality of the nano-scale world, which in a way is the cause of the problem, also comes to rescue. In 1924, the French physicist Louis de Broglie postulated in his PhD thesis that not just photons but all particles have wave-like properties; the heavier the particle is, the shorter the wave. The wave-like behaviour of matter was demonstrated experimentally in 1927; De Broglie got the Nobel Prize in

Living Computers. Alvis Brazma, Oxford University Press. © Alvis Brazma (2023). DOI: 10.1093/oso/9780192871947.003.0005

Physics 'for his discovery of the wave nature of electrons' in 1929. The wave–particle duality opened the possibility that using particles such as electrons might provide a path to an instrument 'sharper' than the light microscope. And indeed, in 1931 the German physicist Ernst Ruska and engineer Max Knoll built the first electron microscope. Ruska had to wait 55 years until 1986, when he was awarded the Nobel Prize in Physics 'for his fundamental work in electron optics, and for the design of the first electron microscope'. Max Knoll died in 1969.

The first electron micrograph of a cell[1] was published in 1945; now, the resolution of fractions of a nanometre has been achieved, sufficient to see the shapes of protein complexes and even the structures of some individual proteins. However, using this technique is neither easy nor straightforward: an electron beam needs vacuum, which is incompatible with life. The cell has to be frozen, moreover frozen so fast that the water in it neither evaporates nor crystalizes.[2] A different 'sharp' instrument is needed to study a live cell. One such an instrument is based on fluorescent labelling. Genomes can be engineered so that florescent labels are attached to specific proteins, which emit photons when excited by a laser pulse, revealing their locations in a living cell. Labels of different colours can be attached to different proteins and using repeated excitation of different labels, the locations of these proteins can be recorded with nanometre precision. This so-called *super-resolution microscopy* makes it possible to look inside a bacterium at the resolution of individual molecules, as shown in Figure 4.1. The high spatial resolution is achieved by trading in the resolution in time—it takes many laser pulses to construct the image.[3]

The enormous advances in high-resolution microscopy have been reflected in recent Nobel Prizes. In 2014, the Nobel Prize in Chemistry was awarded jointly to two US scientists, Eric Betzig and William E. Moerner, and a German-Romanian, Stefan W. Hell, 'for the development of super-resolved fluorescence microscopy'. Just three years later, in 2017, the Chemistry Nobel Prize was awarded to a Swiss scientist, Jacques Dubochet, a German-American, Joachim Frank, and a Scot, Richard Henderson 'for developing cryo-electron microscopy for the high-resolution structure determination

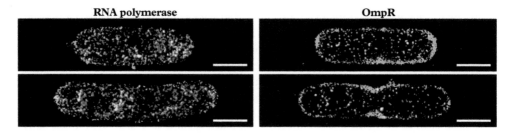

Figure 4.1 *Super-resolution fluorescence microscopy images of E. coli cells (scale bar 1 μm).* **Left.** *RNA-polymerase.* **Right.** *The protein called OmpR, which is a DNA binding protein regulating gene transcription. The polymerase and OmpR are both labelled red; the membrane is stained with green. It is estimated that there are over 3000 RNA polymerases present, while OmpR is clearly less abundant.* (Figure from Foo et al., 2015.)

of biomolecules in solution'. Both techniques can be combined—the so-called correlative light-electron microscopy overlays the images obtained by different technologies to produce a single high-resolution image and to computationally identify particular elements of interest.[4] A complementary way to probe the function of a particular protein is to disrupt the respective gene and look how the cell reacts. Such gene 'knockouts' or 'knockdowns' combined with different modes of microscopy, sequencing a sample of the RNA molecules in a cell, mass-spectrometry, and computational modelling have given scientists quite a good understanding of what different molecules in a cell do, how a cell functions, and how it makes its replica. The times indeed have changed.

Model bacteria

Until the 1970s it was thought that prokaryotes and eukaryotes form the two principal domains of life on Earth. This changed in 1977, when Carl Woese at the University of Illinois undertook a systematic comparison of available nucleotide sequences from organisms sequenced at the time.[5] He found that the prokaryotic sequences clustered into two very distinct groups, at least as distinct from each other, as from the sequences of eukaryotic organisms (Figure 4.2). This earned Woese the Nobel Prize 'for the discovery of a third domain of life' in 2003. The third domain was named *archaea*, as at the time it was thought that archaea were evolutionarily older than bacteria. Many species of archaea live in extreme environments, such as geysers, thermal vents deep at the ocean floor, or the salty water of the Dead Sea, and one can hypothesize that they originated during the early period of formation of Earth, when it was hot and saline. However, later it became clear that this is not what distinguishes archaea: some of them lead very tame lifestyles, while many bacteria or eukaryotic microbes are extremophiles.[6]

Defining the boundaries of a prokaryotic species is difficult, but applying some reasonable assumptions it can be estimated that there are between a hundred thousand and a billion different prokaryotic species on Earth, only a minority of which are known.[7] Amongst the most abundant are the small marine bacteria called *Prochlorococcus marinus* (*P. marinus*) and *Pelagibacter ubique* (*P. ubique*), which have rounded shapes of less than 1 μm in diameter.[8] These organisms survive on rather simple food and are candidates for being the simplest autonomous replicators (as autonomy was defined in Chapter I) existing on Earth.

P. marinus and *P. ubique* have significantly less than 2000 genes; unfortunately, not much about their replication mechanisms is known. Most of our current understanding of the prokaryotic cell cycle comes from larger and more complex bacteria, *E. coli*, which probably is the most studied prokaryote of them all, and from *Caulobacter crescentus* (*C. crescentus*), which is the scientists' favourite species for studies of the bacterial cell cycle. *Crescentus* is Latin for crescent-shaped, which, as shown in Figure 4.3, describes this bacteria well. *C. crescentus* typically lives in freshwater lakes or streams; its replication time there is between 1.5 and 2 hours.[9]

The complexity of *E. coli* and *C. crescentus* is quite similar, they both have around 4000 genes, and thus around 4000 different proteins. If we add the other molecules, we can estimate the number of different 'elementary parts' in these bacteria to be on

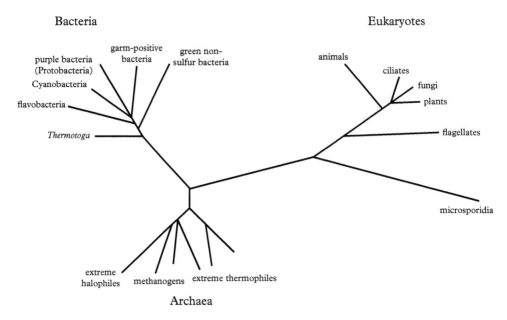

Figure 4.2 *Universal phylogenetic tree estimated from the comparison of the ribosomal RNA sequences from 17 species (Woese, 1987). Using a significant degree of simplification, we can say that the relative evolutionary distance between each two species is calculated from the alignments of their small ribosomal RNA sequences by counting the differences in nucleotides in the respective positions. The phylogenetic 'tree' is then constructed by pairwise joining of the closest species or the estimated consensus sequences of the group. The length of each branch is proportional to the estimated evolutionary distance. The classification shown in the figure has since changed, for instance, what Woese called Purple bacteria is now called Protobacteria. Three of the model bacteria in Table 4.1, E. coli, C. crescentus, and P. ubique, all belong to Protobacteria. P. marinus belong to Cyanobacteria, while the archaeon* M. jannaschii *belongs to (thermophilic) Methanogens. With sequencing the genomes of an increasing number of different species, the hypothetical phylogeny has been since revised and, as noted in Chapter VI, recently it has been suggested that eukaryotes are not a separate domain but a part of archaea.*

the order of 10,000, which makes their complexity quite average in the prokaryotic domain (Table 4.1). The smallest known genome of an organism living outside another live cell belongs to the bacterium *Mycoplasma genitalium* (*M. genitalium*), which has about 500 genes.[10] *M. genitalium* however is a parasitic microbe, living in the human urinary tract and benefiting from many complex molecules and the controlled environment of the host. *P. ubique*, with its 1300 genes, is thought to be the simplest *free-living*[11] organism on Earth. On the other end of the spectrum, one of the largest known bacterial genomes belongs to the species *Sorangium cellulosum* and has over 13,000 genes.[12] Amongst eukaryotes, plants have the largest genomes, which may have up to 50,000 protein coding genes.[13]

The number of protein coding genes is not the only measure of a cell's complexity. The simplest known eukaryotic microbes that live outside another cell can have as few

Table 4.1 *Approximate gene numbers and organism sizes of some representatives of the three domains of life. Some of the bacteria in the list have widely variable genomes depending on the strain (subspecies); for instance, the size of the* E. coli *gene count varies from about 4000 to well over 5000; the reported 4300 belong to one of the most studied strains, K12. The largest bacteria,* Thiomargarita namibiensis, *found in the ocean sediments of the continental shelf of Namibia, can be half a millimetre long (Schulz and Schulz, 2005) and it has multiple genome copies, possibly over a 100,000. Although not all its genome copies are identical, the size of one genome copy is estimated to contain around 3.8 million letters, suggesting in the order of 4000 different genes. The cell sizes given in the table represent the cell's largest dimension in typical growth conditions. In multicellular organisms, cells have highly variable sizes depending on the cell type; for instance a neuron can be over a metre long. The so-called* obligatory intracellular *parasites, such as* Carsonella ruddii *(bacteria, 182 genes) and* Encephalitozoon cuniculi *(eukaryote, 2000 genes), which live inside other live cells, are not included in the table.*

		Species	No. of genes	Habitat	Cell Size in μm
Unicellular (microbes)	Bacteria	*Mycoplasma genitalium*	500	urinary tract	0.3
		Pelagibacter ubique	1300	ocean	<1
		Prochlorococcus marinus	1800	ocean	<1
		Caulobacter crescentus	3700	rivers and ponds	5
		Mycobacterium tuberculosis	4000	human lung	4
		Escherichia coli (K12 strain)	4300	human gut	5
		Thiomargarita namibiensis	~ 10,000	ocean sediments	500
		Sorangium cellulosum	> 10,000	soil	5
	Archaea	*Methanocaldococcus jannaschii*	1700	oceanic thermal vents	1
		Haloferax volcanii	3000	hypersaline water	3
		Haloa rcula marismortui	4200	hypersaline water	3
		Haloterrigena turkmenica	5300	saline soil	2

Continued

Table 4.1 *Continued*

	Species	No. of genes	Habitat	Cell Size in μm
Eukaryotes	*Cryptosporidium parvum*	3800	human intestine	5
	Plasmodium falciparum	5200	human pathogen	2
	Cyanidioschyzon merolae (red algae)	5300	host-springs	2
	Saccharomyces cerevisiae (baker's yeast)	6000	various	10
	Saccharomyces pombe (fission yeast)	5000	various	10
	Chlamydomonas reinhardtii	14,500		10
Multicellular	*Volvox carteri*	14,500		10
	Drosophila melanogaster (fruit fly)	13,600	various	
	Mus musculus (mouse)	20,000	various	
	Homo sapiens (human)	20,000	various	
	Populus trichocarpa (black cottonwood)	45,000	various	

as 3800 genes, which is fewer than for many bacteria, but eukaryotic cells are much larger and have complex inner membrane structure (Chapter VI). Eukaryotes may also have a large number of genes that code for putatively functional RNA molecules in addition to protein coding genes. The average length of a bacterial gene is about a thousand nucleotides and, given that in bacteria most of the genome codes for proteins, the conversion from the genome size in nucleotides to the number of genes in the genome, or back, is easy: divide by a thousand or multiply by a thousand. In *E. coli* the more precise ratio is 1070. Since small bacteria contain on the order of thousand genes, their genome size is on the order of a million nucleotides. In *E. coli* this protein coding part is about 90%. The rest is either coding for functional RNA, including transfer RNA molecules, or consists of promoter sequences regulating the expression of genes.

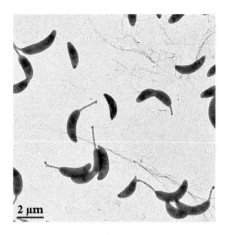
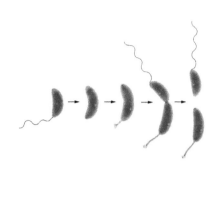

Figure 4.3 *Left. C. cresentus bacteria. Transmission electron micrograph by Cécile Berne from the laboratory of Yves Brun.* **Right:** *C. crescentus bacteria divide asymmetrically producing two different cells—a* stalked cell *and a* swarmer cell. *Only the stalked cells divide. The swarmer cells have flagella (pili), which they use to move. After a while, the swarmer cell changes into a stalked cell, which then can divide again. This is a strategy how these replicating bacteria spread around the environment. Transmission electron micrographs by Yves Brun.*

Are viruses microbes? It does not matter what terminology we use and what exact meaning we attribute to the word 'microbe', but as already noted, viruses do not fit our definition of autonomous self-replicating systems.

Inside a bacterium

As shown in Figures 4.1 and 4.3–4.6, both *E. coli* and *C. crescentus* have an elongated shape and they are typically somewhere between 2 μm and 5 μm long and roughly 1 μm in the smallest dimension, which is about a hundred times smaller than the smallest dot visible by the naked eye. How can we picture the inner architecture and complexity of a bacterium? In a million-fold magnification, a nanometre becomes a millimetre (1 nm → 1 mm), a micrometre becomes a metre (1 μm → 1 m), and a millimetre becomes a kilometre (1 mm → 1 km). Such a magnified *E. coli* would look like a 2–4 m-long cistern with hemispheric ends, 1 m in diameter. An average sized globular protein of 5 nm becomes 0.5 cm large, while a water molecule becomes a tiny, a just barely visible 0.1 mm dot. The DNA becomes a 1.5 km long circular cable of two 1 mm diameter wires wound around each other, packed in a bottlebrush-like structure, as shown in Figure 4.4, maintained by specific DNA-binding proteins that work like clippers.[14] In addition to the main *bacterial chromosome*, bacteria may also have a number of smaller DNA molecules, known as *plasmids*.

The dimensions of a protein complex—the basic functional unit of a living cell—is of the order of 10 nm, which is around 100 times larger than these of an atom.

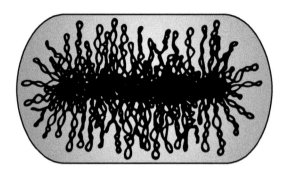

Figure 4.4 *Topological organization of a bacterial chromosome—a schematic representation of the bottlebrush model of DNA packed inside a bacterium.*
(Adapted from Wang et al., 2013).

The characteristic size of a bacterium itself is of the order of 100 times larger than that, that is of the order 1000 nm = 1 μm. Thus, if we arrange 100 atoms on a line next to each other, we get the dimension of a protein complex; if we then similarly arrange 100 proteins complexes, we get the dimension of a small bacterium.[15] The last means that there is space for about a million protein complexes in a 1 μm³ volume of a bacterium. In our million-times magnified model, we can think of several million protein complexes, each taking ~1 cm³ volume, though obviously these complexes are not organized on a grid and some complexes, such as ribosomes, are much larger, some are smaller.

In addition to proteins, there will be many other molecules such as RNA, and smaller molecules, such as ATP. As already noted, 70% of the cell mass is taken by water, amounting to on the order of a hundred billion molecules: around a hundred thousand water molecules per protein. Although proteins are thus 'swimming' in water, *cytoplasm*—the inside of a cell—is a crowded place where proteins are tightly packed. The average distance between protein complexes is less than their size, and thus, for a protein complex to move, other proteins have to be displaced. Like people on a crowded train, they can move when they need to, but only by 'negotiating' ones way with others. Proteins are not hard objects; they are more like jelly toys adjusting their shape as they slide past each other. And most proteins are sticky—some are 'choosy' about which proteins to stick to, some are promiscuous. Finally, we should not forget that a bacterium also contains the 'bottlebrush' chromosome of the genomic DNA.

The environment inside a bacterium is not only crowded but also highly dynamic. Recall that at room temperature proteins jiggle at a velocity around 10–20 metres per second. Despite this crowded environment, for an average protein it takes only about a second to diffuse from one end of a bacterium to the other.[16] The molecular arrangements making up a bacterium are constantly forming and dissolving; one way of thinking of a cell's interior is as being pervaded by transient protein scaffolding, which is constantly assembling and disassembling in various configurations.

The shape of a bacterium is maintained by the *cell's wall*, which is a 'mesh' made of ~ 5 nm long, relatively rigid molecular 'sticks' (sugar molecules and short amino acid polymers—polypeptides);[17] an electron micrograph of an *E. coli* cell wall is shown in

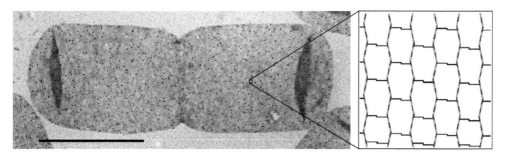

Figure 4.5 *Left. Electron micrograph of a bacterial cell wall from a dividing E. coli cell. The scale bar 1 μm. **Right**. A model of a layer in a section of approximately 30 × 30 nm. The glycan strands—small molecules (vertical zigzag lines) are perpendicular to the long axis, and the peptides—short chains of amino acids (horizontal lines) are in the direction of the long axis of the bacterium.*
(From Vollmer and Bertsche, 2008.)

Figure 4.5. These molecular sticks are linked to each other via the strong covalent bonds (Chapter II), thus the entire mesh can be regarded as one enormous hollow molecule. It has elasticity comparable to rubber—it bends under pressure but springs back to its original shape once pressure is removed.[18] The mesh is porous, but it is 'sealed' from the inside, and for many bacterial species (including *E. coli* and *C. crescentus*) also from the outside, by a lipid membrane—the inner and the outer membranes. In our million-times magnified model these membranes can be thought of as plastic films used in wrapping. In *E. coli*, the wall is around 5 nm thick, in *C. crescentus* is just slightly thicker,[19] thus in our million-times magnification model, the cell wall would be roughly half a centimetre thick.[20]

Some smaller bacteria, called *Mycoplasma* (including *M. genitalium*), do not have cell wall and have only a single membrane. In all bacteria, there are many proteins and protein complexes protruding through their membranes, some spanning the wall and both membranes. Some of these protein complexes have holes, which they can open and close by use of energy from ATP or GTP, taking in food or excreting rubbish. In *E. coli*, more than half a million proteins sit in its membrane.

Just as electronic computers have parts different from Babbage's cogwheels (Chapter I), nano-scale molecular replicators, bacterial cells, have elementary parts, proteins and protein complexes, different from the rigid parts of the kinematic von Neumann's kinematic replicator. Nevertheless, one can find in them elements analogous to parts of macroscopic machines: spring-like hinges of FtsZ, rotating cogwheels of bacterial flagella, and rotating power-generators charging ATP (rotating ATP-ases), among others.[21]

A bacterium replicating

In a favourable environment, bacteria continuously grow and periodically divide in two (Figure 4.6).

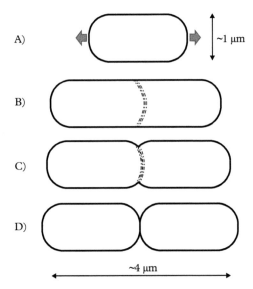

Figure 4.6 *A bacterium growing and dividing. E. coli (and to an extent also C. crescentus) can be viewed as cylinders of around 1 μm in diameter and somewhere between 2 μm to 5 μm in length. A) A bacterium right after its 'birth' (after the previous division). B) As the bacterium is growing in length, FtsZ proteins (Chapter III) are self-assembling in filaments, that further self-assemble into the so-called Z-ring near the 'equator' of the bacterium. C) When the bacterium has reached a certain length, the Z-ring starts constricting, and eventually the bacterium divides into two 'daughter' bacteria D).*

In a fast-growing *E. coli* cell, some 3000–5000 RNA polymerases (Figure 1.1 left) are working in parallel transcribing the genes into mRNA. As the velocity of the RNA polymerases is around 80 nucleotides per second, fully transcribing an average gene takes around 12 seconds. Ribosomes are already 'sitting' on the emerging mRNA molecules as they are being synthesized. It takes about 20 seconds for a ribosome to read and translate a 1000 nucleotide-long RNA to make an average size protein. This is about the same time it takes for an average protein to fold and for a chaperone to complete its cycle (Chapter III). Proteins not only have to fold, they also have to find their partners to form functional molecular complexes.[22] In a fast-growing *E. coli* cell, over 20,000 ribosomes are working in parallel, with the joint 'industrial' output of about 1000 protein molecules per second. As about 40% of cell's proteins are in ribosomes, we can estimate that about eight new ribosomes are made every second. One can make calculations that with this rate it takes close to half an hour for a bacterium to double in size, as observed.

The molecular mesh of the bacterial cell wall grows with the bacterium maintaining its shape. In cylinder-shaped bacteria, patches of a protein named MreB rotate around the cylinder, guiding a complex of enzymes inserting new molecular sticks into this mesh.[23] Mathematical modelling shows that stochastic processes consistent with the laws of physics are sufficient to maintain the cylindric shape,[24] nevertheless it is quite likely that there is regulatory feedback from the cell's shape regulating how fast different parts of the wall grow.[25]

To replicate, the cell also needs to make a new copy of its circular genome. At a specific time, the DnaA protein (Chapter III), changes its state and binds to the genome at a well-defined location to kickstart the genome replication.[26] Two replisomes (Chapter III) assemble there and start moving in opposite directions; when they meet, the genome has replicated. Unlike most cellular processes, which are highly parallel, genome replication is largely a sequential process,[27] therefore, to keep up with the rest of the growing cell, it has to be fast. And indeed, in our million-times magnified model, each replisome covers about 750 metres in 40 minutes, which is about 30 cm in a second, while avoiding collisions with thousands of RNA polymerases transcribing the genes at the same time.

After the genome has replicated, the two DNA molecules need to be separated. In smaller bacteria, maximization of the conformational entropy of the DNA polymers (Figure 2.13), may be sufficient to achieve this.[28] In E. coli and C. crescentus, and probably in most larger bacteria, there are additional active chromosome separation mechanisms at work. One such mechanism, called a burnt-bridge Brownian ratchet, uses energy from ATP to pull the two genome copies apart.[29] In C. crescentus, two proteins, ParA and ParB, are at the centre of this mechanism. The first is a two-state protein, which can be in a high-energy state ParA-ATP or a low-energy state ParA-ADP. In the high-energy state it self-assembles in filaments, one end of which attaches to the genome, the other to the far end of the cell. We can think about this filament as a door chain, one end of which is attached to the door (genome), the other end to the frame (one of the poles of the cell). Removing a link from such a chain and reconnecting the adjacent links generates a pull force, making the door close a little. This essentially is what the cell does. Concretely, the ParA-ATP filament is attached to one pole of the cell via the ParB. This protein is an enzyme changing the state of ParA-ATP to the low-energy state ParA-ADP. In doing so, ParB removes ParA proteins from the filament one by one, and uses the energy released from the ATP to pull and reconnect the remaining part of the filament, thus pulling the genome.

Once the bacterium has doubled in size, the genome has been copied, and the copies have been separated, to complete the replication cycle the bacterium needs to divide in two. Recall from Chapter III that proteins called FtsZ self-assembles in filaments forming a ring around the bacterium. There are around 5000–7000 copies of FtsZ protein E. coli cell. When energized, FtsZ changes its confirmation, making the ring constrict[30] (Figure 4.6). How does the Z-ring 'know' when and where to form? In C. crescentus the positioning exploits the concentration gradient of a protein called MipZ, which runs from the poles to the cell's 'equator', and thus the concentration minimum is achieved near the equator,[31] as shown in Figure 4.7. In E. coli there is a different positioning mechanism, exploiting the bottlebrush genome structure. A protein called SlmA, binds to DNA inhibiting FtsZ from assembling.[32] As the DNA 'bottlebrush' is never far from the cell's inner membrane, only when the cell has grown long enough and the genome has replicated, there is sufficient space free of SlmA near the 'equator' of the long cell. This is where the Z-ring assembles. An additional positioning mechanism in E. coli is based on the already mentioned MinCDE system (see Box 4.1).

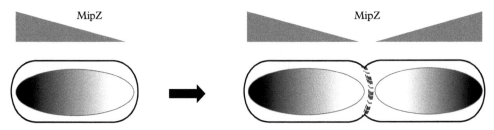

Figure 4.7 *The gradient of MipZ protein guiding the position of Z-ring in C. crescentus. MipZ is a two-state protein, which in one of the states forms dimers (two-unit complexes). A different, protein ParB, which is located near the poles of the cell, is an enzyme that changes the state of MipZ to dimeric. The dimers are not stable and as they move away from the poles, they tend to split into single proteins. Thus, the dimer concentration decreases in the direction from the poles to the 'equator', forming a dimer–singleton gradient, shown dark to light in the figure. MipZ dimers are the Z-ring inhibitors. Only when the cell has grown long enough to be ready for division, near the 'equator' the MipZ dimer concentration is sufficiently low so that the Z-ring can form (Kiekebusch et al., 2012; Monahan and Harry, 2013).*

Box 4.1 Does a cell 'know' its size?

Is there a feedback from the size of the cell at the particular moment to the molecular processes making the cell grow? How can such feedback be achieved? At any particular time, each protein or protein complex 'sees' or 'feels' only its neighbours, but a protein can 'gauge' the concentration of different molecules by how frequently it 'meets' them. Indeed, this is how *E. coli* 'knows' that lactose is present in the environment, as was discussed in Chapter III. However, the overall size of a cell cannot be measured this way, the cell size is a global property of the entire cell. If you are in a large dark cave, it is not possible to tell how big the cave is by just feeling around one's immediate neighbourhood.

One possible way to measure the dimensions of a dark cave is with a radar-like mechanism—clap your hands and count how many seconds it takes for the echo to arrive. Does a cell have such a mechanism? Is there anything that 'tells' an *E. coli* cell, how long it has grown? Indeed, possibly the already mentioned MinCDE system does this. As described in Chapter III, the MinCDE system generates longitudinal waves going back and forth the length of the bacterium. It is possible that by making these 'waves', the Min system works like a 'radar' enabling the cell to 'measure' its size.[33]

But what does it mean that the cell 'knows'? No single protein or molecular complex knows the cell size. This 'knowledge' resides in the self-organizational processes of the cell, which depends on the size of the cell. This is what can be called an *emergent property* of the system—something that emerges only when individual parts interact in a wider whole and which cannot be derived from the properties of any of the parts taken separately.

The growth of the cell, the replication of DNA, the 'snapping' of the cell, and other processes need to be coordinated. In *C. crescentus*, there is an interesting molecular mechanism that tells the cell when it has just completed its division and can start the next cycle[34] described in Figure 4.8.

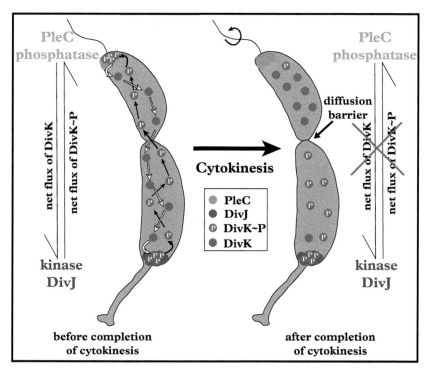

Figure 4.8 *How does a bacterium know that it is the time to start the genome replication? In* C. crescentus *the cell 'learns' that the previous cell division cycle has been completed via an interesting feedback loop mechanism exploiting protein called DivK. This protein has two states: DivK-P (phosphorylated) and DivK (dephosphorylated). Two different enzymes, PleC and DivJ are located at the opposite poles of the dividing cell. These proteins each change the state of DivK to the opposite; concretely, DivJ phosphorylates DivK, while PleC dephosphorylates it. As long as both parts of the dividing cell are connected, proteins in both states diffuse freely throughout the cell and thus DivK and DivK-P are distributed through the cell about evenly. However, once the two parts of a dividing cell have separated, DivK and DivK-P start to accumulate each in just one of the daughter cells. When DivK-P has accumulated in a sufficient abundance, it initiates a cascade of protein state change events, which leads to DnaA changing its phosphorylation state, binding to DNA, and initiating the replication, as mentioned in Chapter III. A series of other processes, including a change in DNA methylation state, ensures that there is only a small window of opportunity for the replication to the start (Biondi et al., 2006; Goley et al., 2007; Laub et al., 2007; Shen et al., 2008).*
(Figure from Matroule et al., 2004.)

To help to coordinate different cell cycle processes, about 15% of *C. crescentus* genes are transcribed in periodic waves, with the progression of the cell cycle.[35] As discussed in Chapter III, gene transcription is regulated by proteins called transcription factors, which bind to DNA upstream of the genes they regulate. In *C. crescentus* there is a 'master regulator' transcription factor appropriately called CtrA, which, on the one hand, regulates its own cyclical expression via a negative feedback loop (when there is too much of it, it switches itself off), and on the other hand, it regulates the expression of other genes, including the expression of FtsZ.[36]

All these cellular processes are highly stochastic; given that on a nano-scale everything is constantly jiggling, it is hard to achieve much determinism. Nevertheless, because different molecular processes are backing each other up, cell cycle progresses quite deterministically; cyclical gene transcription is only one of several redundant mechanisms regulating how cell functions.[37] A living cell is a reliable device made of unreliable components.

Autonomy and ecosystems

In nature no organism and no species are isolated: they all live in symbiotic relationships. Can we actually talk about any organism as an autonomous replicator? In the general sense of this word almost certainly not, however, *autonomous self-replication* defined in Chapter I is a term with a particular meaning. To be considered *autonomous*, the replicator has to be able to make its own approximate copies using parts simpler than itself, for many replication cycles, and without active structured help from the environment. In this way we define a subsystem, which though inevitably is a part of a larger system, can be studied scientifically. Identifying the simplest possible system or subsystem that allows one to ask the questions one is interested in is at the core of any scientific approach, be it called the reductionist or the systems approach.

Organisms can be classified in three groups based on the nutrients they need. The first group includes the organisms that need a range of relatively complex organic molecules, such as glucose, lactose, or amino acids—molecules that are typically produced by living organisms. *E. coli* belongs to this group, which is referred to as *oligotrophs*. On the other end of the spectrum there are *autotrophs*—organisms that only need inorganic molecules. An example of these is *P. marinus*, which, with its ~ 1800 genes, is the simplest known autotrophic bacteria. *P. marinus* obtains its energy from photons of the sunlight. Whether *P. marinus* can be considered as a true autotroph can be contested as it is not able to use (to *fix*) nitrogen N_2 from the atmosphere. *P. marinus* obtains nitrogen from ammonia (NH_3) or ammonia-like simple molecules, which can be considered as inorganic; for instance, ammonia is found in places of volcanic activity. In reality though, most ammonia on Earth is produced by other microbes, or in factories that make fertilizers. One of the simplest photosynthetic fully autotropic marine bacteria that can utilize atmospheric nitrogen N_2 are *Trichodesmium thiebautii*, which have ~ 3400 genes.

In between these two extremes, there are *heterotrophs*, such as *P. ubique*, which need only a small set of rather simple organic molecules, such as pyruvic acid ($C_3H_4O_3$). Pyruvic acid can be considered as 'dead' organic matter, plenty of which is dissolved in the ocean. With its 1300 genes, *P. ubique* is the smallest known free-living bacterium and it is thought to be the most abundant organism in the ocean and possibly on Earth—there are an estimated 10^{29} of these bacteria.[38]

Can any of these bacteria be considered as autonomous according to our definition? Whether inorganic molecules, small organic molecules, or even biological macromolecules are considered as the elementary parts, they all are much simpler than bacterium itself, thus this requirement seems to be satisfied. It is a trickier question to what extent the environments that bacteria need to be able to grow and divide can be regarded as inactive. The favourite environment of *E. coli* is the human gut, which certainly is not inactive or unstructured. *P. marinus* and *P. ubique* live in the ocean, which is not inactive either, however it can be argued that these bacteria are replicating not because of a structured help from to ocean. Bacteria indeed are swimming in their environment like von Neuman's hypothetical kinematic replicators in their container.

But our definition of autonomy does not depend on the nature of their natural or optimal environment, but rather on what is the simplest possible environment in which the bacteria are still able to replicate for a significant number of generations. One of the reasons why *E. coli* is such a popular model bacterium is because it is easy to grow it—to cultivate it—in a laboratory. This however is not true generally; most bacteria are rather difficult if not impossible to cultivate. In some cases, possibly this is because not all the molecules that these bacteria need in the environment, or their concentrations are known, or what other molecules are harmful. But there may also be more complex factors. The symbiotic communities in which bacteria live may not only provide bacteria each other with the necessary molecules, but may also form intertwined regulatory feedback loops between different species, maintaining the necessary balance of concentrations of different molecules. If so, we cannot really isolate these bacteria from their ecosystems, nor consider their environment simple. The field of molecular ecology, which studies such ecosystems on the molecular level, is amongst the most interesting branches of modern life sciences.

That said, some bacteria can be cultivated in quite simple environments, where for significant periods of time no complex feedback loops are required. This is true for *E. coli* and *C. crescentus*, as well as for *P. ubique* and *P. marinus*. All these can be cultivated in a laboratory on a small set of simple nutrients.[39] One could argue that maintaining a laboratory is not simple, however to satisfy the requirements of our definition we only need to prepare the environment in which the organism, when left alone, lives and replicates for a good number of generations. This is true for quite a few microbial species. At the same time, no virus can claim this.

This, however, does not mean that these bacteria are autonomous in their natural environments. For instance, the *Prochlorococcus* genus consists of at least 10 species living side by side and benefiting from each other and from other species. None of the *Prochlorococcus* species have genes for catalase—the enzyme that breaks down hydrogen peroxide H_2O_2. This molecule is dangerous for bacteria but it is abundantly present in the ocean. *Prochlorococcus* depend on other species for catalase to make their environment inhabitable.[40]

How similar are the clones?

How similar are the daughter cells to each other and to their bacterial 'mother'? Can they be considered *clones*? The molecules that are present in the cell in many copies, say a thousand or more, will distribute evenly amongst the dividing cells to maximize the mixing entropy. However, what if the mother cell has, say, five copies of a protein?

It is not inconceivable that a fluctuation could result in all of them ending up in just one daughter. Molecular imaging has made it possible to measure such fluctuations of molecular abundances, and we now know that differences in protein abundances can accumulate and be inherited for several generations.[41] Nevertheless, unless the division goes so wrong that one of the cells die, over the generations these differences will even out.

As noted by Jacques Monod in his famous book *Chance and Necessity*,[42] the main invariant of any living system is its genome. Bacteria are stochastic physical systems, the replicas will tend to vary from each other somewhat, but the information in the genome remains invariant. As we will discuss in the next chapter, for a replicator to be viable, DNA copying has to be extremely accurate and if the DNA replication error rate is above a certain limit, the replicators will die out.

Nevertheless, errors do occur during genome replication. These, the so-called genetic *mutations*, will be passed on to future generations as long as the replicator survives.[43] In *E. coli* one error occurs in several billions of letters per replication, a recent estimate reports one error per 11 billion letters copied.[44] Given that the *E. coli* genome is about 4.5 million letters long, there is about one error per every 2000–3000 replications. For most bacterial species, the mutation rate per replication per genome is about the same— in the range of one error in 1000–10,000 replications.[45] But given that $2^{10} = 1024$, this means that starting from just one bacterium, a genetic error can be expected to occur in at least one of its 'clones' after 10–12 replication cycles.

The simplest possible living cell

How complex does the simplest possible living cell have to be? How many different genes does it need to be able to grow and replicate? Amongst some of the simplest known bacteria, *M. genitalium* has around 500 genes, the simplest free-living bacteria *P. ubique* over 1000 genes. But these bacteria have survived for millions of years and as we saw, have a degree of redundancy built into them. Are all their genes essential? Can bacteria live on a smaller number of genes?

To account for different molecules that are needed to implement the processes of the central dogma, we need around 10 different proteins for the DNA replicase, four for an RNA transcriptase, 54 proteins and three RNA molecules for a ribosome, plus genes for at least 20 different tRNAs and as many enzymes for attaching amino acids to these. This sums up to well over 100, to which we still need to add genes for proteins that make up the machinery synthesizing the necessary small molecules, such as ATP or phospholipids. When describing the bacterial cell cycle, we introduced another two dozen genes. And even this leaves out genes for other essential proteins, such as the trans-membrane transporters maintaining the intracellular environment. Thus, it is hard to imagine a growing and dividing bacterial cell without at least 200–300 genes.

Although this is a rather rough estimate, more rigorous approaches give similar numbers. In 2000, Eugene Koonin at the National Center for Biotechnology Information in Bethesda, Maryland compared the genomes of bacteria sequenced at that time to find which genes were common to all bacteria.[46] His study suggested about 250 genes. In a different approach, George Church's group at Harvard University, designed the 'minimal genome' to cover all necessary functions of the cell, consisting of 151 genes contained in a genome of ~ 130,000 of nucleotides.[47] Looking at the gene list proposed by Church, it seems that their model is implementing a 'biochemical commune' (see next section) rather than an 'individualistic cell'. If we add the genes needed for cell membrane and division, the number will easily increase to 250.

Can these estimates be confirmed experimentally? One way to go about finding the minimal genome is to take a bacterium, knock out its genes one by one, and see if the cell is still growing and dividing. In the minimalistic *M. genitalium*, about 350 of the 482 protein coding genes turned out to be essential in this sense. In the larger *C. crescentus* and *E. coli*, the proportion of essential genes re smaller: 480 out of the 3700 in the first, and 620 out of 4300 in the second.[48] However, this does not mean that the respective bacteria can survive on 350, 480, or 620 genes, as the experiments only tell us what happens if genes are knocked out one at a time. Knocking out gene combinations is cumbersome and there are many combinations to try. In a landmark experiment Craig Venter and colleagues took a more radical approach. They designed their own minimal genome, synthesized it, and then implanted it into a bacterial cell from which the original genome had been removed. They used *Mycoplasma mycoides*, which originally had around 900 genes, as the host.[49] After several attempts to find the minimal working genome, they came up with a combination of 438 protein- and 35 RNA-coding genes. This is the smallest known working genome at the time of writing this book.

Unsurprisingly, it turned out that almost half of the essential genes were involved in making proteins, that is, in transcription and translation. The second largest group, consisting of over 80 genes, were involved in making cell's membrane. Given that a cell's membrane needs quite complex phospholipid molecules, this is not unexpected. The third largest group was about 80 genes involved in metabolism, such as synthesis of ATP and other small molecules. In comparison, 36 genes were involved in DNA replication. (Quite a few genes for essential auxiliary machinery, such as topoisomerase, need to be added to the 10 genes for DNA replicase.) The DNA replication is a process simpler than gene expression, just like in von Neumann's replicator, component R (the tape replicator) is simpler than component U (the Universal Constructor).

The essentiality of the gene is not a yes/no concept. Suppose that in the absence of a certain gene, the cells keep growing and dividing, but at a much slower pace, and their replication is often failing. Is this gene essential? And suppose we have two such 'semi-essential' genes, which when both disrupted made the cell unviable. Thus, essentiality is either a quantitative property, or a threshold needs to be defined to decide one way or the other. This is a consequence of bacteria being reliable machines made of unreliable components. Venter introduced graduation—the truly essential genes, the other genes in the minimal genome, and the dispensable ones. Only 20 genes turned out to be truly essential. Strikingly, the functions of 13 of the 20 were unknown. Somewhat surprisingly, FtsZ gene turned out not to be essential (see Figure 4.9 and Box 4.2).

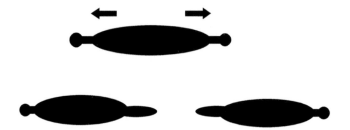

Figure 4.9 *The wall-less* Mycoplasma pneumoniae *cells contain large protein complexes called* terminal organelles. *The cells can divide by terminal organelles 'crawling' in the opposite directions until the wall-less cell raptures.*
(Figure adapted from Krause and Balish, 2001.)

Box 4.2 Dividing without a ring

The FtsZ gene is present in almost all bacterial species,[50] even in the wall-less *M. genitalium*. This gene, however, was not amongst the essential genes in Venter's experiment. Bacteria of Mycoplasma genus do not have a cell wall, and therefore it is possible that their division can be achieved via simpler means. Indeed, some Mycoplasma species are known to divide in a different way, using large internal protein complexes, called terminal organelles.[51] If such a bacterium is placed on a solid surface, the terminal organelle attaches itself to the surface and the bacterium may start crawling in a particular direction. Often these cells have two such organelles and the hypothetical mechanism of Mycoplasma division in the absence of FtsZ is that two terminal organelles crawl in opposite directions, stretching the cell, until the membrane ruptures as shown in Figure 4.9. Although this may be a dangerous strategy, it seems to work well enough.[52]

Terminal organelles are large and complex molecular arrangements, and it is unlikely that Venter's minimal bacteria used this mechanism. Is something simpler possible? Recall the growing lipid vesicles discussed in the Chapter II, Figure 2.12. The volume of a round bacterium is proportional to the cube of the radius, while the membrane surface, to the square. Therefore, both the cytoplasm and the membrane are synthesized at a constant rate, the bacterium cannot remain spherical. It possibly keeps elongating until it eventually snaps. There is a known phenomenon where some bacteria can temporarily lose their cell wall—these are known as L-forms—and indeed, L-forms can divide simply by synthesizing 'excess' membrane.[53]

Thus, the minimal gene set needed for a bacterium to live seems to be somewhere between 400 and 500 genes. Note, however, that *Mycoplasma* are oligotrophs, which benefit from quite complex organic molecules synthesized by their hosts and quite a controlled environment. It has been hypothesized that the necessary minimum for a free-living, self-sufficient heterotrophic bacterium is closer to 1000 genes.[54]

Venter created a synthetic genome, which was implanted in a pre-existing cell, the original genome of which was removed. The genome was synthetic, the cell was natural. Can a truly synthetic cell—a truly artificial molecular replicator—be assembled from non-living molecules? Assembling a living cell from molecules is still beyond the realms of current biotechnology, but there are no compelling reasons why this could not be achieved in principle.[55]

Biochemical communism

Although the number of genome copies in bacteria are strictly controlled and usually there is just one copy, in some archaeal species, the number of genome copies may vary.[56] For instance, *Methanocaldococcus jannaschii* can contain from one to five genomes, which can increase to 15 at rapid growth in optimal conditions. In a different one, *Haloferax volcanii*, 18 genome copies have been counted. The most extreme case is the giant bacterium mentioned in Table 4.1, which may contain thousands of genomes. Having many genome copies may simplify the bacterial division, as then the genome separation does not have to be so strictly controlled.

Can we go further and discard an individual cell entirely? Can genomes be floating in a cytoplasmic 'soup' without any membranes? Moreover, why do we need genomes? Can genes simply float in the environment, make proteins, and keep replicating? Would such biochemical communism work? In some cases, for a limited period, this does seem to happen in nature. However, in the long run, biochemical communism fails for the same reason as the social communism—parasitic elements emerge and take over. In biological communism, the parasites are DNA elements that are very good at promoting their own replication, but not contributing to the environment. It is possible that such a soup was how life emerged, but as we will see in Chapter V and VI, in the long run, at least some degree of individualism seems to be necessary.

According to Maynard Smith and Szathmary's *Major Transitions*,[57] the linkage of individual genes in genomes (or, more generally, in chromosomes) and the emergence of an individual cell are the first major evolutionary transitions after the emergence of the first replicators—the transition from a soup of genes to genes organized in genomes and individual cells. The American evolutionary ecologist Egbert Leigh characterized a genome as a parliament of genes: '*Each [gene] acts in its own self-interest, but if its acts hurt the others, they will combine together to suppress it*'.[58] This is a democratic alternative to communism.

The RNA world

The physical processes at the core of the central dogma link the essentially one-dimensional world of information in DNA, to the three-dimensional world of molecular machinery, which is largely based on proteins. This link is established by RNA, which has stakes in both worlds. As noted in Chapter III, RNA molecules mostly remain single stranded, however complementary fragments of the same strand can form bonds with each other, as shown in Figures 4.10 and 4.11. In this way an RNA molecule may fold in quite complex three-dimensional structures.

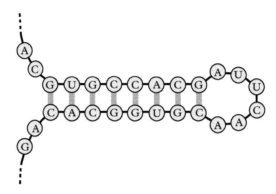

Figure 4.10 *Two RNA sequences* complementary *to each other (nucleotide A is complementary to U, and C to G) tend to form so-called* hairpin *or* stem loops.

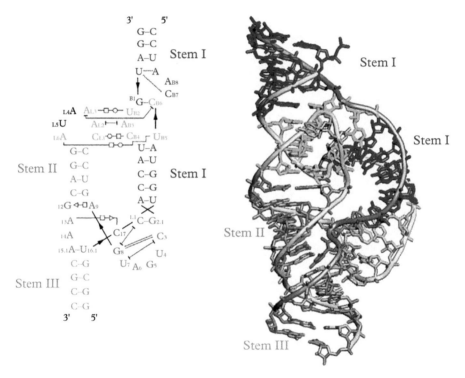

Figure 4.11 *Hammerhead ribozyme.* **Left.** *the nucleotide pairing of RNA, that defines the three-dimensional structure of the molecule.* **Right.** *A model of the actual molecular structure of this ribozyme. The black lines show interactions other than complementary nucleotide pairings.*
(Figure from Przybilski and Hammann, 2006).

RNA is quite a special molecule. On the one hand, RNA can store information that can be easily read; as already mentioned, there are viruses whose genomes are based on RNA rather than DNA. On the other hand, RNA can function as three-dimensional molecular machinery, implementing molecular processes and catalysing chemical reactions. RNA-based enzymes, called *ribozymes*, were first discovered in the 1980s, by the US scientist Thomas Cech at the University of Colorado, Boulder, and the Canadian-American scientist Sidney Altman[59] at Yale University. This earned them the Nobel Prize in Chemistry in 1989 'for their discovery of catalytic properties of RNA'. In contemporary living 'matter', ribozymes are not as widespread as enzymes, nevertheless they do play essential roles. RNA molecules are major components of a ribosome and detailed studies show that the main catalytic function of this multi-unit molecular complex is provided by RNA rather than proteins. One example of a 'pure' ribozyme, acting without proteins, is the so-called *hammerhead* ribozyme, shown in Figure 4.11, which catalyses cleavage and joining of other RNA molecules.[60]

Is it possible to build a replicator without DNA and proteins, where RNA assumes the roles of both—information carrier and interpreter of this information? In support of such an idea, in 1986, Walter Gilbert and colleagues proposed the concept of the 'RNA world'.[61] If one could replace every enzyme in the minimal cell with a ribozyme, this would free up almost 200 genes involved in translation, thus almost halving the complexity of the minimal cell.

The possible existence of such a ribozyme-based living cell is still a speculation; it is not known if all necessary biochemical reactions can be catalysed by ribozymes. This system would have to synthesize the ribonucleotides, phospholipids, or other lipid molecules for the cell's membrane and its other components, as well as to make some sort of energy-rich molecules like ATP. The progress towards understanding if such an RNA-based cell can indeed exist is slow.[62]

For the RNA world to exist, we need RNA-based enzymes that can synthesize other RNA molecules, that is, a ribozyme that can play the role of an RNA polymerase.[63] Moreover, this ribozyme needs to be general enough that it can synthesize a wide variety of different RNA molecules from their complementary templates including itself. No such general RNA polymerase ribozymes are known to exist in nature; however they have been made in a laboratory. In 1991 Szostack and colleagues at Massachusetts General Hospital showed that an RNA-molecular complex can have a ribozyme activity and can serve as a catalyst to synthesize itself from RNA templates. But to have a self-contained replicating system, these different RNA molecules need to be held together as a unit, otherwise communist parasites are likely to take over, as discussed above. We will return to a discussion in Chapter VI, but to finish this discussion for now, note that if there were a single self-replicating RNA-molecule, the parasite problem could potentially be avoided, see Box 4.3.

Box 4.3 RNA quine

Does a single RNA molecule exist that can synthesize itself? In 2001, details of a ribozyme called 'Round 18', just over 160 nucleotides long, was published,[64] which was able to synthesize RNA molecules up to 14 letters long from their templates. Ten years later, a ribozyme[65] about 200 ribonucleotide long, named tC19Z, could synthesize RNA molecules about half its own size, making about one mistake in a hundred ribonucleotides. In particular, tC19Z could synthesize the hammerhead ribozyme. Perhaps one day a ribozyme-polymerase able to synthesize itself will be found.

Let us denote an RNA polymerase ribozyme sequence by R+, and its reverse complement by R−. Then, given that R− is a template for the synthesis of R+, the pair [R+, R−] would synthesize a new ribozyme R+. If R+ was able to use its own copy as a template, then the pair [R+, R+] would produce a new molecule R−. Thus, the process of self-replication could kick off: [R+, R−] would produce another R+, and [R+, R+] another R−. Thus [R+, R−, R+, R+] would make [R+, R−, R+, R+, R+, R−] and so on.

However, we also need to consider that copying errors can happen, recall that the error rate of tC19Z was about 1%. Thus, R+ is also likely to make imperfect copies, let us denote them by R'+ and R'−. These too will be multiplied. Experiments show that the replication rate depends not only on the replicator (the ribozyme), but also on the template—some templates replicate faster than others. If a replication mistake creates a template that is good at presenting itself for replication but is not good as a replicator—it becomes a parasite. This is the problem of the biochemical communism. To fight it, we need R+ and R− be kept together as a unit.

Theoretically this can be achieved by a palindromic RNA polymerase. A palindrome is a sequence of letters that read backwards has the same meaning. A popular example is the phrase: 'Eva, can I stab bats in a cave?' In the context of RNA or DNA, when we read the word backwards we have to replace the nucleotide letters with the complementary ones. An example of such a polydromic RNA sequence is AUCCGCGGAU. More generally, sequence R+R− (the sequence of R+ followed by its complement R−) is a palindrome.

Now, suppose R+R− has polymerase activity and is able to synthesize itself. Since this sequence equals its own (reverse) complement, we have just one molecule that self-replicates by working on its own copy as a template. If it now makes a copying mistake that disrupts its polymerase function, it stops replicating and thus parasitism does not emerge. If such a molecule existed, it would be an ultimate self-replicating physical object containing its own description, in other words, a molecular *quine* (Chapter I).

A living cell and the Universal Constructor

Let us return to the analogy between a living cell and von Neumann's self-reproducing automaton. The most non-trivial part of von Neumann's device was the Universal Constructor. Can a living cell be viewed as a universal constructor? The Universal Constructor has to be able to execute arbitrary instructions given to it on a tape. For a living cell, the tape is DNA. Inserting a new DNA fragments into the genome would be analogous to adding code(D) to the tape in von Neumann's replicator, as discussed in Box 1.1.

This, for instance, can instruct the Universal Constructor to produce a bicycle in every replication cycle. And indeed, additional genes can be inserted in a genome, making the cells produce substances that they do not produce naturally. Such genetic engineering is used for making medicines, such as insulin. And it is possible to go even further. Venter and colleagues replaced the entire genome of the cell with a genome from a different (though rather close) species. In several generations, the cell got converted into the species encoded in the new genome; the cell did what the instruction 'tape' told it to do.[66] Thus, an analogy between a living cell and the Universal Constructor seems to be there.[67]

Nevertheless, there are apparent differences between von Neumann's kinematic replicator and a living cell. Von Neumann's replicator is a deterministic device, while a living cell is a highly stochastic system, which achieves its reliability via a high degree of redundancy. Von Neumann's replicator builds its copy sequentially, step by step, while the processes in a living cell are highly parallel. The copy of von Neumann's replicator is made stillborn and then brought alive by an 'ignition' signal, while a living cell replicates by growing and dividing, and all instances are alive all the time. A central concept in biology is that of a gene, but in von Neumann's automaton we did not have its direct analogue.

On a closer look however, none of these differences are as clear-cut as they may appear. We have already discussed the concept of stochastic automata, how to build a reliable devices from unreliable components, and how this relates to genes (Box 1.1). The self-replicating configuration in the *Game of Life* uses a high degree of parallelism—the states of the cells on the grid are recomputed in parallel. Moreover, a self-replicating configuration in *Life* would be a continuously 'living' structure.

Still, there is one fundamental difference—a living cell exploits the molecular processes of self-organization. Von Neumann did not consider the possibility that his Universal Constructor would only need to assemble linear elementary part 'chains', which would then fold into the right three-dimensional structures simply to 'comply' with the laws of physics. But this is how proteins fold. In a living cell, the link between the one-dimensional sequence of monomers in DNA and the three-dimensional structure of the protein is 'provided' by the tendency of matter to minimize its free energy. The folded proteins further self-assemble in functional complexes, spread around the cell, and build increasingly larger and more complex structures. The same is true in the hypothetical RNA world. Thus, in a living cell, the conversion of the one-dimensional information into three-dimensional structures is 'outsourced' to physics. A living cell can be viewed as von Neumann's kinematic replicator that uses the processes of self-organization and self-assembly.

This strategy of exploiting the processes of self-organization and self-assembly works because only a tiny minority of all theoretically possible amino acid sequences have been carefully selected for the specific structures into which they fold. Where does the information about which few of these over 20^{300} possible sequences to be used come from? This is the topic of Chapters V and VI.

V

Evolving Replicators

Life is what replicates and evolves.

(Martin Nowak and Hisashi Ohtsuki, 2008)

Without deviation from the norm, progress is not possible.

(Frank Zappa, 1971)

In 1988, Richard Lenski, who at the time was an assistant professor at the University of California, Irvine, started an experiment which more than three decades later, at the time of writing this book, continues and is still producing results. Lenski took a culture of *E. coli* bacteria called B-strain out of a freezer, defrosted it, split it into 12 parts, and placed each into a flask filled with identical media. The medium was unusual for these bacteria; there was only small amount of glucose, the favourite food of *E. coli*, but there was plenty of sodium citrate, which *E. coli* was unable to utilize under the aerobic conditions of the experiment. Would the bacteria placed in such an unusual medium change to adapt to their new environment? Would the changes be the same in different flasks, and would the molecular mechanism of any changes be explainable?

While in glucose-rich media *E. coli* replicates in less than in half an hour, in this glucose-poor medium, the average replication time was around three and a half hours. Nevertheless, in the comfortable 37 °C temperature the bacteria grew and multiplied, depleting the little glucose that was there.[1] At the end of every day, 1% of the bacteria from each flask were transferred to a new flask containing the same glucose-poor and citrate-rich medium. The other 99% were discarded, except that every 75 days some of them were put in a freezer where they became revivable 'fossils'.

Every day the bacteria underwent the so-called lag phase of waking up, followed by exponential growth, which later in the day slowed down due to the diminishing food, followed by hours of starvation. Thus, the conditions were seasonal; seasons of plenty were periodically followed by seasons of starvation. After each transfer, every flask contained around 5 million cells. In a day the bacteria underwent about 6.6 replication cycles, growing in numbers about $2^{6.6} \approx 100$ times, reaching half a billion. As replication leads to the population doubling in size at each cycle, even slow growth can lead to enormous numbers fast (see Box 5.1). As the experiment was continuing, the bacteria underwent some 2,400 division cycles per year and 24,000 in 10 years.

Living Computers. Alvis Brazma, Oxford University Press. © Alvis Brazma (2023). DOI: 10.1093/oso/9780192871947.003.0006

For humans, 24,000 generations would take over half a million years. When, on 19 February 2018, Lenski gave the Inaugural Saunders Genetics Lecture[2] at Cambridge University, England, his experiment had been going on for 68,000 generations—almost 1.5 million years if bacteria were humans.

Box 5.1 Exponential growth and large numbers

There is an ancient anecdote about a chessboard, grain of rice, and an emperor, in one version of which a man persuades the emperor to pay for something in rice by putting one piece of grain on the first square of the board, two on the second, four on the third, eight on the fourth, and so on, doubling the amount on each next square, until reaching the last—the 64th square. Can anybody blame the emperor for his foolishness if even in the computer age many of us may not realize that this is unaffordable—one would need about 1000 times more grain than the current global rice production. The total number of pieces on the board comes to 18,446,744,073,709,551,615.

Such doubling, or more generally, multiplying at every step by a constant factor is called *exponential growth*. Not very much seems to be happening at the beginning. The initial numbers are 1, 2, 4, 8, 16, ..., and even on the tenth square there will be only $2^{10} = 1024$ pieces, just over 1000. Assuming that one piece weights 30 milligrams, on the twentieth square, there will be about 30 kilograms. This will take a large chessboard, but it is not a major problem. The emperor's problems really begins only when he gets well into the second half of the board.

Charles Darwin realized the implications. In 1859 in *On the Origin of Species*—the book that changed our understanding of living 'matter' forever, he wrote:

> The elephant is reckoned to be the slowest breeder of all known animals, and I have taken some pains to estimate its probable minimum rate of natural increase: it will be under the mark to assume that it breeds when thirty years old, and goes on breeding till ninety years old, bringing forth three pairs of young in this interval; if this be so, at the end of the fifth century there would be alive fifteen million elephants, descended from this first pair.

Even if there are many deaths occurring during replication, as long as the birth rate exceeds the death rate, even marginally, the growth is still exponential. The value of the expression $(1 + s)^n$ will get out of hand sooner or later, no matter how small s is, as long as $s > 0$. Whether it is 1.1^n or 1.01^n or 1.001^n, for large enough n we will obtain exploding numbers. To keep the size of the population constrained, from some point after each cycle the 'excess' of individuals will have to be culled. If the population is large, say a billion replicators, then even 0.1% advantage of replication over death in each step provides us with a million 'excess individuals'. In Lenski's experiment almost half a billion cells were discarded every day. But these excess individuals can also be used to experiment if changing something in them improves their fitness. And if it does not, we have lost nothing.

If we have a population of several different types of individuals, one fitter than the others, the fittest one takes over. Although all populations grow exponentially, the ratio between the number of the fitter type to the number of the less fit grows exponentially too. Relatively, after sufficient time, the fraction of copies of the slower replicators will be negligible. Those who have experienced the COVID-19 pandemics in the early 2020s have seen this—every SARS-CoV-2 variant with a minor advantage over the previous one soon took over. And, if the population is constrained, then the faster replicator will eventually squeeze out the slower one entirely. Any fitness advantage, say 0.01% or 0.001%, leads to the fitter population taking over after enough time has passed.[3] The things become more interesting if the competing populations interact in a way that their abundances change their relative fitness values.

The enormity of the numbers that result from exponential growth is one of the reasons why some computable information processing tasks, may not be computable in practice. This is also one of the reasons why predicting the behaviour of a complex biological system is so difficult, often impossible.

Lenski monitored two key features of his bacteria—the average size of the cells and their *relative fitness*. The fitness of an individual replicator quantifies a combination of two factors: how fast it replicates over a number of cycles and how robust it is—what is the probability of a successful replication compared to death. At population level, fitness can be interpreted as the rate at which a population grows or shrinks. If we have several populations, we can measure their fitness relative to each other. If we use one of them as a reference, assign it fitness $f = 1$, then, relative fitness $f = 2$ means that the particular population grows twice as fast as the reference, while $f = 1/2$ that it grows at half the reference speed.

To measure the fitness of each of his 12 experimental populations relative to the original B-strain, Lenski had found a strain largely identical to B-strain with the exception of colour. The coloured strain was frozen before the start of the experiment and, therefore, some of it could be defrosted whenever needed. By growing a population mixed with the defrosted coloured strain and monitoring the changes in the colour, relative fitness could be estimated. For instance, if the fitness of the bacteria in one of the flasks had increased, then after mixing it with the coloured strain and growing the mixture for some time, the colour of the mixed population would gradually get paler.

By 1994, the bacteria in the experiment had undergone 10,000 generations and Lenski could test if the bacteria had changed.[4] It turned out that in all the 12 populations, both the fitness and cell size had been continuously increasing, but at slowing rates, as shown in Figure 5.1. If one looked at the measurements more closely, it appeared that the cell size had been increasing in steps, indicating that the increase was due to discrete events. Later it became clear that the fitness had continued to increase with gradually diminishing gains through the entire long-term experiment.[5] It was likely therefore that the bacteria were continuously adapting to the new, glucose-poor and citrate-rich environment, but at decreasing rates.

Cell size is just one of the phenotypic features or *phenotypes*—observable characteristics of an organism. For larger organisms, such as animals or plants, there are a plenty of phenotypic features that can be described or measured; for example, the length of the neck, the colour of eyes, the number of limbs, or the shape of a flower. Bacteria too may differ from each other in shape and appearance, however arguably the most important phenotypic feature for bacteria is how fit they are in various media; that is, what food they like. This can be measured by placing the bacteria in a medium with particular nutrients and monitoring how well they grow. Throughout his long-term experiment, Lenski was

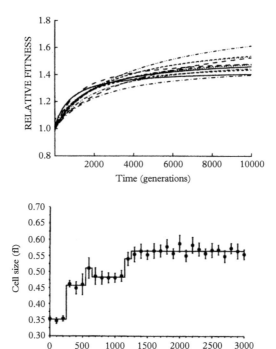

Figure 5.1 *Top. The* E. coli *fitness growth in Lenski's experiment over the first 10,000 generations. After initial rapid increase, in all 12 populations the fitness growth slowed.* **Bottom.** *Cell size was increasing during the initial 3,000 generations in discrete steps. Each point is the mean of the 12 populations.*
(Top: From Lenski and Travisano, 1994. Bottom: From Elena et al., 1996.)

measuring the fitness of his populations in the glucose-poor/citrate-rich medium. However, after a few thousand generations, he also checked their fitness in a range of other media. It turned out that the bacteria had been changing their dietary preferences— having not seen some types of food for generations, they had been losing their appetite for these. Some of the 12 lineages could gradually regain their ability to utilize the old food again, but each of the 12 populations behaved somewhat differently.[6] Clearly, the bacteria were changing, and not exactly the same way in all 12 lineages.

After the experiment had been going on for about 15 years, one morning, Lenski discovered that in one of the flasks the medium had grown turbid.[7] He defrosted some of the earlier samples and 'restarted' them. Three weeks later the turbidity reappeared, the phenomenon was reproducible. Moreover, Lenski found that it was not caused by contamination. More investigations showed that after 30,000 generations at some point, *E. coli* had started acquiring a new nutritional habit—they had 'learnt' to consume citrate. When, after considerable additional research Lenski finally published his findings, he was much too modest to claim that a new bacterial species had evolved in his laboratory, but this would not have been a massive overstatement.[8] One of the defining phenotypic features of *E. coli* as a species is their inability to utilize citrate in the presence of oxygen. One of the 12 populations in the experiment had broken this trait. This was like a carnivore animal becoming an herbivore after being starved of meat but having plenty of vegetables on offer. Although it took genome sequencing to find out what exactly had happened to the cells, it was already clear from the experiments that these bacterial populations were evolving.

Changing genomes

Possibly it was a coincidence, probably an inevitability, or perhaps it was Lenski's foresight, that he began his long-term experiment right at the onset of the genome age. The first *E. coli* genome was sequenced in 1997; another decade later genome sequencing had become so straightforward and affordable that one could sequence a bacterial genome at will. Lenski teamed up with a small international group of scientists and sequenced several genomes from his evolving bacteria at different generations; at first, from just one of the 12 lineages. By that time the ancestral B-strain sequence was already available, therefore, by aligning the respective genomes sequences and comparing the letters, Lenski could see if anything had changed.[9]

As expected, Lenski and colleagues found that as *E. coli* were adapting to the new environment, its genome was changing. In the first 20,000 generations, 45 changes had occurred, on average one change in every 400–500 generations. Of these, 29 were so-called *point-mutations*, where one nucleotide in the over 4 million nucleotide genome DNA had changed to a different one. The others were either deletions, insertions, or inversions of pieces of DNA of different sizes. Of the 29 point-mutations, 26 had happened in genes coding for protein. Given that 26 out of 29 is about 90%, and that protein-coding genes take about the same percentage of the *E. coli* genome, this was about what could be expected by chance.

The locations of the 26 mutations, however, did not appear random. As noted in Chapter III, the genetic triplet code is redundant: more than one triplet can code for the same amino acid. For instance, ATT, ATC, and ATA all code for the amino acid *isoleucine*, therefore mutations that interchange T, C, or A in the third position of these triplets will not change the coded protein. Such mutations can be viewed as *synonymous*. In contrast, as the triplet ATG codes for a different amino acid, *methionine*, a mutation that, for instance, changes C in ATC to G will change the protein. It can be calculated that slightly over 20% of all possible nucleotide substitutions in *E. coli* genes are synonymous. Therefore, by chance, about five of the 26 observed mutations would be expected to be synonymous. However, none was. The probability that this could be a coincidence can be calculated to be less than 1%. Not impossible, but highly unlikely. The mutations that had spread and established themselves in the population, seemed to be somehow selected to impact the proteins. Even more improbably by chance, the first 26 point-mutations had occurred in only 14 out of the ~ 4300 *E. coli* genes. Later Lenski found that by generation 68,000, half of all mutations had occurred in just 57 genes, which is less than 2% of all *E. coli* genes. Why would mutations occur mostly in particular genes, moreover in positions likely to affect the coded protein?

Sequencing of more clones showed that synonymous mutations were sometimes occurring, but at a slow constant rate of about one mutation per 20,000 generations.[10] The rate of non-synonymous mutations was much higher, in particular initially, but were gradually slowing down. As noted above (Figure 5.1 Top), a similar slowdown was observed in the increase in the fitness of the bacteria, suggesting a potential link. The stepwise nature of the increase in the cell size (Figure 5.1 Bottom) too seemed to indicate discreate events, such as mutations.

To understand why the detected mutations did not seem to be random, one only needs to realize that many more mutations were likely to have occurred during the experiment than the ones that had become established in the population. As mentioned in Chapter IV, the mutation rate in *E. coli* is around one mutation per genome per 1000 to 10,000 generations. The size of the evolving populations in Lenski's experiment was never less than 5 million bacteria, therefore in 20,000 generations we should expect that well over 10 million mutations had occurred. But the genome sequencing found less than 50. Less than one mutation in 200,000 had spread through the population and was detected. This is not surprising; as 99% of bacteria were discarded every day, most mutants would have been discarded with them. Moreover, as the bacteria were discarded after their population had grown to half a billion, we could expect that the absolute majority of the mutants were indeed discarded before they had a chance to spread widely. Probably most readers will agree that it is quite intuitive to hypothesize that the mutations that had spread were the ones that helped the bacteria to adapt to the new environment. But can this be proven?

A death for every birth

Although the number of bacteria in each of the 12 populations fluctuated enormously every 24 hours, overall the populations were constrained. To keep a population constrained, for every bacterium that replicates, on average, there must be another one that dies (or is discarded). Such a process—one death for every birth—was studied mathematically by the Australian statistician Patrick Moran in 1957 and is now called *Moran process*.[11] In the Moran process, pairs of bacteria are repeatedly randomly picked: one for replication, the other one for death. The probability at which a bacterium is picked for replication rather than death is proportional to its fitness. In this way the Moran process is a simple model of evolution where two types of replicators are competing while the total size of the population stays constant.

What is likely to happen to a new mutant that has just appeared in the population? Initially there is only one copy, therefore, unless its fitness is very different from that of the original bacteria, the mutant has about 50:50 chance to disappear before it had replicated. If the mutant does replicate, there are two copies now, but each has 50% probability of dying in the next step. Consequently, the probability that a new mutation will survive for two cycles is less than 50%. As we move forward in replication cycles, for a while the probability that the mutant will still be around is decreasing. But if the mutant is lucky to survive until it has spread through more than half of the population, its chances of survival start improving. It can be proved mathematically that if a new mutant is *neutral*—if its fitness is the same as that of the ancestral bacteria—then the probability that in the Moran process it will spread through the entire population of N bacteria equals $1/N$. For proof see the excellent book by Martin Nowak, *Evolutionary Dynamics*.[12] As the population size was never smaller than half a million bacteria, no more than two neutral mutants in a million were likely to get established. And, given that synonymous mutations do not change the protein and thus are largely neutral, the probability that such a mutation would spread was rather small.

Unsurprisingly, mathematical analysis of the Moran process also shows that if a mutation has significantly increased the fitness of the bacterium, then such a *positive mutant* has a better chance of survival, while the opposite is true for a *negative mutant*. Concretely, the probability of a positive mutant spreading through the entire population is roughly proportionally to the *gain* in the relative fitness value.[13] For instance, if the fitness gain is 2% or 0.02—that is, if the mutant's relative fitness is $f = 1.02$ (assuming the fitness of the ancestral strain as $f = 1$)—then its probability of becoming established also equals 2%. In other words, on average, two in 100 such mutants are likely to spread through the entire population. Consequently, if the size of the population is half a million, a positive mutant with 2% fitness gain has more than 10,000 times a better chance to become established than a neutral one. Thus, under the assumptions of the Moran process, the selection of the fittest is a mathematical fact.

For a negative mutant, calculations show that if its fitness has dropped by significantly more than by $1/N$, this mutant will almost certainly disappear. On the other hand, if the mutant's fitness changes by less than by $1/N$ in either direction, the mutant behaves as if it was neutral. To reflect such an impact of the population size N, biologists talk about the *selection horizon*: if the mutation changes the relative fitness of the individual by less than the inverse of the number of individuals in the population, the selection of the fittest cannot 'see' this change and 'treats' the mutation as if it was neutral. This is an important observation as it implies that if for some reason the population has shrunk, negative mutations can spread. Accumulation of such invisible to selection mutations can play a role later if the population starts growing again. For large organisms, such as animals, particularly in bad times the population size may drop so much that the selection of the fittest at these times is weak. It is thought that in human evolution at some point in time, our population size dropped well below 15,000 individuals.[14] Various mutations might have been accumulating at that time, later possibly playing an important role in the evolution of human language. When a population of replicators is large and 'happy', most changes get suppressed and cannot spread. When the population is in trouble, changes can happen. This seems to be true both in biological, as well as in cultural evolution.

The impact of the population size should not be too surprising. If we toss a fair coin 1000 times, the fraction of heads and tails is likely to be close to 500 each. On the other hand, if we toss the coin just 10 times, we can easily get, say, eight heads and two tails. The greater the number of tosses, the closer the result is likely to be to the expected probability of 50/50. Similarly, if we have a slightly bent coin which, for instance, has a probability 55% of landing on heads and 45% on tails, then after tossing it 1000 times we are likely to have close to 550 heads and 450 tails. On the other hand, it is not likely that we will be able to infer any difference in the preference of heads over tails after tossing it just 10 times. When a mutant with a small fitness advantage has occurred and begins spreading, at each replication cycle it has some probability to replicate or die. This is like tossing a coin.

If we have a stochastic process that generates random mutations at a constant rate, then it can be shown that the neutral mutations are working like a clock; even though each individual neutral mutation has only a minuscule chance to spread, cumulatively they spread at a rate at which they are generated regardless of the size of the population. This fundamental mathematical result was discovered by the Japanese biologist Motoo Kimura in the first half of twentieth century and is known as theory of *neutral evolution*.[15] This has two important implications. First, at least on the molecular level, not all evolution has to be adaptive; evolutionary changes can be neutral drift. Second, the clockwise nature of neutral mutations provides us with a tool to perform genome archaeology— if two genomes have a common ancestor, comparing the number of synonymous and non-synonymous changes between them allows us to estimate when they began diverging. This is how, as will be discussed in Chapter VIII, we know when the last common ancestor of all present-day humans lived.

As any model, the Moran model is a simplified description of real-world processes and thus it does not explain everthing that can be observed. In fact, one of the purposes of building mathematical models is to try to refute them which then points to more subtle effects to be included in the next model. An important phenomenon not captured in the Moran model is *clonal interference*: when two different positive mutations compete, for instance, positive mutations outcompeting neutral ones. It can also happen that a neutral mutation occurs in the same clone as a strongly positive one and is spreading 'on the back' of this positive one.

Can we now explain Lenski's observations? After the first 20,000 generations, 26 point-mutations were detected in protein-coding genes, none of them synonymous. These mutations affected the coded proteins and through this, possibly, the fitness of the bacteria. As the bacterial fitness was increasing, we can hypothesize that the mutations that spread were likely to be positive. To test this, Lenski and collaborators synthetically reintroduced in the genome of the ancestral B-strain the first five of the mutations that had been detected in his long-term experiment. As expected, all of them turned out to be positive. Together, the five mutations increased the fitness of the population by the factor of $f = 1.35$. In other words, after these mutations had occurred, the bacteria were growing 1.35 times faster than the ancestral strain. Apparently, after *E. coli* was placed into a new environment, mutations increasing the fitness of the bacteria were 'wating to happen'. The selection of the fittest were finding and selecting them amongst many millions of mutations occurring at random. Possibly the mutations giving the highest fitness increase were spreading first, squeezing out those giving less advantage as well as the neutral ones via the clonal interference. However, as the bacteria were adjusting to the new conditions, there were fewer and fewer mutations that could give bacteria a large fitness advantage in one step. This is known in evolutionary biology as *the law of diminishing returns*.

Thus, extended with the empiric laws of diminishing returns and clonal interference, the Moran model seems to explain Lenski's observations quite well. Nevertheless, this is not yet a direct proof. Can we find a direct, 'mechanistic' link between a particular genome variant—*genotype*—and the respective phenotype?

From genotype to phenotype

More genome sequencing revealed more interesting phenomena. In the very first lineage that Lenski sequenced, something dramatic had happened after the first 20,000 generations. Sequencing this population at generation 40,000 revealed 653 point-mutations uncovered; most were specific changes of A to C or T to G. The mutation rate had increased 15-fold. Sequencing other clones later showed similar phenomena in several other evolving populations too. Moreover, all these *hyper-mutating* populations contained a mutation in the gene called mutT. We already came across mutT in Chapter III; this gene codes for a protein that is a part of the DNA proof-reading and error correction machinery. Specifically, this protein helps to prevent the formation of erroneous A–G nucleotide pairs. In DNA, the nucleotide A pairs with T, but if mutT gene is disrupted, some erroneous A–G pairs tend to form, which after the genome replication may lead to permanent erroneous C–G pairs.

Mutations in mutT gene fundamentally changed the evolutionary dynamics, creating what can be described as a *hyper-mutator phenotype*. Both synonymous and non-synonymous mutations began appearing at a fast rate. Initially, the faster mutation rate also correlated with a somewhat faster increase in bacterial fitness; later though this stopped, and moreover, in some of these populations the mutation rate had slowed down again.[16] Apparently, having such a high mutation rate for a long time was bad for bacteria; possibly further mutations had partly reversed the hyper-mutator effect.

The link between the mutT gene mutation and its effect is clear: by disrupting the DNA error correction machinery, this mutation leads to other mutations accumulating at a much higher rate. This hyper-mutator phenotype however is not a 'classic' phenotype: it is not directly observable without genome sequencing. Can we find an explanatory link between a mutation and a 'classic' bacterial phenotype?

Probably the most studied trend of bacterial evolution of them all is the emerging bacterial resistance to antibiotics. Antibiotics are poisons for bacteria, but at the same time they are relatively harmless for eukaryotic organisms, and thus can be used for treating bacterial infections. The first antibiotic was discovered in 1928 by the Scottish scientist Alexander Fleming, when he accidentally contaminated his bacterial colonies with mould—fungi of the genus *Penicillium*. Fleming noticed that this eukaryotic microorganism was killing bacteria. He isolated the chemical substance causing this and named it *penicillin*. In 1945, Fleming shared the Nobel Prize in Physiology or Medicine with the Australian Howard Florey and German Ernst Boris Chain, who were instrumental in converting this discovery into an effective medicine. There are few scientific discoveries throughout history that can claim to have changed the world as dramatically as antibiotics: before antibiotic drugs were developed, even a small infection could kill a patient and every surgical operation carried a deadly risk. Tellingly, in his Nobel lecture, Fleming predicted that bacteria would evolve resistance to antibiotics. We now know that there is a constant evolutionary race going on between various eukaryotic microbes evolving to produce antibiotic substances and bacteria evolving genes to resist them. Human enterprise has accelerated this race, but the race is not new. The more antibiotics are around, the more resistant bacteria become. Since the discovery of penicillin, over 150 different antibiotics have been found in nature or synthesized artificially.

To study the mechanisms by which bacteria acquire antibiotic resistance in a laboratory, bacteria can be grown in the presence of the particular antibiotics in low enough concentration not to kill the bacteria. A robotic device monitoring the growth of bacteria and automatically increasing the antibiotic concentration as the bacteria are developing the resistance can be used to speed up the evolutionary process. In one such experiment, bacterial adaptation to an antibiotics called trimethoprim was tested.[17] Trimethoprim works by inhibiting the synthesis of a bacterial enzyme called DHFR. The experiment demonstrated that the resistance to trimethoprim was developing in step with mutations, all of which were in the part of the genome coding for the catalytic centre (Box 3.1) of the DHFR enzyme. Thus, there was a direct link between each mutation and the evolving phenotype; moreover the biochemical mechanism of what was happening was clear too.

In Lenski's long-term experiment, the strongest example of a mechanistic link between a specific mutation and its phenotype was the already mentioned emergence of the ability of *E. coli* to utilize citrate. *E. coli* bacteria have all the enzymes needed for citrate utilization and can use citrate in anaerobic conditions when oxygen is absent. Moreover, spontaneous switching to citrate utilization in aerobic conditions has been observed in *E. coli* in nature.[18] The showstopper of why these bacteria normally do not use citrate when oxygen is present is that in these conditions a gene called citT is switched off. Apparently, in the wild, it is advantageous for *E. coli* not to express this gene in an aerobic environment. In Lenski's environment however, the ability to utilize citrate gave the cells an advantage.

The crucial mutation leading to citrate utilization turned out to be a duplication of an about 3000 nucleotide-long genomic region, which joined the citT gene with a promoter (Chapter III) of a different gene[19], thus activating citT. This was followed by subsequent amplifications and additional mutations that increased the expression of the citT gene further. Thus, the crucial event on the road to citrate utilization was linking an existing gene to a new promoter and thus creating a new gene regulation circuit.

Microbes in the wild

Until now we have only been discussing observations done in an artificial, laboratory environment. Do bacteria in the wild evolve in a similar way? One of the most widely known manifestations of bacterial evolution in nature is the emerging resistance to every new antibiotic drug that is invented. An example is the emergence of drug-resistant strains of *Mycobacterium tuberculosis* (*M. tuberculosis*)—bacteria causing tuberculosis. This bacterium was first discovered by the German physician and microbiologist Robert Koch in 1882, for which he received a Nobel Prize in 1905. In the eighteenth and nineteenth centuries, tuberculosis was epidemic in Europe and North America. The first effective treatment of tuberculosis became available only when *streptomycin*—an antibiotic effective against tuberculosis bacteria—was discovered in 1944. Other effective antibiotics were discovered later, and by the 1980s tuberculosis was in decline. However, in the 1990s the decline rate slowed down. The first drug-resistant tuberculosis case was detected in 1993 in New York and another one in Beijing two years later. The Beijing strain has 53 different mutations, mostly linked to genes associated with the resistance to antibiotics. By now a thousand different antibiotic resistance-causing mutations affecting some 40 genes have been discovered in tuberculosis microbes.[20]

Natural evolution has been observed in many bacteria, including *E. coli*. In 2016, my colleagues at the European Molecular Biology Laboratory obtained genomes from almost 900 strains of *E. coli*, some of which were already available but most of which they sequenced afresh.[21] Over 500 of these were isolates from the wild; the others came from various experiments. It is probable this is still only a small sample of the existing *E. coli* genome diversity, nevertheless it provides a glimpse into the evolution of this species.

Alignments of the available genome sequences revealed nucleotides differed between the strains. Large stretches of their genomes were identical, thus it was a reasonable hypothesis that these bacteria had evolved from a common ancestor. If so, the differences between the genomes were likely to represent mutations that had happened after the strains started diverging. Overall, my colleagues found over 600,000 point-mutations, most present only in one of the genomes, the other containing the same nucleotide in the particular position. On average, one nucleotide in seven was different in at least one strain from the respective nucleotides in the genomes of other stains. Thus, even though the genomes of the species were quite similar, by no means they were identical.

About 80% of the mutations that had happened inside protein coding genes were synonymous—not altering the coded protein. Recall that by chance only about 20% of mutations are supposed to be synonymous. This is in stark contrast to what happened in Lenski's experiment. It is possible that in the wild, *E. coli* bacteria do not often have to cope with such sudden dramatic changes in the environment as in Lenski's experiment and thus there are few positive mutations possible, while the negative ones are 'purified' by Darwinian selection.

Can we estimate how long it has taken for the sequenced *E. coli* bacteria strains to diverge from their common ancestor? Consistent with what we know about mutation rates in bacteria, let us assume that there has been one mutation per 5000 generations. If about half of these mutations were neutral—synonymous or not changing the structure of the protein significantly enough to change its function—we have about one neutral mutation every 10,000 generations.[22] Kimura's theory tells us that neutral mutations are spreading in the population at the same rate as they are occurring in the individuals. Consequently, for 600,000 neutral mutations to get established, about 6 billion generations were needed. It is thought that in the wild, on average, *E. coli* replication cycle is one to two days.[23] Thus, on the order of 10 billion days or roughly 20–30 million years would be needed to generate the observed genome diversity.[24]

Although these calculations are rather simplistic, the conclusion is consistent with more elaborate estimates. The bacterial species closest to *E. coli* is *Salmonella enterica* (*S. enterica*), which shares with *E. coli* roughly two-thirds of all genes.[25] Although it is difficult to compare the entire genome sequences of such different species, we can compare sequences of the genes they have in common. As ribosomal genes are shared by all known organisms, they can be used universally. By this approach, it was estimated that *E. coli* started diverging from *S. enterica* around 100 million years ago, which is consistent with our estimate above.[26]

By aligning and comparing gene and genome sequences, we can investigate the relationships between any group of species and try to reconstruct their putative evolutionary history.[27] If we align and compare every pair of ribosomal genes, we can find which species are the closest to each other. Moreover, from these alignments we can reconstruct the putative ancestral genes—the nucleotides that are identical in two related genes are also likely to be the same in the ancestor. By repeatedly joining up the most similar sequences (including these of the derived putative ancestors), a putative *phylogenetic tree* of these species can be reconstructed. Although the real phylogeny reconstruction algorithms are somewhat more complex, this is roughly how Woese discovered archaea (Figure 4.2). However, comparing ribosomal RNA does not tell the full story and deep evolutionary history cannot always be represented by a tree-like structure because species tend to exchange genes with one another as discussed below.

Pangenomes and exchange of information

Finding that all *E. coli* genomes were similar but not identical was hardly unexpected; it was more striking to find that from the about 4300 genes in a typical *E. coli* bacteria, only around 2000 were present in all strains.[28] On the other hand, in total, around 30,000 different genes could be counted across the 370 genomes. This total gene pool—*pangenome*—is what defines a bacterial species.[29] *E. coli* may thrive in any particular stable environment with just over 4000 genes; however, when the environment changes, other genes may be needed, while some existing genes may become superfluous. The entire *E. coli* pan-genome thus has more protein-coding genes than humans, but only around 2000 of these form the *core genome* present in every *E. coli* strain. The existence of a pan-genome is not restricted to *E. coli*, for instance, *Procholorococcus* pangenome has at least 6000 genes, while its core genome has fewer than 1300 genes.[30]

As the environment changes, the bacterial clones that have the 'right' genes, expand, the ones that do not will shrink. However, what also seems to play an important role is that genes that give bacteria an advantage in the particular environment can be transferred from one bacterium to another. Information is not only inherited from parent to offspring, but it also is exchanged 'horizontally' between cohabiting bacteria. Such horizontal gene transfer is one of the major ways in which bacteria acquire antibiotic resistance in the wild. As we will see in Chapter VI, eukaryotes exchange genome information regularly by having sex. Bacteria too have sex, though less regularly.[31]

There are several mechanisms by which horizontal gene transfer can happen.[32] When bacteria die and disintegrate, pieces of DNA can get into the environment, and as DNA is quite a stable molecule, other bacteria may pick these pieces up. Viruses may pick up and carry bacterial genes, transferring them from one bacterium to another. As already noted, in addition to their main chromosome, bacteria may contain little circular pieces of DNA, known as *plasmids*. These bits of DNA can integrate into the main chromosome, and vice versa—genes can escape from the main chromosome to form plasmids. Some bacteria harbour specific molecular complexes that facilitate the exchange of plasmids, which is the bacterial equivalent of sex. The B-strain used in Lenski's experiments was intentionally chosen such that it lacked this mechanism.

Gene transfer may happen not only between bacteria of the same species but also between different species, and even between bacteria and archaea.[33] Information exchange may be seen as a part of the symbiotic relationships between different bacteria, allowing them to keep their genomes smaller. Not only are materials and energy a symbiosis 'currency' but information too can be used in a trade. Horizontal gene transfer probably is an ancient mechanism present in all forms of life, and such transfer may have affected every single gene at some point in its evolutionary history.[34] Possibly this mechanism played an essential role early in evolution before the emergence of cells, when boundaries between 'individuals' were established (see Chapter VI).

However, gene import into an organism does not create new genes. Where did the 30,000 genes of the *E. coli* pangenome come from originally? And more generally, where do the billions of different genes present on Earth come from?

Complexity evolving

The mechanism of how life evolves new genes and how genome complexity can increase was proposed by the Japanese-American geneticist Susumu Ohno in 1970, in his influential book *Evolution by Gene Duplication*.[35] Much of what he described there turned out to be true, even though the process may be less straightforward than he depicted.[36] Nevertheless, the main principles are relatively simple: parts of an existing genome get duplicated, and then each mutate and evolve somewhat independently.

There are several ways in which a genomic region can get amplified; we will not discuss these in detail here beyond mentioning that one mechanism is related to the machinery that repairs damaged DNA.[37] Repair can go wrong and as a result part of the genome gets duplicated. Genomic regions that include repetitive or partially repetitive sequences are particularly prone to such duplication, leading to their further expansion and to many repeated or partially repeated regions. This mostly happens in eukaryotic genomes, though, as we saw in Lenski's experiments, genome region amplification happens in bacteria too.[38] As will be discussed in Chapter VI, eukaryotic genomes contain DNA sequences that actively promote their own amplification.

The most frequent fate of a duplicated gene is that one of the copies gets 'silenced' and then is lost.[39] Occasionally, however, gene amplification provides an advantage; for instance, when an antibiotic-resistant gene gets amplified, the abundance of the protein it produces may increase, which may help the bacteria to survive.[40] Importantly, when a duplicated copy of a gene is not lost, the other copy is now 'backed up' and can evolve more freely. Often, both copies tend to evolve faster, as the constraints on their evolution are now different.[41]

Genome-region duplication is also how new gene-regulation circuits evolve; for instance, a new copy of a gene can be joined with a promoter of a different gene, which was how citrate utilization evolved in Lenski's experiment. It is not always new genes that are created but new combinations of genes, or a new combination of genes and promoters regulating their expression, in other words—new gene regulation circuits. An import of a gene from different species may have a similar effect. By creating a material on which evolution can work, genome expansion is a mechanism how small ancient genomes evolved into increasingly larger ones. Given that complexity and information are two sides of the same coin, DNA amplification, followed by mutations and natural selection, is one of the most essential mechanisms by which evolution 'learns' from the environment and concentrates information into genomes (Chapter VII).

To summarize the chapter so far, let us note that all evidence seems to point to the conclusion that bacteria evolve through mutations in DNA, including amplification, deletion, and exchange of pieces of DNA, followed by the selection of the fittest. These observations do not yet prove that something else could not be happening too; as the famous detective Sherlock Holms said: 'absence of evidence is not evidence of absence'. Is there anything else going on in addition? One of the main principles of science is captured in what is known as Occam's razor: *plurality should not be posited without necessity*. This implies that given two competing theories, the simpler one is to be preferred. My personal favourite formulation of this principle is to try to *explain as much as possible with as little as possible*. The observations encountered so far could be explained by mutations and selection, as described above, and thus there is no need to evoke anything additional yet. But we will return to this question later.

Finally, let me note that although we tend to think of evolution as a progressive process creating organisms of increasing complexity, in reality evolution often reduces an organism's complexity. Genes that are not needed are forgotten and dropped.[42]

Fitness landscape and evolvability

As already mentioned, in one of the follow-up experiments, Lenski took the original B-strain and synthetically reintroduced in its genome the five mutations that in his long-term experiment got established first. Recall that these mutations jointly increased the fitness of the bacteria by the factor of $f = 1.35$. But Lenski also measured the fitness of their $2^5 = 32$ different subsets—the fitness effect of each individual mutation, every pair, every triplet, and every quadruplet.[43] The results are shown in Figure 5.2.

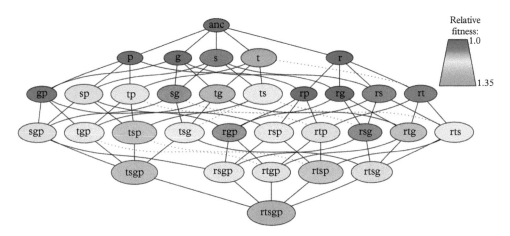

Figure 5.2 *The fitness landscape of all combinations of five mutations. Each letter* r, t, s, g, *and* p *is the first letter of the name of the gene affected by the particular mutation:* r *for rbs;* t *topA;* s *spoT;* g *glmUS; and* p *for pykF. Each node represents one of the 32 possible combinations of five mutations and is labelled by letters representing the mutated genes. The fitness value at each point is the 'altitude'. (The notation—green for high ground and red for low ground—is the reverse of the usual convention in geographical maps but follows the conventions of the stock market instead.) The edges show the possible one-step mutations, together representing the 120 unique paths to go from the 'ancestor' genotype* anc *to five mutations* rtsgp. *Solid and dotted lines indicate whether the mutation along the respective path cause an increase or decrease in the fitness.*
(from Khan et al., 2011)

There are 5! = 120 different orders how the final state of five mutations can be reached navigating through these 32 subsets: any of the five mutations can be the first one, any of the remaining four mutations the second one, etc., and thus we get 5 × 4 × 3 × 2 × 1 different paths. How does the fitness of the bacteria change along each path? Is the fitness continuously increasing with each new mutation, or are there routes along which the fitness at times goes down? In other words, what is the *landscape of fitness* along these paths?

The answer is shown in Figure 5.2. It turns out that the fitness is monotonously increasing along 86 of all possible 120 routes; the other 34 routes have a descending stretch at least somewhere. For instance, the route anc→t→rt→rtg→rtsg→rtsgp first goes up to t, then down to rt, then up again until reaching the top at rtsgp. Moving through this fitness landscape is like trying to reach the top of a hill when hillwalking. Some routes keep ascending until one reaches the top of the hill, but along others sometimes one has to descend before finally reaching the top.

As the selection of the fittest is likely to discard the negative mutations fast, it is not likely that evolution utilizes the routes where fitness is decreasing significantly before it goes up again. Thus, the fitness landscape defines which evolutionary paths are the likely ones leading from one genotype to another—from one point in the landscape to another. The concept of a fitness landscape was first introduced by the US geneticist Sewall Wright in 1932, well before the structure of DNA was known. Wright did not have mutations in mind, he was thinking about genetic variation—different genotypes that are already present in a population. Wright considered fitness landscapes as a visualization of a high-dimensional fitness space in two dimensions, and he saw evolution as 'walks' in this landscape. In the context of DNA mutations, this concept was first introduced in 1970 by John Maynard Smith.[44] Fitness landscape is a network where the nodes represent genomes, the genomes different from each other by one nucleotide are connected by an edge, and each node has a 'height' proportional to the fitness of the genome associated with it (Figure 5.2).

We can view evolution abstractly as a highly parallel stochastic algorithmic process navigating the fitness landscape. The algorithm is sending out agents wandering around in the landscape, trying to move to higher ground. The agents that move on low ground die, the ones on higher ground multiply. The climbers are in the dark, they cannot see ahead. A few steps going down may be fine, but if the path leads downwards for some time, the agent is in trouble. When they exchange information, sometimes the agents are able to make jumps longer than one step. This class of stochastic algorithms is known as *genetic algorithms*; we will discuss this in some more detail in Chapter VII. The question we want to ask now is in what type of landscapes such an algorithm can succeed?

The above algorithm will be able to find the higher ground only if the fitness landscape permits this without making the agents to descend too low. By definition, the fitness landscape supports evolution, or is *evolvable* from a given point, if there exists a (mostly) monotonously ascending path from this point to locations with higher fitness. Conversely, if the genotypes with high fitness are isolated from the rest of the landscape by deep 'ravines' of low fitness, this landscape is not evolvable. We can talk about the degree of evolvability by how many ascending paths exist in the landscape. The possibility of information exchange between the organisms complicates the definition of which landscapes are evolvable, as some longer jumps that potentially may allow evolution to cross 'ravines'.

More traditionally, the concept of evolvability is defined in evolutionary biology as the 'ability of populations to adapt' to environmental changes.[45] In my opinion, this definition is somewhat fuzzy but not inconsistent with the one above if we include environmental factors in the landscape in addition to genotypes. Suppose the environment suddenly changes and our population finds itself on low ground, for instance, as at the start of Lenski's experiment. Does a mostly ascending path towards higher ground exist from there? If yes, the landscape is evolvable. We can also ask how fast and how many of the possible paths go upwards. This is the degree of evolvability. Even more generally, we can ask for how many different environmental changes the landscape is evolvable?

What makes microbes and other life forms evolvable is the redundancy in their genomes. As we discussed in Chapter IV, many genes can be removed, but the microbe still keeps replicating, be it at a slower rate. Microbes are reliable devices made of unreliable components. We can change quite a few amino acids in a protein and that protein will still perform its function. We can knock out a gene and it is quite possible that the cell continues functioning. We can change quite a few nucleotides in the genome without any visible phenotypic effect at all. But this also gives evolution a room for manoeuvre. Life has evolved to be like this, and this is also why it can continue evolving.

Martin Nowak and Hisashi Ohtsuki defined life as '*what replicates and evolves*'.[46] Can an evolvable replicator be built artificially? Does an evolvable replicator need to be based on polymer molecules such as DNA and proteins, or can such a replicator be made of, for instance, steel, copper, and silicon? Or rubber, if we want flexibility. Proposing an evolving replicator made of substances different than DNA and protein like polymers would amount to proposing life based on physical implementation entirely different from the known one. Conceptually there is nothing that would confine evolvability to systems based on polymer molecules (see Figure 5.3), but it is a different question whether such evolvable landscapes exist in nature. In the physical world of atoms and molecules, the fitness landscape and associated evolvability is the property of the system and its environment.

Returning to polymer-based fitness landscapes, the example shown in Figures 5.2 provides a glimpse into a tiny region of a real-world fitness landscape. The 'global fitness landscape' is hierarchical—genomes can be viewed as consisting of discrete modules: genes. We can first look at the landscapes inside a single gene and then combine such intragenic landscapes.[47] For the combined *intergenic* landscape, such as the one Figure 5.2, usually it does not matter which exact nucleotide in the particular genes is mutated.

What is the shape of the global landscape of all possible genomes? Is it one huge mountain or many small hills? It seems quite likely that only a small fraction of all the possible 4^n genomes of length n will allow for functioning machinery of a cell. If so, the

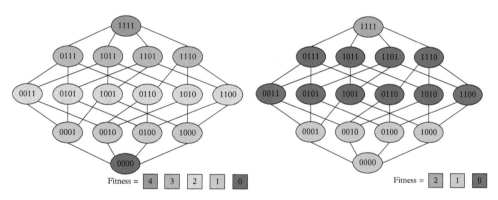

Figure 5.3 *An examples of the fitness landscape that enables evolution (**Left**) and does not (**Right**). As in Figure 5.2, genotypes are represented by nodes, two genotypes are linked if they can be reached from one another by exactly one point mutation. The colour scheme is the same as in Figure 5.2. We can define the distance between two such genotypes as the minimal number of point mutations, that can lead from one genotype to the other. Some readers may note that this is* Hamming *distance: distance* $D_H(X,Y)$ *between two genotypes X and Y equals the length of the shortest path in the network, leading from one node to the other. In evolvable landscape (left), the genotype 1111 with the top fitness 4 can be reached from any other genotype via a monotonously ascending path. In non-evolvable landscape (right), the genotype with the highest fitness 2 cannot be reached from any of the genotypes of the minimal survivable fitness 1 without passing through genotypes with fitness 0, meaning death.*

global landscape is likely to be an almost empty flat desert, with a few small dunes or mountainous regions where life is possible. Are these mountains connected by 'ridges' or are they isolated? Is this landscape rugged or smooth?

It has been argued by theoretical biologists for the last two decades that the fitness landscape is 'rugged'.[48] But what exactly do they mean by this, and how can ruggedness be measured in a multidimensional space? One way to look at this is through the effects of combinations of mutations. If the combined impact of two or more mutations on the organism's fitness roughly equals the sum of their individual impacts, then the landscape can be regarded as smooth—the mutations are either good or bad, and by accumulating good mutations we gradually climb uphill. However, if a combination of mutations can produce effects that are not predictable from the fitness effects of the individual mutations, then the landscape is rugged. For instance, if mutations A and B individually each increase the fitness but jointly they decrease it, then the landscape will have an element of unpredictability and arguably of ruggedness. In physics this is called non-linearity, in biology it is called *epistasis*—a term introduced by the British biologist William Bateson[49] in 1905. Non-linearity can be quantified by comparing the predicted additive fitness of two mutations with the empirically observed effect.

Large-scale systematic studies in *E. coli* and other species have revealed quite a few such epistatic effects,[50] including those where a combination of two mutations has the opposite effect of each of the individual ones, the so-called *sign epistasis*. However, there are also reasons to believe that epistatic effects are not widespread; overall, linear models are quite good at predicting the effects of combined mutations.

How many of the possible mutations can be expected to lead upwards in the fitness landscape, to improved organism's fitness, and how many downwards? The answer depends on whether the organism is in its typical or in a new environment. Various estimates of the proportion of positive mutations in natural environments have been suggested, and all agree that the fraction of positive mutations for bacteria in their 'normal' conditions is tiny.[51] However, when bacteria are placed into a new environment, the fraction of positive mutations can increase dramatically. Depending on how different this new environment is, the estimates range somewhere between 0.0001% and 1% and even higher.[52] Overall, if we look at the effects of changes at different genome positions in bacteria, as the rule of the thumb, about half of mutations are likely to be negative, about half neutral, and only a very small fraction will be positive. But positive mutations do exist, as Lenski's experiment shows.

How much change is too much?

In the world of perfection where mistakes are never made, nothing new can emerge. '*Without deviation from the norm, progress is not possible*', as the American musician and film-maker Frank Zappa once said. Errors are a necessary ingredient of evolution. But how many errors are permissible? Earlier we discussed the hyper-mutator mutation, which temporarily accelerated adaptation to a new environment. Does this mean that a higher rate of mutations is bound to lead to faster evolution, or is there a limit on how many mutations can be tolerated? Is there an optimal mutation rate? These are fundamental questions. There is a similar question in cultural evolution—does conservatism have any role in a progressive society? Can too much disruption be bad for progress?

In this chapter we will constrain the discussion to biological evolution. Intuitively it feels that there must be a limit on the maximal acceptable mutation rate, that too many mutations will kill the organism. For instance, X-rays are dangerous to life because they cause mutations. But what is this limit? In 1979, the German scientist Manfred Eigen and his colleagues at the Max Planck Institute for Physical Chemistry in Göttingen proved a fundamental theorem providing the answer.[53] At the time Eigen already had a Nobel Prize awarded for his work on measuring fast chemical reactions some 20 years earlier. The fundamental theory of replication error threshold was later simplified and developed further by John Maynard Smith.[54]

Consider the fitness landscape, where the organism with one particular genome, the *master genome*, has the highest fitness, while the fitness of other organisms decreases as the distance between their genome and the master genome increases. Thus, we have a simple fitness landscape, with one hill and the master genome on the top of it (as in the example in Figure 5.3 Left). Eigen proved mathematically that in such a fitness landscape, if the mutation rate is above a certain threshold, the replicators will 'slide down the fitness slope' and eventually the population will die out. If the error rate is below this threshold, then a structure in the replicator population will emerge: a certain proportion of replicators will have the master genome, plus there will be a decreasing number of replicators with any particular genome that is one, two, three, etc., mutations away from the master genome. The particular population structure will depend on the rate of mutations and on how steep the fitness slope is. If the error rate is low or the fitness slope is steep, this distribution is narrow: most replicators will either have the very fittest genome or just a few mutations. If the error rate is high or the fitness decreases only slightly with every step away from the top, then as long as the error rate is below the permissible threshold, the distribution is wide: the population will be a large 'cloud' of many different mutants centred around the fittest genome.

Thus, Eigen's result shows that *there exists a maximum error threshold above which a population of stochastic replicators disintegrates*. More specifically, Eigen proved that for the population to survive, the mutation rate must not be higher than one letter per genome per replication. For a sketch of proof see Box 5.2 and Figure 5.4. Denoting the probability of the copying error of an individual letter by u, and the length of the genome by L, this relationship is $u < 1/L$. This puts constraints on the length of the longest possible genome given this error rate: The longer the genome, the higher replication fidelity is needed for an organism to exist and to evolve. For instance, for a copying error probability $u = 1\%$, the maximum length of the genome is about 100 letters. And vice versa, for a genome length of 100 letters, the maximum permissible copying error per letter is 1%. Eigen called this *the error meltdown threshold*; sometimes it is referred to as the *error catastrophe threshold*.

Box 5.2 Eigen's model simplified

I will largely follow John Maynard Smith here, while avoiding the use of differential equations. As before, we consider a population of replicators with different genomes; the fitness of the replicator is defined by its genome. Maynard Smith's model assumes that there is one particular, so-called *master genome*, which has fitness equal to f, where $f > 1$, and which is the highest point in the fitness landscape. The model assumes a population of $x + x'$ replicators, where x replicators have reached this highest point f, while the other x' replicators have the mean fitness $f' < f$. We ask what is the maximal rate of mutations per replication cycle under which the proportion of the fittest replicators does not decrease?

The model assumes that all genomes have the same length L, and that all genome positions have the same probability of mutating equal to u, where u is small ($0 \leq u \ll 1$). The probability that any given nucleotide will be copied correctly is $1 - u$. The probability that two nucleotides are copied correctly is $(1 - u) \times (1 - u) = (1 - u)^2$, that three are copied correctly is $(1 - u)^3$, and the probability that the whole genome is replicated without errors equals $Q = (1 - u)^L$. The quantity Q can be called the replication fidelity—the probability that a replication of the entire genome happens without an error.

Let us first look at the case when there is no death, that is, we assume that the fitness values f and f' are defined by the number of replications per time unit, in other words, f and f' are the birth-rates. Figure 5.4 shows how each of the populations x and x' change in a short time interval Δt, say in 1 minute. If there are no errors in replication, that is if $u = 0$ and thus $Q = 1$, and if initially we have x replicators, then the number of replicators that are born in time interval Δt, equals $f \times x \times \Delta t = fx\Delta t$. In the general case, when the fidelity of replication is $Q < 1$, the increase in the number of replicators with the master genome will be $\Delta x = Qfx\Delta t$. Thus, after the time interval Δt, the total number of the replicators that have the master genome will be $x + Qfx\Delta t$.

Following similar arguments we get the increase in the number of sub-optimal genomes is $\Delta x' = [f'x' + (1 - Q)fx] \times \Delta t$. Indeed, $f'x'$ is the number of new replicators with sub-optimal genomes that have been added by their own replication (recall that their birth rate is f') and $(1 - Q)fx$ is the number of new replicators with suboptimal genomes that have been added by erroneous replication of the master replicators. The probability that a replicator from population x' can change to the one specific genome x is negligible, thus $f'x'$ represents the number of the replicators of population x' that have replicated with or without an error.

The expressions that we just derived imply $\Delta x > 0$ and $\Delta x' > 0$, which as is consistent with our assumption that there are no deaths, means that the total population $x + x'$ is growing. Now we make one final assumption, namely that the size of the population of replicators is stable, that is $\Delta x + \Delta x' = 0$. To ensure this, we need to introduce terms representing death in our equations. Consistent with the probabilistic interpretation of fitness, we choose individuals from both populations for death with a certain rate c, proportionally to the size of each population.[55] Thus, the death-adjusted expressions are

$$\Delta x = [Qfx - cx] \times \Delta t$$

and

$$\Delta x' = [f'x' + (1 - Q)fx - cx'] \times \Delta t$$

Using straightforward high-school algebra and setting $\Delta x + \Delta x' = 0$, we get $c = (fx + f'x')/(x + x')$. Finally, to ensure that evolution does not lead to decline of the population of the fittest replicators, we have.

$$\Delta x \geq 0$$

The opposite, $\Delta x < 0$ would mean that the proportion of the fittest replicators in the population had declined. Relatively straightforward calculations show[56] that $\Delta x \geq 0$ implies $Q > f'/f$. We can always scale the relative fitness values so that $f' = 1$, in which case the inequality simply becomes: $Q > 1/f$. This is the necessary condition that allows the population of replicators not to slide down from the fitness hill.

Finally, given that $Q = (1 - u)^L$, and that for small u, $\log(1 - u) \approx -u$, taking the logarithm[57] and dividing both sides of the inequality by L, we obtain

$$u < \frac{\log f}{L}$$

This inequality links the error rate u, the relative fitness of the best replicator over the population average, and the length of the genome. It shows, that the longer the genome and the lower the relative fitness of the best replicator, the more reliable the replication has to be to avoid sliding down the fitness slope.

By assuming that on the logarithmic scale, the global fitness landscape is relatively flat (as was described by Maynard Smith: 'f is not too large and not too small'),[58] we can further simplify the above relationship as

$$u < 1/L$$

Eigen did a more detailed analysis and came to a similar conclusion. Looking at this relationship from the other side, we conclude that *the genomic mutation rate must not exceed 1 mutation per genome per replication*.[59] In my opinion this is one of the most fundamental, perhaps *the fundamental theorem of biology*. It shows what is needed for the very simplest life to exist.

The simplest viruses have genomes of some 1000 nucleotides, therefore the nucleotide-copying error rate has to be less than 0.1%. Observations show that many RNA-based viruses indeed 'operate' close to this meltdown threshold. Possibly in this way they achieve a faster evolution rate, and thus can evolve to avoid the host's immune system. But this also means that many of them die. On the contrary, bacteria seem to operate at a safe distance from the catastrophe threshold. For instance, given the *E. coli* genome length of about 4.6×10^6 letters, its mutation rate of just over 10^{-10} nucleotides per generation[60] is more than 1000 times below the error meltdown threshold. In Lenski's experiment, the hyper-mutator mutation increased this rate 15-fold, but this is still on the order of 100 times less than the theoretical maximum. *E. coli* seems to play it quite safe. Possibly this is why *E. coli* has been around as a species for some 100 million years and no longer has a need for extreme evolutionary experiments.

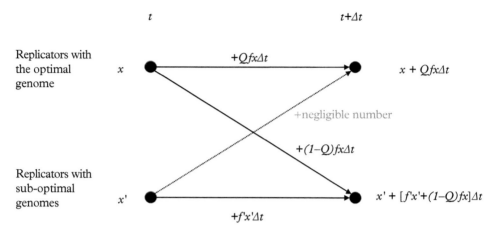

Figure 5.4 *We split all replicating individuals in two groups. The first group consists of the individuals with the single optimal genome; their fitness is f and at time point t their number is x. The second group consists of the individuals with any other genome; at time point t their average fitness is $f' < f$ and their total number is x'. The replicating sequences with the optimal genome replicate perfectly with probability Q, thus they make a replication error and 'move' to the second group with probability 1–Q. The sequences of the second group replicate with the rate f', but they are also 'supplemented' from the sequences of the first group whenever a replication error happens.*

Or possibly, the more complex fitness landscapes of more complex organisms impose additional constraints not captured in Eigen's model.

Recall the RNA-world from the previous chapter and the hypothetical self-replicating RNA molecule (Box 4.3). Assuming for a moment that there exists a 200 nucleotide-long RNA molecule that can copy RNA of its own size, for this molecule to form an evolving population of self-replicating molecules, the error rate u would have to be below $1/200 = 0.5\%$. Moreover, for such a self-replicating molecule to grow in length and complexity, its copying fidelity would have to grow in parallel. And vice versa, if such a sequence evolves towards more reliable copying, it can grow in length. As argued in Chapter I, there is a certain minimal complexity below which autonomous self-replicator complexity is not possible and thus there is a minimum genome length that can support a replicator. This implies that the first evolving replicator had to be quite reliable from the very beginning. As we will discuss in Chapter VI, how such a replicator could have emerged is still a puzzle.

Eigen and Maynard Smith's arguments show that there is a minimum copying fidelity necessary for evolution to be possible, regardless of how this hereditary information is encoded—in our reasoning we never assumed that the information is encoded in a polymer. If *n bits* of information need to be passed on, then for the evolution to happen, the maximum permissible error rate is $1/n$. This observation somewhat limits the possible role of transgenerational epigenetic inheritance in evolution, because, as will be discussed in Chapter VII, the fidelity of epigenetic inheritance is quite limited.[61] Also, there are indications that there is a minimum information transmission fidelity that is needed to ensure cultural evolution,[62] which may explain why cultural evolution accelerated dramatically with the emergence of human language.

Darwinian and Lamarckian evolution

In 1859 Charles Darwin wrote in his *On the Origin of Species*:

> We see nothing of these slow changes in progress, until the hand of time has marked the long lapses of ages, and then so imperfect is our view into the long past geological ages that we only see that the forms of life are now different from what they were.

Nowadays we know that not only microbes but many fast-reproducing species, such as flies and even some birds, do evolve on a timescale that allows for direct observation and laboratory experiments.[63] The idea that all biological species are evolving first took root in the nineteenth century. One of the first to point this out was the French naturalist Jean-Baptiste Lamarck, who thought that animals acquire features that help them adapt to the environment and then pass on these features acquired in their lifetime to descendants. Now we know that this is the way cultural evolution works, but not biological evolution.

The first scientists to realize that different principles 'govern' biological evolution were Charles Darwin and Alfred Russel Wallace. In 1859, in his book *On the Origins of Species*, Darwin outlined how the variability of individuals in a population and preferential selection of the fittest lead to adaptive evolution. In an example that is something of a caricature, the individuals of an animal species whose main food is leaves on tall trees have a comparative advantage if they have longer necks, and thus they are likely to produce more offspring. If the long neck phenotype is inheritable—that is, if the information encoding this phenotype is passed on to future generations—the average length of the neck in the population is likely to grow. Wallace had arrived at a similar idea at roughly the same time, though Darwin not only published these ideas first but he had also gathered more evidence.

What makes Darwin's and Lamarck's explanations different from each other is that according to Lamarck, the changes acquired in an organism's lifetime are inherited; giraffes' necks may be stretching during their lifetime and then the longer necks are passed on to their offspring. Trying to find elements of Lamarckian evolution in biology, where acquired phenotypes are inherited epigenetically, has recently become a popular activity, particularly amongst science writers. We will come back to this in Chapter VII. Even if occasional examples of acquired transgenerational inheritance can potentially be argued (which is not proven), this does not change the basic facts. Unlike cultural evolution, biological evolution follows the Darwinian rather than the Lamarckian model.

A prerequisite to either Darwinian or Lamarckian evolution is that information is passed on from generation to generation. In Lamarckian evolution this includes the information acquired during an individual's lifetime. In Darwinian evolution this is information inherited from the individual's predecessors, possibly changed by mutations. Only the rediscovery of genetics and the discovery of DNA structure in the twentieth century revealed the mechanisms of how the genetic information is coded, how it can change, and how it is passed on.

VI

Life on Earth

Complexity is hard to define or to measure, but there is surely some sense in which elephants and oak trees are more complex than bacteria, and bacteria than the first replicating molecules. Our thesis is that the increase has depended on a small number of major transitions in the way in which genetic information is transmitted between generations.

(John Maynard Smith and Eörs Szathmáry, 1995)

How does the over 60,000 generations of bacterial evolution discussed in the previous chapter, a period of time equivalent to well over a million years for humans, compares to the time life has existed on Earth? When did the first life emerge on Earth? When did the first animals and when did humans emerge? How did this happen? Can evolution explain the present-day complexity of life? But to begin with, how old is the Earth itself?

In 1492 Columbus discovered America; in 1543 Copernicus dethroned Earth from the centre of the Universe; in 1610 Galileo found that not only Earth but Jupiter too had moons, and in 1687 Newton published his *Philosophiæ Naturalis Principia Mathematica*. By the end of the seventeenth century the modern scientific method was firmly adopted by the natural philosophers.[1] Nevertheless, for over a hundred years that followed, the same natural philosophers believed that the Earth was only 6000 years old! From the contemporary perspective this is hard to comprehend: how could, for more than a century, Newton's *Principia* coexist with the belief that the Earth had existed for only about 300 human generations?

Probably this was because nothing that would indisputably contradict the biblical beliefs was yet discovered. Newton himself, being a religious person, regarded his celestial mechanics as laws describing the solar system created by God. Once set in motion, the planets moved around the Sun according to the laws of mechanics and gravity, but this did not say anything about how the solar system began. Direct contradictions between physics and the 6000-year-old Earth became apparent only later, at the onset of the twentieth century, with the discovery of radioactivity.

Nevertheless, doubts about the Earth's young age started growing long before that. In 1670, the Italian painter Agostino Scilla described large fossilized teeth found on the Mediterranean island of Malta[2] (Figure 6.1). He was not the only person who noted that some fossils, known at the time as 'tongue-stones', show their internal structure, and used this to argue that these were not 'jokes of Nature' that grow in the ground to test

Living Computers. Alvis Brazma, Oxford University Press. © Alvis Brazma (2023). DOI: 10.1093/oso/9780192871947.003.0007

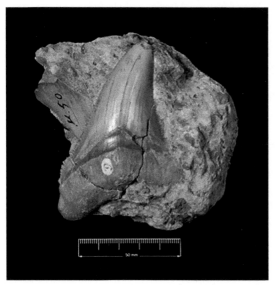 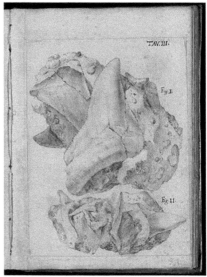

Figure 6.1 *Left. Shark teeth over 5 million years old found in Malta, currently in the Woodwardian collection in the Sedgwick Museum in Cambridge, UK (object CAMSM E-27-6).* **Right.** *Illustration from Scilla's book* La vana speculazione disingannata dal senso (Vain Speculation Undeceived by Sense, 1670)*, where he argues for a scientific explanation for such fossils, as opposed to them being of fantastic origin or a test of faith to God.*
Reproduced with permission from the Sedgwick Museum.

one's faith in God but are the remains of animals that lived a long time ago. Doubts in Earth's young age came also from geology. In 1785, the Scottish Enlightenment figure, James Hutton, gave two lectures at the Royal Society of Edinburgh, suggesting that the formation of rocks must have been a very slow process.[3] Defending the belief in the 6000-year-old Earth was becoming increasingly difficult, and at the beginning of the nineteenth century, two great rival French palaeontologists, Georges Cuvier and Jean-Batiste Lamarck, both agreed on one thing: more time than the chronologies in the Book of Genesis would allow for was required to explain the fossil record. By 1859, when Darwin published his *On the Origin of Species*, most scientists had accepted that the Earth must be tens, if not hundreds of millions of years old.

The quantitative estimates of how old Earth was, perhaps unsurprisingly, fell to physicists, who were used to working with numbers. In 1862, just three years after the publication of Darwin's *Origin*, the Scottish physicist William Thomson, who later became Lord Kelvin, calculated the time it would take for the surface of Earth to cool to its present temperature, assuming that the Earth had formed as a molten object.[4] He estimated this to be between 20 million and 100 million years.[5] Others, using different assumptions, came to similar conclusions; amongst them, George H. Darwin, the son of Charles Darwin, gave the age of Earth at 56 million years.

By that time, the geologists already felt that even 100 million years were not enough. However, there was one big problem—the age of the Sun. Straightforward and indisputable calculations showed that if the Sun was fuelled by chemical reactions, then radiating light at its present rate it should have burned out in just a few thousand years. A temporary relief was provided by gravitational energy. Gravity pulls objects towards each other, thus converting the potential energy into kinetic energy, and eventually, when the objects collide, into heat. So, might the energy have come, for instance, from meteorite bombardment? The meteorite hypothesis was soon discarded, but it evolved into the idea that the Sun could be powered by gravitationally driven contraction. Knowing the planetary orbits, Newton's mechanics could be used to estimate how much mass was available. In 1856 the German physicist Hermann von Helmholtz calculated the time it would take for the Sun to condense down to its current diameter and brightness starting from a nebula of gas to be 18 million years. Kelvin himself made similar calculations and came to a similar estimate. The age of the Sun became generally accepted to be around 20 million years, which was the upper limit of the age of the solar system consistent with classical physics.

Newton's physics was in contradiction with geology and palaeontology for the entire second half of the nineteenth century; for half a century, the physicists were the reactionaries. But the classical theory of physics itself was increasingly coming into conflict with experimental observations. It took the discovery of radioactivity by Henri Becquerel in 1896 and a major paradigm shift in physics to resolve these conflicts.

The implications of Becquerel's discovery on geology did not become immediately apparent, but when in 1903, the French physicist Pierre Curie and his assistant Albert Laborde announced that radium produced enough heat to melt its own weight in ice in less than an hour, geologists quickly realized what this meant for them. In 1903 Pierre Curie shared the Nobel Prize in Physics with his Polish wife, Marie Skłodowska-Curie, and Henri Becquerel 'in recognition of the extraordinary services they have rendered by their joint research on the radiation phenomena discovered by Professor Henri Becquerel'. This was the start of a new scientific revolution, which led to the development of nuclear physics, quantum mechanics, and the theory of relativity.

The discovery of the radioactive elements, such as radium and uranium, did not automatically provide the answer to what makes the Sun shine, but it showed that there was a previously unknown enormous source of energy. Twenty million years were no longer a limit on the Earth's age. In 1905 Einstein published his famous equation $E=mc^2$, linking mass and energy; it took another 34 years before the nuclear reactions fuelling the Sun were worked out in detail by the German physicist Hans Albrecht Bethe. Bethe got his Nobel Prize in Physics in 1967 'for his contributions to the theory of nuclear reactions, especially his discoveries concerning the energy production in stars'.

Nuclear energy, however, not only made it possible for the Sun to keep going for billions of years; radioactivity also provided scientists with the means to measure the age of rocks. In 1906, studying helium trapped in radioactive rocks, Ernest Rutherford and colleagues estimated the age of Earth to be about 400 million years. Revised estimates by the British physicist Robert John Strutt, the 4th Baron Rayleigh, and several others put the age of Earth at 2 billion years. Although still less than half of what detailed calculations found eventually, this was of the right order of magnitude. In about a hundred years the Earth had aged almost a million times. But radioactivity also provided a means of dating the age of fossils.

Atoms come in different flavours—chemical elements have different *isotopes*. As we mentioned in Chapter II, an atom can be viewed as a positively charged nucleus surrounded by a cloud of negatively charged electrons. The particular chemical element is defined by the number of protons in the nucleus, however in addition to protons, the nucleus may also contain chargeless particles called *neutrons*. For instance, the carbon atom C always has six protons, but the number of neutrons vary between six and eight. The number of neutrons is what defines the isotope. Carbon has three isotopes ^{12}C, ^{13}C, and ^{14}C, where the upper index shows the sum of the number of protons and neutrons (for instance, 6 protons + 6 neutrons = 12). Almost 99% of carbon that is present in the Earth's atmosphere is the isotope ^{12}C, and about 1% is ^{13}C. About one atom per trillion is the isotope ^{14}C. The last may seem very few but given that 12 grams of carbon contain about 6×10^{23} atoms, on average, there will be 600 billion ^{14}C isotope atoms per gram of atmospheric carbon.

Some isotopes are unstable and decay into other types of atoms, releasing energy in the process. For instance, ^{14}C decays into nitrogen ^{14}N by emitting an electron and another particle called an antineutrino. The opposite can happen too: when hit by a high energy neutron, ^{14}N may get converted into ^{14}C plus a free proton. Indeed, the ^{14}C isotope is continuously produced in the upper layers of the Earth's atmosphere by cosmic rays; the ratio of one ^{14}C atom per trillion is a dynamic equilibrium where ^{14}C atoms are constantly born and decay.

Importantly, the decay rate of an isotope is constant; usually, we talk about the *half-life* of a radioactive isotope. For instance, the half-life of ^{14}C is 5700 years. This means that if we initially had, say, 2 billion atoms of ^{14}C isotope in a piece of a bone, then after 5700 years, there will be only 1 billion left. After 10,400 years there will be only half a billion left, and another 5700 years later, only a quarter of a billion. As the atmospheric carbon atoms get incorporated into photosynthetic plants and microbes, which are then consumed by other organisms, the constant ratio of ^{14}C to ^{12}C in the atmosphere is reflected in most living organisms. But once the animal dies, the ratio of ^{14}C to stable isotopes will start decreasing. We can determine how long ago the animal died by measuring how far has this ratio decreased. This is known as carbon dating.

In fact, the ratios between different carbon isotopes in a sample can also be used to look for evidence of life more directly. Isotopes of the same element have very similar, but not identical, chemical properties, and enzymes can tell isotopes apart. In particular, photosynthesis enzymes, which extract carbon from the atmosphere, tend to prefer the ^{12}C over the other stable isotope ^{13}C. Although about 1% of the carbon in atmosphere is in the form ^{13}C, in living organisms, this percentage is smaller. Thus, in some cases, we can tell if a fossil has a biological origin by measuring the ratio of ^{12}C to ^{13}C.

As ^{14}C decays quite fast, carbon dating can only be used to measure the age of relatively young fossils. A radioactive element that provides means to look back millions to billions of years is uranium (U). All uranium isotopes are unstable; in fact, the radioactivity of uranium is what Becquerel discovered. The most stable uranium isotope ^{238}U decays with a half-life of 4.5 billion years into the lead isotope ^{206}Pb, while the faster decaying isotope ^{235}U has a half-life of 700 million years decaying into lead ^{207}Pb. All these isotopes can be found in volcanic rock, but there is an important difference in their origins. Uranium atoms incorporate into growing mineral crystals, but lead is rejected because of very different chemical properties. Thus, we can be certain that most lead in a mineral is a product of radioactive decay that happened after the crystal had already formed. Consequently, the ratios between uranium and lead can be used to determine the age of the rock. Such a radiometric dating has shown that the oldest rocks on Earth are about 4.4 billion years old, which is consistent with other evidence indicating that the Earth formed around 4.54 billion years ago.[6]

When did the first life appear on Earth? Organic matter does not normally survive billions of years, but minerals carried by water seeping into pores of a dead organism fill the spaces within the tissue or around it, and crystals begin forming. Dating the oldest mineralized fossils of microscopic size is, however, challenging. Different parts of the rock might have formed at different geological times, and it is not always clear which part the microfossil belongs to and if indeed it represents a once living organism. The prevailing view amongst palaeontologists is that the first signs of life can be dated to at least 3.5 billion years ago. In the light of recent discoveries, it is increasingly beginning to look that life on Earth may have already existed 4 billion years ago, though this is not a full scientific consensus.[7] What is certain is that for the first half a billion years the Earth was hot, violent, and almost certainly uninhabitable to any form of life, while on the other hand, no later than 2.5 billion years ago life on Earth was well established.

It does not make a huge difference whether it took life hundred million or a billion years to emerge after Earth became habitable. Human civilization is barely 10,000 years old, which is 0.01% of 100 million or 0.001% of a billion years. Our current knowledge is insufficient to say that if life could have emerged in a billion years, 100 million years were not enough. A different possibility is that life arrived on Earth from somewhere else, for instance, brought by a meteorite (or by aliens), which would have given life more time to form—the universe is believed to be about 14 billion years old. But for this to have occurred, life would have to travel enormous distances.

What is clear is that after the first life emerged on Earth, biological evolution had several billion years to 'experiment' with different life forms before arriving to forms inhabiting the Earth presently. In the remaining part of this chapter, we will describe more than 2 billion-year-long history of life on Earth in less than 14,000 words—less than one word per 100,000 years. As we will see in Chapter VIII, human language is thought to have originated somewhere between 100,000 and 200,000 years ago, or even more recently. If we spread the words in this chapter out evenly, a single word stands for the entire period since human language exists.

From pre-life to life

But let us begin at the beginning—how did life start? How did the first system of evolving replicators emerge, kick-starting biological evolution? The origin of life itself is not a part of biological evolution: this was pre-biological or chemical evolution. In his *Origin*,

Darwin consciously avoided discussing the origin of life, though some of his correspondence suggests that this question was on his mind.[8]

The first scientific hypothesis of the origin of life was formulated by the Russian scientist Aleksandr Oparin in 1924 and independently, verified by the British scientist John Haldane in 1929. They suggested that synthesis of a wide range of organic molecules in the early oxygen-poor atmosphere was possible, given a sufficient supply of energy, which could be provided, for instance, by lightning or ultraviolet light. Later, in the 1950s, Stanley Miller and Harold Urey at the University of Chicago demonstrated this experimentally—in the presence of water and simple molecules, such as CH_4 and NH_3, electric discharge produced a rich mixture of more complex organic molecules, including amino acids.[9] Thus, most likely, a rich mixture of small organic molecules and some larger ones were present on the early Earth.

However, there is an enormous gap to bridge from even complex organic molecules to a self-sustaining system of evolving replicators. This is often glossed over in the popular science and even scientific literature, probably to avoid giving ammunition to creationists. This book is primarily aimed at those who believe in science, therefore I am not concerned in stating that it is still very unclear how this gap can be bridged. The spontaneous synthesis of any of amino acids is possible in the right conditions quite easily; this happens. Spontaneous assembly of just 10 amino acids in a particular order without having information guiding this assembly seems to be much less likely—there are 20^{10} > 10^{13} different orders from which to choose. As discussed in Chapter V, the shortest known RNA molecule that works as a polymerase-ribozyme is about 200 nucleotides long. Even assuming the existence of a polymerase-ribozyme that is half this length, there are $4^{100} \approx 10^{60}$ polynucleotides to choose from. And even if many different combinations of many different polymers can provide the basis for a replicator, and even if there are many 'experiments' to find them performed in parallel, it is still hard to imagine that a random search can produce a system of such complexity. Some way of gradually 'climbing' towards a set of sequences that jointly are able to catalyse their own synthesis seems to be needed. Once we do have an information-carrying replicator, selection of the fittest gives us evolution, but how the selection of the fittest or any sort of gradual evolutionary process can happen without self-replication already in place? Is this a chicken-and-egg problem?

Establishing a possible path from organic molecules to the first evolving system of replicators is one of the major unsolved problems of modern science. Recognizing this, one of the authorities on evolutionary molecular biology, Eugene Koonin, proposed an explanation based on the anthropic principle and the multiple Universe hypothesis— if life had not emerged, we would not be here to talk about it; we happen to live in a Universe where life did emerge.[10] Most scientists, though, will not accept this as an explanation.

Others working in the field are more optimistic. For instance, the Nobel Laureate Jack Szostak and his co-author Itay Budin wrote in 2010: '*Recent synthetic approaches to understanding the origin of life have yielded insights into plausible pathways for the emergence of the first cells*', mentioning '*the ability of unexpected physical processes to facilitate the self-assembly and self-replication of the first biological systems*'.[11] But by mentioning 'unexpected', this optimistic statement hints at something presently missing.

As discussed in the previous chapters, all known life on Earth is based on template-guided copying of polymer molecules (Chapter II). The theory of the origin of life has to explain how an evolving system of such polymers could have emerged in geochemical processes. Is it possible to find a gradual process leading from monomer or short polymer molecules to a system of self-replicating polymers, or are we stuck with waiting for a lucky fluctuation kick-starting life? This could then be a long wait.

Various hand-waving arguments and speculations about the possibility of selection of the fittest without replication have been made before, however in my opinion, the very first clear model of how this potentially could happen was proposed in 2008 by Martin Nowak and colleagues.[12] More concretely, Nowak and Ohtsuki proposed a model of what they called *pre-life*—a growing set of binary sequences (representing polymer molecules) in a hypothetical fitness landscape (see Box 6.1 for more detail). There is no replication in *pre-life*, but there is selection of the fittest (defined by the fitness landscape), as some sequences grow faster than others. Nowak and Ohtsuki showed that in such *pre-life*, template-guided copying is needed for sequences to grow not only in length but also in complexity. Importantly, they also showed that in *pre-life*, a minor fitness advantage gives a minor advantage in growth. The last is in contrast to a system of replicators where even a minor advantage in fitness leads to the fittest completely taking over (Box 5.1). Thus, if a set of (polymer) sequences able to catalyse their own replication emerges in *pre-life*, the self-replicating system soon outcompetes all other sequences. In other words, there is a phase transition from *pre-life* to life. The dividing line between *pre-life* and life is clear, there is no gradation between the two.

Box 6.1 Growing sequences

Nowak's model of *pre-life* assumes that there is a mixture of monomers of different types and their polymer sequences in the environment.[13] We can think of the polymers as RNA molecules, and monomers as nucleotides. There is a stochastic process in which monomers are added to the polymers one by one, and in this way the polymers grow in length. Importantly, the probability by which a monomer can be added to a particular polymer, and thus the rate at which each particular polymer grows, depends on the particular sequence of monomers. Conceptually, each sequence has a defined fitness; the fitter the polymer, the faster it grows. In the physical world, this fitness landscape would be defined by the chemical properties of the respective polymers, but in Nowak's model, the fitness landscape is an *a priori* given mathematical function assigning a value to each sequence. Finally, it is also assumed that the polymers can stochastically break in two with a probability independent of the specific sequence.

In its purest conceptual form, the polymers can be represented by binary sequences, just like information on von Neumann's tape. Thus, the symbols 1 and 0 represent two different monomers. Simplifying Nowak's model somewhat, in this binary 'world', given a polymer sequence p, there are two 'chemical reactions' possible:

$$p + 0 \rightarrow p0$$

continued

Box 6.1 *continued*

---

or

$$p + 1 \rightarrow p1$$

For instance, $p = 0010110$ can give rise to 00101100 or to 00101101. Any of these sequences can split into two; for instance, 0010110 can split into 0010 and 110, each of which then grow again. As noted above, the rate at which each such sequence is growing depends on the actual sequence of 0s and 1s. Unless the probability of splitting is too high, these rules result in a particular dynamic mixture of binary sequences. The composition of this population will depend on the particular fitness landscape, just like in Eigen's model in Chapter V. Thus, there is a selection of the fittest going on, even though there is no replication.[14]

Nowak and Ohtsuki showed that for many fitness landscapes, the sequences generated this way were of a limited complexity (for instance, typically the sequences included many repetitions and therefore their Kolmogorov complexity was low) and the variety of the emerging sequences was limited. But this model did not make use of template copying of information, which is an important property of nucleotide sequences. Nowak introduced an additional operation—template-based joining of two polymers: if a sequence obtained by two polymers joined together was complementary to an existing polymer sequence, then (with a certain probability) the first two polymers are concatenated.[15] For binary sequences, 0 and 1 are complementary to each other; thus, sequence 0010110 is complementary to 1101001. An example of the described rule, if we have sequences 1101, 001, and 0010110, then with a certain probability the first two can be joined up into 1101001. Adding this operation to the rules resulted in sequences gradually increasing not only in length but also in complexity.

Nowak's collaborators performed an experiment using DNA polymers demonstrating that their theoretical predictions were indeed true. A slightly different model was studied by another group of scientists in Munich with similar results.[16] They introduced the rule that two complementary polymer sequences that form a duplex, for instance a double helix of DNA, are more stable and less prone to degradation than each separate strand.

But there is a principal difference between the phase transition in cooling water and what happens in transition from *pre-life* to life. If liquid water loses heat at a constant rate, at some point freezing is inevitable. It cannot however be guaranteed that pre-biological polymer evolution necessarily leads to the emergence of a replicator. For life to emerge, it is not sufficient that the *pre-life* polymers grow in length and complexity; it is also necessary that in this process a mixture of polymers can emerge such that it catalyses its own replication. Whether this can happen depends on whether the particular fitness landscapes for the non-replicating and replicating sequences are 'joined up'. There is no obvious logical reason why we could not construct such a fitness landscape mathematically, but whether such a landscape exists in the physical world of atoms and molecules is the key question. Nevertheless, the proposed model clearly pinpoints to how prebiotic evolution can happen and what is logically needed for life to emerge.

Nowak noted that there could have been many attempts of pre-biological evolution to reach life before one succeeded. We could think of this situation as living in a two-storey building, where the upstairs is inhabited by life and the downstairs by *pre-life*. The floors are not connected by a staircase, and it is virtually impossible to jump to the higher flour unaided. Initially, only downstairs was inhabited but the inhabitants were trying to build ladders to reach the ceiling and break through it. The ladder not only had to be high enough, it also had to be positioned so that a hidden hatch between the levels could be opened. Many attempts to get upstairs might have failed, but given that the upstairs is now inhabited, apparently at least one attempt succeeded (recall Koonin's argument). Once the upstairs got inhabited, it soon got crowded. By now we may have lost the knowledge where exactly the successful ladder was positioned.

The chemistry of early life

Could the growing polymers of *pre-life* have been RNA? Was it the RNA-world (Chapter IV) where the transition from *pre-life* to life happened? This indeed is the leading hypothesis pursued by several scientists, including the Nobel laureate Jack Szostak.[17] There are two alternatives: RNA at the basis of the very first replicator, or RNA as an intermediate transitional stage from an even simpler form of replication.[18] Each of these alternatives provide answers to some problems but leave unresolved others. For instance, the monomers making up RNA themselves are quite complex molecules, as shown in Figure 6.2. It is still not known how these molecules can be synthesized without the use of enzymes. Several possible chemical paths to life are discussed in some more detail in Box 6.2.

An alternative to RNA-life hypothesis, the so-called proteins-first hypothesis, was proposed independently by Stuart Kauffman and Freeman Dyson[19] in the 1990s. This hypothesis suggests that life started as a set of polypeptides—small proteins—which

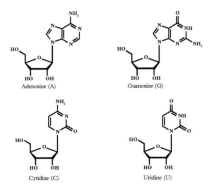

Figure 6.2 *Ribonucleotides are quite complex organic molecules; it is not known if they can be synthesized abiotically (that is, without help of enzymes).*

Image reused from © StudioMolekuul/Shutterstock.

catalysed each other's synthesis and jointly formed a set of reactions implementing a self-sustaining autocatalytic cycle. Polypeptides can indeed form in chemical reactions mimicking prebiotic conditions like these in Urey–Miller experiments. Moreover, it is known that some polypeptides show enzymatic activities and can catalyse reactions joining other peptides into longer ones. However, Kauffman and Dyson made several unrealistic assumptions and, tellingly, they came to opposite conclusions difficult to reconcile.[20] But most importantly, by itself self-replication is not sufficient—it needs to be capable of evolution. An updated proteins-first model has been proposed more recently,[21] however the authors still conclude that it is not clear how such autocatalytic peptide networks can evolve.

Whether the transition from *pre-life* to life indeed happened in the RNA-world or in some other physical media, life needs to be able to harvest energy from the environment to power its replicators. As we discussed in Chapter III, a sophisticated energy management system operates in a contemporary living cell, but where did the early life get its energy? For instance, a significant amount of energy is needed to attach a nucleotide to a growing RNA polymer. Contemporary cells do this by first energizing the nucleotide with ATP. However, making ATP is a complex biochemical process. How were ribonucleotides energized in the early cells?

Box 6.2 RNA first?

One of the difficulties for 'RNA first' hypothesis is that all the versions of RNA life that we can currently think of are still rather complex. There are at least three sets of problems. First, the individual ribonucleotides—the nucleotide monomers from which the RNA polymer is made—are complex molecules. We still do not know how to synthesize either ribonucleotides or even their simpler relatives abiotically in sufficiently large quantities.[22] Moreover, as already noted, joining up monomers into a polymer requires significant energy.

Second, even if we did manage to solve the problem of where the ribonucleotides came from and how to join them up, there is a problem that even the simplest RNA molecule that can catalyse RNA-template-based copying are complex. There are two possibilities. First, a single RNA molecule able to copy itself, and second, a multi-molecular complex. Regarding the first possibility, recall from Chapter IV that the best currently known RNA-polymerase ribozyme (RNA-based enzyme) tC19Z is around 200 ribonucleotides long and able to copy RNA molecules half its length. Even if we assume that there exists a single reliable enough self-replicating RNA molecule of a quarter of the length of tC19Z, it would still be a sequence of over 40 nucleotides. To specify the order of 40 RNA letters, we need around 80 bits of information, though less if the sequence contains regularities (for instance, to specify a palindromic sequence will require only half of this—40 bits of information). Could *pre-life* 'guess' such a molecule in the absence of pre-existing information guiding its synthesis? Describing a complex of several shorter RNA molecules requires less information than one long polymer.[23] Could RNA polymers, growing as proposed by Nowak's *pre-life* model, gradually evolve into the first self-replicating complex of RNA molecules? It has been shown that in the RNA-world, Eigen's error meltdown threshold could be avoided by a replicating system with up to 7000 ribonucleotides, which would allow for up to 100 short genes.[24] But if the system is based on a set of molecules, then to form an evolving self-replicating unit, these molecules need to be somehow kept together as a unit. Assuming similarity to current life, possibly they were enclosed in lipid membrane bubbles. We saw in Chapter II that indeed such membrane bubbles can form by self-organization and can grow and divide spontaneously, provided that there are enough lipid molecules in the environment.

This points to the third problem. For the known RNA-based polymerases to work efficiently, magnesium ions Mg++ (magnesium atoms with two electrons lost) are needed in the environment. These ions, however, are disruptive to membranes consisting of simple lipid molecules.[25] Thus, either the membrane has to be made of more complex molecules, for instance, phospholipids, which are the material of the contemporary cell membranes, or an RNA-based polymerase needs to be found that can work without the presence of magnesium ions. Phospholipids are rather complex molecules; recall from Chapter IV that even a minimalistic cell has over 80 genes for membrane synthesis, thus it is not likely that such complex membranes were available to *pre-life*.

Or perhaps, in early life, the replicating molecules were kept into units by means other than membranes? For instance, the replicating molecules could have been enclosed in bubbles of water floating in oil, or was early life evolving in microscopic water droplets in atmosphere? The currently prevailing hypothesis, described already in 1995 by John Maynard Smith and Eors Szathmary,[26] is that the first replicating molecular systems formed inside porous minerals such as hydrothermal vents on the ocean floor. The chemical reactions leading to self-replication were possibly happening on the surface of such minerals, which was serving as an inorganic catalyst. Thus, possibly the first life was two-dimensional—bound to the surface of a mineral.

A different version of RNA-life hypothesis is that the RNA-based life was only an intermediate stage. The so-called genetic takeover hypothesis states that in a self-replicating system based on entirely different principles, a mechanism for synthesizing and polymerizing the components of RNA evolved for its own selective advantage (for instance, as 'selfish' genome elements do) and then took over.[27] One such system, already mentioned in Chapter II, is clay crystals.[28] However, we neither know whether such clay crystal-based evolving replicators are actually possible, nor what is a possible path from these to an RNA-based replicator.

A hypothetical solution to early life's energy problems comes from hydrothermal vents deep in the ocean,[29] popularized by Nick Lane at University College London, in his book The Vital Question.[30] Hydrothermal vents are porous chimney-like rocks on the ocean floor mixing hot geothermal and cold ocean water, in this way creating 'steep' thermal and chemical gradients and energy flows. Additionally, the pores offer natural 'cells', while their surface can inorganically catalyse various chemical reactions. Thus, possibly the thermal and chemical gradients within these structures could have provided pre-life with energy. One such proposed nurture site of early life is the 'Lost City Hydrothermal Field' in the Atlantic Ocean. Recently, it has been reported that possible signs of life more than 4 billion years old have been discovered in hydrothermal vents in Quebec, Canada.[31] This still quite speculative hypothesis, does not explain though how the harvested free energy enabled the growth of information-carrying polymers.

How long did life exist as a symbiotic complex of molecules, possibly living in porous minerals, if indeed this is where life started, before the first membrane surrounded cell emerged? The contemporary bacterial and archaeal cell membranes (Chapter IV) are made of phospholipid molecules, each of rather different type. Which of these types is closer to what was used in the early cells? Maybe the first cells contained membranes that included both types of phospholipids. Did the RNA life (if it existed) evolve to membrane-surrounded cells, or did the membrane evolve only after DNA/protein-based life had emerged? We do not know.

After all that has been said, the reader might appreciate that the emergence of life is a difficult unsolved scientific problem. Can it be solved at all? It is possible that the 'chemical ladders' that scientists are exploring currently will not reach the ceiling. Trying to reconstruct the chemical evolution that led to the emergence of life is like trying to put together an enormous puzzle where one only has some of the pieces at hand.

From RNA to DNA and proteins

Let us assume that the first evolving self-replicating system has emerged and that the replicating units consist of RNA and smaller molecules enclosed in a membrane. In other words, it is a cell where both the information carriers and interpreters are based on RNA. How did DNA and proteins emerge in this system? For this, a system for translating information from nucleic-acid to amino-acid polymers needed to evolve. Recall that in a minimal cell, discussed in Chapter IV, there were almost 200 genes devoted to translation. How could such a complex system emerge? Just like we do not know how *pre-life* evolved into life, we do not know how RNA life evolved into RNA/DNA/protein-based life—this happened over a billion years ago. Nevertheless, this problem is simpler, as we already have an evolving replicator.

The current leading hypothesis builds upon an observation that amino acids and short polypeptides can serve as 'enzyme assistants', conventionally called *co-enzymes*. A co-enzyme is a small molecule that increases the efficiency of an enzyme or potentially a ribozyme (Chapter IV). Short polypeptides can be synthesized without the machinery translating DNA, possibly by ribozymes or even via inorganic catalysis. The presence of a particular amino acid or polypeptide may have increased the efficiency of a ribozyme, and in this way, a preferential linkage of amino acids or polypeptides to RNA could have evolved. Possibly, this is how transfer-RNAs evolved, eventually providing a link between DNA and proteins.

RNA/DNA/protein-based life was likely to outcompete the pure RNA-life: DNA is a more stable information carrier than RNA, while proteins have more diverse three-dimensional structures and thus have more diverse and efficient enzymatic activities. We do not know how exactly this transition happened, but once we had the first replicator, there was exponential growth, and many 'experiments' could be performed to look for improved efficiency. We know that currently there are over 10^{30} bacteria living in the ocean, on average replicating once a day. Assuming that RNA-based life was a thousand times slower, and that the number of replicators that were supported by the ocean at that time was a billion times smaller, we still get 10^{18} replications a day. In 100 million years over 10^{27} evolutionary 'experiments' would have been performed.[32]

Is this much? On the one hand, obviously this is an enormous number; on the other hand it allows for an exhaustive search of sequences only up to about 45 nucleotides long.[33] Thus, it isn't much really. But evolution did not search the entire sequence space—it only explored the fitness landscape travelling along the higher 'ridges' (Chapter V). Any replicators that wandered away from the ridge died and the exploration in these directions stopped. Evolution does not do exhaustive search; it uses genetic algorithms (Chapter VII) and explores only the tiny part of the global fitness landscape that is 'inhabitable'.[34]

Given the universality of the genetic code, we can be sure that all contemporary existing life has evolved from a common ancestor, which is sometimes referred to as the Last Universal Common Ancestor (LUCA). What was LUCA like? Was it similar to a contemporary prokaryotic cell, or have prokaryotes changed in billions of years beyond recognition? How many genes did LUCA have?

Given that contemporary prokaryotic cells belong to two distinct domains of life—archaea and bacteria—LUCA must have had elements common to both. Beyond sharing the genetic code, bacteria and archaea share only about 100 genes and almost all these common genes are coding for parts of translation and transcription machinery.[35] Surprisingly, the genes that code for the DNA replication machinery in bacteria and in archaea are rather different. This would seem to indicate that the DNA-replication machinery evolved after LUCA was already established. Can this be true?

One hypothesis is that LUCA had only one of the two different DNA-replication machineries—either the archaeal or the bacterial one—while the other one evolved independently in viruses and was later picked up by one of the domains of life. A different hypothesis is that LUCA had a small genome, therefore its replication machinery might have been much simpler and very different from that in contemporary organisms. It is also possible that a lot has changed since contemporary cells diverged from LUCA and that intensive horizontal gene transfer has mixed things up beyond recognition.

Whatever LUCA was like, it did already have all the most fundamental features of contemporary life; it was a membrane-enclosed DNA/RNA/protein-based evolving replicator. The evolution of life had achieved a major milestone. Information was encoded in DNA and was interpreted by proteins.

A complex cell

After bacteria and archaea had diverged, at some later point they came back together again, but now in a different sense. Possibly, archaea were grazing on a bacterial mat on the ocean floor and occasionally swallowed bacterial cells alive. On one such occasion, rather than getting digested, a bacterium got 'enslaved' by the archaeal predator, or possibly it was a friendly symbiotic relationship of a bacterium entering and living inside an archaeon. Whichever way it happened, we know almost for certain that somehow an ancestor of a contemporary *alpha-proto bacterium* became internalized inside an ancestor of a contemporary archaeal cell and became an endosymbiont—a symbiotic organism living inside another one.[36] Arguably this was the defining event in the birth of the eukaryotic cell—the principal building block of all complex multicellular organisms, such as animals and plants. And arguably, the emergence of the eukaryotic cell was one of a very few most important evolutionary transitions in the path from the emergence of life to present-day complexity.

We do not know whether attempts to evolve a cell of a eukaryotic type happened only one or many times, but we do know that only one lineage succeeded and eventually became the ancestor of all contemporary eukaryotic cells. Most scientists estimate the time of eukaryotic origin as somewhere between 1.2 and 1.4 billion years ago, which is long after the emergence of life itself. The scientific consensus window is much wider though: from 2.7 billion years ago—eukaryotes early hypothesis, to 800 million years ago—eukaryotes late hypothesis.[37] Whether there was one defining event or many, the evolution of the eukaryotic cell was a long and gradual process that introduced at least 60 major evolutionary innovations, both in respect to how information is stored in a cell, and how it is passed on to descendants.[38]

A typical eukaryotic cell has hundreds to thousands of times larger volume than a typical prokaryotic cell. Even more importantly, eukaryotic cells have complex inner membrane structures known as *organelles*, as shown in Figure 6.3. Such membrane-enclosed cellular compartments allow for different concentrations of molecules at different locations of the same cell, opening enormous possibilities for a high degree of organization inside one cell. Another important difference is that eukaryotic cells do not have a cell wall; the shape of the cell is defined by intracellular scaffolding, which can change dynamically.

In prokaryotic cells, molecules find their partners and locations mostly by diffusion. Eukaryotic cells have evolved active, energy-driven molecular transport mechanisms moving around not only individual molecules but entire structures. In a particular type of such trafficking, a membrane-surrounded 'bag', known as a *vesicle*, is brought to the cell's external membrane, merged with it, and as a result its contents are released into the external environment. As cells also can have receptors that can sense various molecules in the environment, such a mechanism can be used for sending signals between cells. Possibly such vesicle-based intercellular signalling is what enabled the evolution of nerve cells interconnected in networks forming the brain—the main information processor of an animal.[39]

The innovations in how information is stored and processed inside a eukaryotic cell are just as spectacular. Bacterial and archaeal genomes usually do not contain more than 5000 to 6000 genes, rarely more than 10,000. Multicellular eukaryotic organisms almost always have well over 10,000 protein-coding genes; some have as many as 50,000. As a rule, eukaryotic cells split their genomes into several chromosomes, which are complex structures where DNA is 'wrapped around' specific proteins and packed in hierarchically structured folds. This gives eukaryotes an additional means of gene regulation—some parts of the chromosome may be open, accessible to transcription, while others may be closed. Thus, different cells can have different *epigenetic states*, defining different cell types, such as muscle cells, skin cells, or neurons—they each have a different epigenetic state of the same genome. Eukaryotic cells have evolved

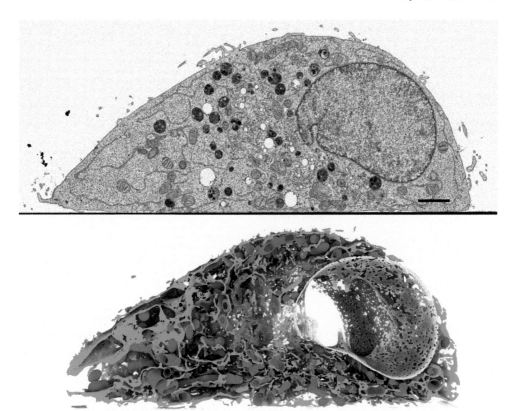

Figure 6.3 *Eukaryotic cell.* **Top.** *Focused Ion Beam Scanning Electron Microscopy generated image of an 8 nm-thick slice of a human (HeLa cell line) cell. The scale bar (black) represents 2 µm (2000 nm), which is in the order of the size of a prokaryotic cell. For details, see Hennies et al., 2020.* **Bottom.** *A picture reconstructed by computationally aligning 1,700 such slices in a stack and identifying and rendering different inner membrane structures (so-called* organelles). *More specifically, a machine learning algorithm was trained to recognize* nuclear envelope *(yellow),* mitochondria *(magenta),* Golgi apparatus *(blue), and* endoplasmic reticulum *(green).*
Figure by Julian Hennies, Anna M. Steyer, Nicole L. Schieber, and Yannick Schwab at EMBL.

mechanisms for how such epigenetic states can be inherited through a number of generations and in this way, a multicellular organism maintains its cell types. Obviously, the division of a eukaryotic cell is complex; for instance, the chromosomes have to be correctly distributed amongst the daughter cells.

Eukaryotic genes themselves are more complex too. Typically, eukaryotic genes are split; they consist of stretches of genomic DNA that code for parts of a protein—*exons*, interrupted with usually longer stretches of DNA—*introns*, which do not code for amino acids (Figure 6.4). In eukaryotic cells transcription and translation are physically

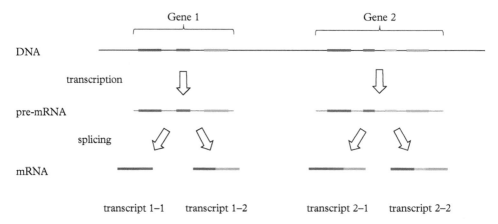

Figure 6.4 *In eukaryotic organisms, genes are typically split: the protein coding parts—exons, which are shown in colour—alternate with introns, which are spliced out before mature mRNA is translated into proteins. Given that different exons of the same gene can be used in the mature transcript, the same gene can code for several different proteins.*

separated—transcription happens inside a membrane-enclosed *nucleus* (Figure 6.3), which is another eukaryotic invention, and the translation occurs after the RNA has been transported outside the nucleus into the cytoplasm. The introns need to be *spliced* out before the genes are translated into proteins. As transcripts can be spliced in different alternative ways, more than one type of a protein can be produced from a single gene. Thus, in eukaryotes the simple one-to-one relationship between genes and proteins—one gene, one protein—is not valid.

But arguably the most important eukaryotic innovation in respect to information representation is that a eukaryotic cell usually contains two similar but not identical copies of its genome. More specifically, chromosomes in a eukaryotic cell come in pairs; each two *homologous chromosomes* have very similar but not identical DNA sequences. The human genome, for instance, is packaged in 23 chromosome pairs (46 chromosomes in total) and the homologous chromosomes differ from each other by one letter in about a thousand.[40] Having such a *diploid* genome not only provides a eukaryotic cell with information redundancy, but as we will discuss shortly, diploidy is an essential part of the mechanism how eukaryotes exchange genetic information on a regular basis.

The key epimutation

The event of the internalization of a bacterial cell by an archaeal one was not a genetic mutation. We know about this because one of the most important organelles of the eukaryotic cell, *mitochondria* (see Box 6.3), which is present in all eukaryotic cells, contains its own small genome that is evolutionary related to that of alpha-proto bacteria. A series of genetic mutations had to happen too in order to enable this internalization, but

Box 6.3 How many domains of life are there?

As noted, the internalization of a bacterium into an archaeal cell and the emergence of the symbiotic relationship between these two arguably was the defining event that led to the emergence of a eukaryotic cell. One of the most important membrane-enclosed structures present in all eukaryotic cells is mitochondria.[41] These structures have their own small genomes and are thought to be remnants of the original alpha-proto-bacterium internalized by the archaeal cell. Mitochondria are the power-plants of eukaryotic cells where ATP molecules are made.[42] The importance of mitochondria notwithstanding, there is growing evidence that many of the eukaryotic innovations, including membrane-vesicle trafficking, might already have been in place in the archaeal ancestor before this internalization.[43]

New discoveries suggest that there might be only two domains of life—bacteria and archaea—and that eukaryotes are a sub-domain of archaea. This is also supported by the recent discovery of a new archaeal group called *Lokiarchaeota*, the genome of which contains several proteins that were thought to belong exclusively to the eukaryotic domain.[44] Interestingly, these archaea include proteins that are essential for membrane-vesicle trafficking. It is fascinating that these archaea live deep in the Arctic ocean at sites of hydrothermal activity at over 3 kilometres deep near the active venting site known as Loki's Castle—a possible cradle of early life.[45]

The way these organisms were discovered was by sequencing the DNA in water sampled from the particular location in the ocean, and then computationally reconstructing the genomes of the organisms present there. It became clear from this *metagenome* analysis that these archaea were not contemporary eukaryotes, even though they contained eukaryotic genes and genome features. Thus, possibly a significant stretch of evolution from prokaryotes to eukaryotes had already happened in the archaeal domain before symbiosis with bacteria led to the evolution of mitochondria. Maybe this is why the time window of emergence of eukaryotes appears so wide—perhaps indeed it was a slow evolutionary process and there was no single point when we can say that the eukaryotic cell had emerged. Possibly, the Last Eukaryotic Common Ancestor (LECA) before the eukaryotic cells began diversifying was quite a complex cell of over 4000 genes.[46] Compare this to LUCA.

this important evolutionary event itself was not one of them. It was an epigenetic event, a mega-*epimutation*, if we can call it so. Similar events happened when green plant cells acquired photosynthetic bacteria enabling the plants to utilize solar energy.

Cellular membrane structures normally are inherited from a mother cell by the daughter cell and then replicated via division in a process in a way like the cell division itself. Eukaryotic cells can grow and multiply various membrane structures, but whether entirely new membrane structures are created by a cell *de novo* is not clear. Thus, the maintenance of a cell's membrane structures is as a form of epigenetic inheritance.

Are there any other epigenetic evolutionary events of a different type as important as mitochondrial internalization? More generally, do epigenetic structural mutations play a regular evolutionary role, or are they rare? This is a hotly debated topic in evolutionary biology;[47] we will come back to this in Chapter VII.

From cells to organisms

A eukaryotic cell, though much larger and more complex than a bacterium, still is microscopic in size. Animals and plants may consist of trillions of cells of hundreds of types forming complex tissues and functional organs, all working in concert. The simple small worm *Caenorhabditis elegans* (*C. elegans*) consists of about 1000 cells, while a human body contains over 3×10^{13} cells of hundreds of different types.[48] How did the cells that originally competed for resources and with each other evolve to cooperate to form complex organisms containing trillions of cells all working in harmony?

Not all multicellular organisms are as complex as animals or plants. Measuring an organism's complexity is difficult, but one possible contributor to an organism's complexity is the number of different types of cells it has. Animals and plants almost always have over 30 different types of cells; algae, such as sea kelp, have around 10 while multicellular fungi, such as mushrooms, typically less than 10.[49] But much simpler multicellular organisms exist, some even made of prokaryotic cells. Photosynthetic bacteria, called cyanobacteria can form multicellular filaments that have at least three different cell types: first, photosynthetic cells, which usually form under favourable growing conditions; second, climate-resistant spores, which form when environmental conditions become harsh; and third, heterocysts, which are cells that fix nitrogen when oxygen is absent. Bacterial biofilms or fruiting bodies formed by myxobacteria, such as those shown in Figure 6.5, can also be argued to be simple multicellular organism.[50] Although it can be debated whether these simple multicellular structures are true multicellular organisms or are communities of symbiotic cells, regardless of where we draw the line there is a continuum from single cell organisms to complex multicellular ones. This continuity may help us to understand how multicellular organisms emerged.[51]

Figure 6.5 *Electron micrograph of the* Stigmatella aurantiaca *(a member of myxobacteria group) fruiting body. The size of the fruiting body is about 80 μm. Myxobacteria form such multicellular fruiting bodies upon nutrient depletion.*

Photograph by Heinrich Lünsdorf, GBF Braunschweig, Germany (Plaga and Ulrich, 1999).

Unlike the eukaryotic cells, which all have the same ancestor, different multicellular groups of organisms, including algae, mushrooms, land plants, and animals, each evolved from unicellular ancestors independently at different times. In fact, *fungi* apparently have evolved multicellularity from single cellular ancestors twice.[52] Overall, multicellularity has evolved on Earth at least a dozen times and possibly first as early as 2.5 billion years ago.[53] It is believed that the first animals evolved from their common unicellular ancestor somewhere between 800 million and 600 million years ago, perhaps even earlier. Fungi evolved at about the same time, while land plants evolved quite a bit later—about 400 million years ago.[54] A simple multicellular green alga, known as volvocine alga *Volvox carteri* (containing only two cell types), evolved from a single cellular ancestor as recently as 200 million years ago,[55] which is around the same time the first mammals appeared. It seems that on a geological timescale, transitions from single to multicellular life have been quite a regular event, even though only few of the attempts have led to the emergence of complex organisms.

Cooperation between cells is at the basis of all single-to-multicellular transitions, but the point at which a cooperative community of cells becomes an organism is not always clear cut and defining one transition point may not be possible. For instance, an important element in this transition is the emergence of cell types that 'commit suicide' to allow other cells to replicate more efficiently. However, something similar also happens in ant colonies: the worker ants die without ever reproducing.[56]

Arguably the most important feature of an organism is its *individuality*—being the unit of natural selection. Darwinian evolution happens to populations of individuals and to species but selection applies to individuals—these are individuals that multiply or die. However, as we will see in a moment, the last statement is not entirely true either.

Genomes vs individuals

Here we are mostly interested in a particular group of multicellular organisms, animals, therefore we do not need to worry about where communities of cells stop and multicellular organisms begin. As every pet owner will confirm, animals are individuals.

In complex multicellular organisms, and in animals in particular, only a small minority of cell lineages, the so-called *germline cells*, can go on replicating potentially indefinitely; most cells, the so-called *somatic cells*, are programmed to die after a finite number of divisions. One can say that the somatic cells are there only to serve the germline cells. Muscle cells and brain cells are somatic, they will die when the animal dies, the germline, however, is potentially passed on to the animal's descendants. Is there a paradox—how did the cells, which originally competed for resources to produce as many descendants as possible, evolve to accept a certain death for the benefit of others?

Just like similar conflicts in cultural life, this conflict is resolved via sharing information. Once we realize that first, all cells in an organism 'share' the genome, and second, that ultimately it is the genome that holds the key to what the cell does, the paradox resolves. By saying that 'the genome holds the key', I do not mean to take the position of genetic determinism. What the cell does depends on the interactions between its genome, the rest of the cell (*epigenome*), and the environment. The statement is true under the assumption that the environment and the epigenomes of the cells under consideration are similar enough.

Let us see how sharing the genome can lead to cooperation. Consider the following *gedankenexperiment* (thought experiment). Suppose we have a mixture of two types of bacteria, one with genome G, the other with genome G', and suppose the fitness of the cells that have genome G is higher. Suppose initially there are two bacterial cells with genome G, say cell c_1 and cell c_2, and many cells with genome G'. Further, suppose that after 100 generations there are a hundred thousand cells with genome G, while the cells with genome G' have disappeared. Does it matter whether all hundred thousand successful cells are the descendants of c_1 (or c_2), or alternatively, about half of the successful bacteria are the descendants of c_1 and the other half of c_2? How would we tell the difference? Although these were the cells—the individuals—that were fighting for existence, in the end it was the genome that succeeded. The reason is that, assuming the same epigenetic states, the cells are 'slaves' of their genomes, the same way as the Universal Constructor is the 'slave' of the information given to it on the tape. Thus, effectively it was the fittest genome, rather than the fittest individuals, that were selected.

This type of an argument was first explicitly formulated by George Williams in 1966, in his seminal book *Adaptation and Natural Selection*, and developed further and popularized by Richard Dawkins in his many books, such as *The Selfish Gene* and *The Extended Phenotype*.[57] The arguments presented by Williams and Dawkins were more sophisticated than my reasoning above. Williams and Dawkins were considering organisms that reproduce sexually, and in that setting they concluded that these were individual genes, rather than whole genomes, that were the objects of selection. We will come to sexual reproduction in a moment, but for now, staying with bacteria, we can assume that the genome, or to be precise, the information encoded in the genomic DNA, is the main object of Darwinian selection.

Let us consider an example of how this can lead to cooperation of individuals. To digest food as efficiently as possible, some bacterial species release enzymes into the extracellular environment to partly digest the food before it is swallowed and utilized by the bacteria internally. For each individual bacterium, this comes at a cost—producing an enzyme takes energy and materials. On the other hand, a cell benefits not only from its own enzymes but also from the enzymes released by its neighbours. For the community as a whole, the cost–benefit balance may be positive.[58] If the bacteria in the colony all have the same genome, then at the end of the day there may be more copies of the genome that make the bacteria to release the enzyme than if the bacteria were 'selfish'. Thus, the genome benefits. Consequently, a genomic mutation that makes the bacteria cooperate—to release such a digestive enzyme—improves the fitness of the genome. Note also that the bacteria that are closer to the centre of the colony are likely to gain more; the ones that are at the periphery are relative losers. Thus, the bacteria at the periphery of the colony are made to accept their own sacrifice for the benefit of their genome.[59]

The sacrifice for the benefit of the genome can go further. If there is a shortage of food, then rather than replicating as much as possible, a better strategy for the genome may be to provide some of the individuals with additional nutrients, to make them robust to the unfavourable environment, disperse them, and let them 'sleep' and wait for better times. This is what the genomes are trying to achieve via fruiting bodies of mycobacterial colonies. And indeed, fruiting bodies can be induced by starvation. Genomes of many different bacterial species harbour genes that make the cells to undergo such preprogrammed death.[60] Evolving a gene that induces the death of a cell is a step towards the evolution of somatic cells.

If we have two cells with identical genomes, what tells them which one is to die and which one will survive—which one will become the somatic cell and which the germline one? Given that all cells have the same genome, this decision either has to be a stochastic event or it has to come from the epigenome or the environment. For instance, the genome may be programmed so that only the cells that reach a size above a certain threshold can become germline cells. Once the decision has been made, for a while the cell's fate can be passed on from cell to cell via epigenetic inheritance. The amount of information needed to define whether a cell belongs to the germline or to the somatic lineage is 1 *bit*—this is a choice between two options. And indeed, small amounts of information can be encoded and transmitted epigenetically.[61]

Returning to the example of sharing the extracellular enzyme, note that to maximize the benefits, the bacterial colony, or rather its genome, may want to keep individual cells close together. One way to achieve this is by producing cell adhesion proteins—proteins in the cell's membrane that stick to similar proteins in the adjacent cells. And indeed, if we look at the most recently evolved multicellular organism, the already mentioned Volvox algae, and compare it to its closest single cellular relative *Chlamydomanas reinhardtii*, we see that they have about the same number of genes (~ 14,500), the main difference between the two relatives being that the multicellular one has genes coding for adhesion proteins.[62] The comparison of Volvox algae and its unicellular relative also shows that proteins that already exist in the unicellular ancestor can evolve to assume such new roles. It does not have to be a huge evolutionary step for a simple multicellular organism to emerge from a single cellular relative.[63]

Cooperation between cells can be enhanced by somatic cells specializing further. For instance, the colony of cells described above, and therefore their genome, might benefit from forming a structure where the outer layer consists of cells that protect the colony against dangerous substances and keep the digestive enzyme inside the colony. Such specialization is a step towards evolution of a primitive digestive system. Forming such cell structures is a form of self-organization (Chapter II), though here these are cells, rather than molecules, that self-organize.

If an enzyme-excreting bacterial colony grows beyond a certain size, one strategy may be to bud—to break off parts of it, which then float away. To achieve this, some cells have to be made to stop expressing the adhesion genes, achieving which requires transcription regulation and inter-cellular signalling. And indeed, three types of genes are usually considered to be the hallmarks of a multicellular organism—genes involved in cell adhesion, intercellular signalling, and programmed cell death.[64]

Obviously, there is a long way to go from such simple organisms to mushrooms or animals, where cells differentiate into many different types and self-organize into complex organs and body parts in a highly controlled fashion.[65] Nevertheless, we can see that Darwinian selection of fittest genomes can lead to emergence of cooperating cells and eventually of complex multicellularity. Sharing genetic information is the key.

Sex and information exchange

The Earth is inhabited by over 10^{28} tiny, micron sized *Proclorococus* bacteria, but by fewer than half a million African elephants. Even though in the early part of the twentieth century there may have been up to 5 million African elephants, this is still some 10^{21}–10^{22} times fewer than the *Proclorococus* population. Given the relative sizes of these two organisms, this will probably not be a surprise to anybody. And although these species may represent the extremes in the size distribution of creatures inhabiting the Earth, the observation illustrates a general trend: overall, the larger the average size of the individuals, the less populous the species. Additionally, the larger the individual, the slower the replication cycle.

It is not hard to see the implication for evolution: the larger the individuals, the fewer evolutionary 'experiments' can happen in a given time. In 100 years *Proclorococus* species performs on the order of 10^{32} evolutionary 'experiments', in the same period African elephants less than 10^7. How can the evolution of elephants keep up? Elephant species have not stopped evolving. DNA-sequence comparison shows that the African and Asian elephants diverged from each other less than 8 million years ago, and the woolly mammoth diverged from the Asian elephant around 7 million years ago (about the same time that the lineage leading to humans diverged from that of chimpanzees). The African elephants themselves are of two separate species—the savannah and the forest elephants, which, adapting to their environments, started diverging from each other approximately 4 million years ago.[66]

Evolution has invented an ingenious solution to make up for the reduced numbers: routine exchange of genetic information. As a result, evolutionary discoveries need be made only once after which they can spread through exchange of information. This is similar to how cultural evolution works. In biological evolution such an exchange of information is known as *sex*. Although bacteria in the wild do exchange genetic information, multicellular eukaryotic species have raised this to a new level and many of them exchange genetic information before every replication cycle.

Individuals of most animal species come in two types known as sexes—males and females. Recall that in eukaryotic cells, genomes are typically diploid: chromosomes come in pairs, where the chromosomes within the pair differ from each other slightly. The organism's germline cells can divide in such a way that their descendants inherit only one chromosome per pair, after which these cells no longer have diploid genomes but have so-called *haploid* genomes. Under specific circumstances, a germline cell with a haploid genome searches for a haploid germline cell of the opposite sex. If the search is successful, the pair fuses and in this way a new diploid cell is born. In this new cell, half of its diploid genome comes from the mother (the contributing female), half from the father (the contributing male). In multicellular organisms, this new cell will go through repeated cell divisions, on the way cells differentiating into different cell types and developing into a new multicellular organism, containing somatic and germline cells.

The merger of two haploid cells, as described above, may appear symmetric but normally the female haploid germline cells—the eggs—are much larger than the male ones. The males only contribute their genome, females contribute the genome, the epigenome, and in some multicellular organisms also provide the environment during the early development of the new individual. Biologists say that only females reproduce directly. It is debatable whether sexual reproduction can be called replication, as the new individual is a clone of neither the mother nor the father but is a 'mixture' of both. Nevertheless, a new individual has been added to the population—some sort of replication has happened. Certainly, genes of these individuals have replicated.

Importantly, in sexual reproduction, before the participating cells become haploid, the homologous chromosomes in each of the merging cells exchange pieces in a process called chromosome *crossover* (Figure 6.6). As a result, each of the haploid chromosomes becomes a mixture of the previous mother's and father's chromosomes. This has two

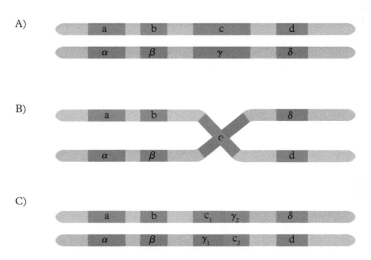

Figure 6.6 *The chromosome crossover. A) The orange chromosome contains genes denoted by a, b, c, and d; the homologous blue chromosome contains the respective homologous genes α, β, γ, and δ. Typically, α, β, γ, δ will differ in no more than a few nucleotides from respective a, b, c, d. Also, typically, there will be hundreds or thousands of genes on each chromosome, while in multicellular organisms the chromosome intragenic regions will be much longer than the genes. Suppose the orange chromosome is inherited from the mother and the blue one from the father. B) The two homologous chromosomes cross over. The figure depicts one crossover point, but several are possible. C) After the crossover, the new chromosome will contain parts from the mother, others from the father. Consequently, in the following reproduction cycles, the new inherited chromosomes will contain information from all grandparents. Also note that crossover does not respect gene boundaries (though if the intragenic regions are much larger than genes, the probability that a gene is split is relatively small).*

important consequences. First, after each cycle of sexual reproduction, every new individual inherits about a quarter of its genetic information from each of its grandparents, about one-eighth from each of the great-grandparents, and so on. Thus, genes from all the ancestors have a chance to contribute; information about genetic 'discoveries' done in previous generations is exchanged. For instance, if a grandparent on the mother's side had a mutation that has made an enzyme A more efficient, while grand-grandparent on the father's side had a more efficient enzyme B, their descendants have a chance to inherit both improved enzymes, without reinventing any of these improvements. Of course, they can also inherit none, but Darwinian selection can now look for the best combinations of the inherited genes.

The second consequence is that in a sexual reproduction, the genomes do not stay intact for more than one generation. Thus, genomes can no longer be the objects of selection—they do not last long enough. Williams and Dawkins concluded that these are neither the individuals nor genomes that are objects of Darwinian selection but genes. This approach too has difficulties, to which we will return shortly.

In germline cell replication, mutations can happen, however, most of the genetic variation, upon which Darwinian selection is working, comes from shuffling the already existing genome variants. In the human germline, there are between one and two mutations per 100 million genomic letters per generation.[67] At the same time, one nucleotide in about 1000 differs between homologous chromosomes. The genome shuffling creates new combinations of these existing variants. The artificial selection of farm animals, which was one of the major sources of data for classical genetics, is almost entirely based on recombining different gene variants already existing in the population; mutations play little role there. This is why the classical genetics at the beginning of the twentieth century did not consider mutations to be important. Sexual reproduction really changes the evolutionary dynamics dramatically.

Is it obvious that sexual reproduction speeds up evolution? Although genome shuffling makes it possible for two positive mutations to come together without each having to be 'rediscovered' independently, the opposite can happen too—two beneficial mutations can get split apart. Moreover, given that only females reproduce directly, the population can grow at only half the speed in comparison to a similar asexual population.[68]

Despite some still persisting counterarguments, it is proven that sexual reproduction speeds up evolution. It can be shown mathematically that if the fitness effects of individual genetic variants are mostly additive (if there is no epistasis, Chapter V), then crossover speeds up evolution by the square root of the length of the coding part of the genome (in comparison with asexual species with a similar size genome). For a species with a genome with the coding part similar in size to that of the human genome, this amounts to about a thousand-fold acceleration.[69]

If there are interactions between genes (if there is epistasis), the mathematical calculations become much trickier and depend on the properties of the particular genome-wide fitness landscape. Nevertheless, it has been shown theoretically that under realistic assumptions, sex does speed up adaptive evolution, in particular in a changing environment.[70] Laboratory experiments with yeast also supports this conclusion—sexually reproducing yeast strains evolve faster than similar asexual ones.[71]

The elusive concept of a gene

Let us return for a moment to asexual selection and to the *gedankenexperiment* described a few pages back. Do the entire genomes need to be identical for the individuals to cooperate? Suppose the two genomes differ in one or a few nucleotides that do not affect the individual's phenotype. Would this change the conclusions from our *gedankenexperiment* that these were the genomes rather than individuals that were selected in the Darwinian selection of the fittest? As the differences do not affect the fitness, Darwinian selection would not see them. Thus, not the entire genome matters, but only the parts that have impact on the phenotype, or more precisely, the parts that if changed, would impact the individual's fitness. Thus, apparently, these are not the entire genomes but only the parts that affect the individual's fitness which are the objects of Darwinian selection. The other sections of the genome can be changed arbitrarily. If the tape of von Neumann's replicator contains sections that are skipped over, it does not matter what bitstrings are encoded there.

One could think that the parts of a genome that do affect individual's fitness are what we call genes. Is this so? In the classical genetics indeed, a 'gene' (a Mendelian unit of inheritance, as defined by Johannsen in 1909) was viewed as 'something' that affected organism's phenotype in a heritable way, for instance determining whether the individual had blue or brown eyes. It was an abstract information-carrying entity, the physical basis of which was unknown. With the advances of molecular biology however, the definition of a gene changed. Nowadays, by the word 'gene' we usually understand a segment of a genomic DNA that code for a protein or a functional RNA molecule. This is why we talk about 20,000 protein coding genes in the human genome.

But are these two concepts of a *gene* the same? Although today we associate genes with functional molecules, in some contexts we also want to associate genes with phenotypic effects, such as a disease. This complicates the situation. For instance, what about promoters (Chapter III)? These are DNA sequences which do not code for functional molecules but regulate the expression of functional molecules, and thus mutations in promoters are likely to have phenotypic effects. So far this is not a problem—in genomics, promoters usually are regarded as parts of the respective genes. However, in eukaryotes, the expression of a gene can also be affected by parts of the genome that in the genomic DNA sequence are located far from the particular gene—by the so-called distal enhancer elements. And moreover, the same enhancer may affect the expression of several genes, while the same gene may be regulated by several enhancers. Should such enhancer elements be regarded as genes too? Conversely, not all changes in coding sequences will have a discernible phenotype, for instance, some synonymous mutations in a gene may have little effect. Thus, defining the concept of a *gene* that pleases everybody is not easy.

Williams and Dawkins, when arguing that genes are the main units of Darwinian selection, stayed within classical genetics. Dawkins defined 'gene' as '*any partition of chromosomal material that potentially lasts for enough generations to serve as a unit of natural selection*' (suggesting that molecular biologists have '*usurped*' terminology—after all, geneticists introduced the term *gene* much earlier; the correct term for a piece of DNA coding for a protein is *cistron*).[72] Dawkins' definition however is somewhat fuzzy, more-over when defining this concept via natural selection while declaring it to be the unit of selection we need to be careful not to fall into the trap of circularity.[73] In Dawkins' defence, some have argued that too rigid terminology 'would only hamper the for-mulation of current ideas and research aims'.[74] This might have been true in 1980s but persistently fuzzy terminology may lead to persistent hand-waving. For better or for worse, with the advances of genomics and bioinformatics, today genes are broadly associated with genome sequences coding for functional molecules.

Let us return to the units of Darwinian selection. In sexual reproduction, the genome changes in every reproductive cycle, thus a genome does not stay intact long enough to serve as the unit of selection. This is essentially why George Williams concluded that genes are the ultimate units of selection. However, these are not different genes coding for very different functional molecules in the organism that directly compete with each other, but rather different variants of the same gene, or as geneticists call them—different *alleles* (for instance, one allele coding for blue eyes, the other for brown). If a gene has different variants (different alleles), one of which gives the organism higher fitness, then the fitter one will be selected over the other variants. Indirectly though, in a longer time period, evolution of one gene can lead to a different gene becoming redundant, and thus disappearing. Moreover, if a species goes extinct, all its genes disappear. Thus, in the long run, different genes do compete with each other too. This is known as the 'gene's-eye view of evolution'.[75]

The gene's-eye view has problems too. One is that chromosome crossover does not respect gene boundaries. Not only genomes but also genes (according to their molecular biology definition) can get scrambled in a cross-over. Thus, in the new individual, a part of a gene may come from one ancestor, a different part from another.[76] In fact, such gene scrambling is a mechanism by which a new gene can be born. But arguably, genes last long enough to be able to serve as objects of Darwinian selection.

There is however a more fundamental difficulty. Williams' and Dawkins' reasoning is straightforward only if the effects of gene variants on the individual's fitness can be added (if there is no epistasis, Chapter V). If gene variants (alleles) are either good or bad, having two genes of a good variant (two good alleles) is better than having one good and one of a bad, which is still better than having two bad ones. The good genes would take over. But if genes interact with each other, their effects cannot be added. As we discussed, in some cases a combination of two individually good gene variants may jointly have a bad effect (sign epistasis). Which of the gene variants will be selected then? Williams and Dawkins address this by measuring the gene fitness in all possible genomes or all existing individuals and averaging it. Although, the validity of this approach is debatable, in many cases the gene contributions to the fitness can indeed by added, and the conclusions of 'gene's-eye view' seem to be largely valid.

As this discussion demonstrates, finding the 'true' unit of Darwinian selection is dif-ficult. It has been argued amongst evolutionary biologists for decades whether groups of individuals, or even entire species, can be objects of Darwinian selection. But species are born and die much less frequently than individuals, thus selection on the level of species is likely to be weak. Nevertheless, Darwinian selection can happen on multiple levels in parallel, different entities contributing to evolution to a different degree. A clear example demonstrating how selection can happen on two levels in parallel is described by Novak.[77] Models of evolution that view evolution as learning from the environment (see below) can also be used to explain *multi-level selection*. Accepting that different level entities can contribute to Darwinian selection makes finding the 'true' unit of selection less important.

Runaway genomes

Between a half and three-quarters of the human genome consists of shorter or longer sequences repeated in many copies, most of which are not believed to have a biological function and thus do not directly affect the phenotype.[78] These repeated genome sequences are mobile (or transposable) DNA elements, which exploit the machinery of a cell to multiply themselves and spread. Or they are 'fossils' of such mobile elements—former mobile elements that are no longer active. Just like genes use organisms as vehicles to propagate themselves, and viruses hijack cell's machinery for their own replication benefit, so do these mobile DNA sequences. In fact, there is no strict distinction between virus genomes and mobile DNA elements; some viruses, including the AIDS-causing HIV virus, multiply by occasionally inserting their sequences into the host genome. Some of the mobile DNA elements are 'fossils' of former viruses.

When evidence for such mobile genomic elements were first discovered in the 1940s by the US biologist Barbara McClintock, the discovery was received with scepticism. McClintock's conclusions were confirmed in the 1960s and 1970s, but it took longer to understand the prevalence and importance of these 'selfish' genomic elements. It was only in 1983, when McClintock received her Nobel Prize in Physiology or Medicine 'for her discovery of mobile genetic elements'.

Mobile elements are non-autonomous replicators. To describe this phenomenon, Koonin introduced the concept of the *fundamental [genomic] unit of evolution*.[79] These are genome elements, either genes or any pieces of DNA that replicate, spread, and mutate, and thus are subjected to Darwinian selection. The dynamics of such 'selfish' genetic elements can be described as multi-level selection. Effectively, a 'selfish' genomic element is a 'cheater' or a parasite, which can successfully spread via the genome for a while, but if it becomes a major burden to its physical host then the host will either have to find a mechanism to stop it or it will die together with the 'selfish' elements. In this way a balance is maintained. For species with larger individuals and thus usually a smaller population, the Darwinian selection horizon (Chapter V) makes it harder for selection to purge such elements, and thus, in such species selfish elements can grow.[80] In human genome, less than 2% of the code for proteins and over half of the genome consist of sequences that are repeated many times and do not seem to have functional roles. Much of the human genome is thought to be 'junk'. But this 'junk' possibly provides material on which evolution can conduct its experiments, thus making the species more evolvable.

Some aspects of genome evolution can be understood by using an analogy with artificial life simulated on a computer. Like artificial live, selfish genome elements are pieces of information that replicate and compete. But unlike in artificial life, selfish genome elements feed back to their physical hosts—if they become too much of burden, together with their physical host organisms they will die. This is what makes the real-world life different from life simulated on a computer. In real life there is constant feedback between the replicating information pieces and the physical entities interpreting this information. And as long as there is no feedback from the computer software to its hardware, we do not need to worry about computers evolving to take over the world. Some readers may have heard of the three 'Asimov's Laws of Robotics': first, 'a robot may not injure a human being or, through inaction, allow a human being to come to harm'; second, 'a robot must obey orders given it by human beings except where such orders would conflict with the First Law'; and third, 'a robot must protect its own existence as long as such protection does not conflict with the First or Second Law'. Perhaps to ensure that computers and robots do not take over the world, we need to add the fourth law: 'No robot shall ever modify itself'. I am afraid, however, that sooner or later this law will be broken, and to protect ourselves we will have to rely on more liberal regulations.

Regardless of the extent to which gene's-eye view explains evolutionary dynamics, it elucidates the role of the interactions between the world of information and the physical world of atoms and molecules. Individuals—cells or organisms—are physical entities, which live or die. Genes and genomes are pieces of information. Williams calls them material and codicil domains respectively.[81] In Dawkins' terminology, the replicators are the genes—thus pieces of information. Cells or organisms are vehicles—they belong to the physical domain. In the terminology of this book, the vehicles are interpreters of information. When we talk about selection of genes or genomes versus selection of individuals we are comparing two different domains. For real-world evolution to happen, both need to interact[82].

New means of information processing

After life emerged on Earth, for billions of years it almost exclusively existed as microscopic, mostly single-cell organisms[83] (Figure 6.7). As mentioned earlier, a single cell already has at its disposal all the means necessary for information processing. On top of this, multicellular organisms evolved neurons, neural nets, and eventually brains, and through this evolved new and much faster means of information processing, and with a larger information storage capacity. How did this happen and when?

The first animals of simple spherical shapes probably appeared around 800–700 million years ago (mya). This is the period of time when macrofossil diversity begins to increase gradually. It is quite likely that the last common unicellular ancestor of the present-day animals was an organism cycling between unicellular and colonial behaviour. Choanoflagellates—motile organisms, shown in Figure 6.8—are believed to be the closest living single cellular colonial relatives to contemporary animals.[84] As discussed earlier, forming spherical colonies, where the food is digested inside the formation, may have made the colony more efficient, leading to increasingly cooperative behaviour.

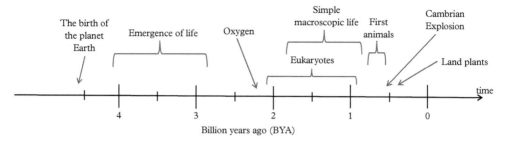

Figure 6.7 *Rough timeline of the evolution of life on Earth. The first life appeared on Earth possibly just half a billion years after the Earth itself formed, however the undisputed fossil evidence is more recent. About 2.3 billion years ago (BYA) significant amounts of oxygen appeared in the atmosphere, which probably was a product of life. Eukaryotes most likely emerged ~1.2–1.4 BYA, though the full scientific consensus puts the window at much wider. The first life forms that were large enough to be seen by naked eye appeared 1.8 BYA, perhaps later. Just over 0.5 BYA, macroscopic forms of life started diversifying rapidly, preceded by another increase of oxygen levels. About 0.4 BYA, or possibly a slightly earlier, macroscopic life came out of the ocean and started colonizing dry land, which is also when the first land plants appeared. (Han and Runnegar, 1992; Knoll and Carroll, 1999; King, 2004; King et al., 2008; Nielsen, 2008; El Albani et al., 2010; Donoghue and Antcliffe, 2010; Knoll, 2011).*

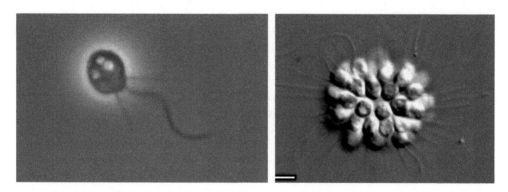

Figure 6.8 *Choanoflagellates are a group of free-living unicellular and colonial eukaryotic organisms considered to be the closest living relatives of the present-day animals. **Left.** The unicellular organism Choanoflagellate* Monosiga brevicollis. ***Right.** Scanning electron microscopy image of a related colonial organism,* Salpingoeca rosetta, *which is thought to be the predecessor of animals. Scale bars = 5 μm. The figure shows a colony oriented in a sphere around a central focus, with their flagella oriented outward. S. rosetta has a sexual life cycle with transitions between haploid and diploid stages. In response to nutrient limitation, haploid cultures of S. rosetta become diploid. S. rosetta can differentiate into at least five distinct cell types, including three solitary cell types and two colonial forms (rosettes and chains). Cells within colonies are held together by a combination of fine intercellular bridges, a shared extracellular matrix, and other means. For comparison, M. brevicollis genome has just under 10,000 genes, while S. rosetta has just over 11,500 genes.*

(Left: phase contrast image of Monosiga brevicollis, by Stephen Fairclough, Wikimedia, CC BY SA 2.5 licence. Right: Figure from Dayel et al., 2011)

It is believed that early animals grazed on algae mats on a shallow ocean floor, where in order to maximize their food intake they needed to be able to move around. For an organism to be able to move, it needs cells that can change their shape; moreover, a multicellular organism also needs to coordinate what different cells do. Eukaryotic cells had the means to evolve these functions efficiently. First, eukaryotic cells can change their shape because of their dynamic inner molecular skeleton, and second, the exocytosis, described earlier, provides these cells with the means to send signals rapidly from cell to cell. It is believed that muscle and nerve cells (or *neurons*) are 'sister' cell types, which had a common evolutionary origin, diverging to specialize for two different, but closely related functions.[85]

The first animals with a bilateral symmetry—animals with head and tail, stomach and back, and by implication a right and left side—probably appeared more than 600 mya.[86] They already had a simple nervous system and simple photoreceptory organs—predecessor of contemporary eyes. Bilaterians gave rise to most animal species dominating the contemporary fauna, including insects and vertebrates. In contrast, sponges, which diverged from the rest of the animals no later than 650 mya, and which do not have neurons, still exist largely unchanged. Whether sponges should be considered as evolutionary very successful species or not depends on one's point of view.

There is ample fossil evidence showing that the diversity and structural complexity of life forms started increasing rapidly from about 540 mya, or slightly earlier. This period of rapid diversification and greater abundance of macroscopic life is known as the *Cambrian explosion*. The extent to which the Cambrian explosion was sudden and what started it is still debated, however this period does seem to be the time when many complex macroscopic forms emerged in a relatively brief period of time.[87] New classes of genes and new interactions between them evolved to enable increasing organism complexity; for instance, genes for a hard skeleton, which was one of the Cambrian inventions. Gene amplification is mostly likely to have played a major role; there is a clear direct link between the amplification of genes coding for some transcription factors and the increase in organisms' complexity.[88] Increasing morphological complexity probably soon led to a more complex behaviour too.[89]

During the Cambrian diversification, the ancestors of contemporary vertebrates (animals with spines) split from invertebrates (such as insects, worms, molluscs, and starfish). The first mammals appeared later, around 200 mya, at the time when Earth was dominated by dinosaurs. Mammals started rapidly diversifying just about 70 mya, when dinosaurs were in decline. Some 55–60 mya, mammals 'gave birth' to primates, the lineage leading to great apes and humans.

To enable the animals to acquire information about the external environment more efficiently, sensory organs evolved. Sensors are present also in unicellular organisms, but multicellularity provided for means to evolve a light-focusing eye and other complex information-acquiring structures.[90] As the animals that were better able to integrate and process information from sensors located in different parts of the body were likely to have an evolutionary advantage, increasingly sophisticated neural networks, the nervous system, and eventually the brain evolved (see Figure 6.9). There is evidence that early bilaterian nerve systems already integrated sensory cells with neurons conducting electric impulses.

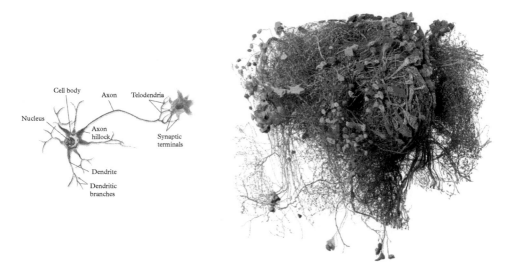

Figure 6.9 *Left. A schematic representation of a neuron. A typical neuron consists of three main components: the body of the cell, dendrites, and axons. Dendrites usually branch, axons can extend for great distances, and usually there is only one axon. The structures that allow neurons to communicate with each other, using chemical/electric signals, are called synapses, typically located at the tips of dendrites and axons (telodendria). Neurons communicate via discrete, though asynchronous, impulses processed by neurons in non-trivial ways. Synaptic signals may be excitatory or inhibitory, and neuron can combine incoming signals in a way analogous to the logical gates (Chapter I), though neurons are much more complex than logical gates. **Right.** A part of* Drosophila *fruit fly brain, specifically 613 neurons with most of their synapses in the so-called superior medial protocerebrum region. The neurons were imaged with electron microscopy, segmented with automated and manual techniques, and rendered with path tracing. The colours are false and chosen in order to maximize the distinction between neurons.*

(Left: Figure is adapted from Wikimedia (author BruceBlaus). Right: Figure courtesy of the FlyEM project at the Janelia Research Campus, Howard Hughes Medical Institute. For more, detail see Scheffer et al., 2020.)

Thus, new means of information processing, based on rapid conduction of chemical and electric signals specific to animals, emerged. Within single cells, information processing is done via chemical reactions; in multicellular organisms, electricity started playing an increasingly important role.

In some evolutionary lineages, the size of the brain and the number of neurons in it were gradually increasing. Apparently, in some ecological niches, larger brain gave animals an advantage in fitness. It is hard to know how many neurons various extinct animal species had, but in contemporary animals this can be estimated. The worm *C. elegans* has exactly 302 neurons, while the human brain contains about 10^{11} (100 billion); these two numbers are probably close to extremes. Note, however, that the information that contains 'instructions' how the brain, like the one shown in Figure 6.9, develops out of a single germ cell, is mostly encoded in the genome of a single fertilized egg (see Chapter VII).

Is evolution learning?

Learning can be defined as using one's past experiences to optimize one's actions in the future. For instance, when one is learning to drive a car, a link gets established between turning the steering wheel clockwise and the car turning right, or pressing on the brakes and the car decelerating. Investigating what exactly happens when an animal is learning is a research subject of neurobiology, but in very general terms, it is believed that the strengths of signal conducting connections between neurons get adjusted or new connections developed.

Artificial systems, such as computers or robots, can be designed to be able to learn. Most artificial learning systems use feedback between trial and outcome, gradually adjusting various parameters of the learning algorithm to make the system behave in a way that would have achieved better results in the past. For instance, an artificial neural network, a schematic example of which is shown in Figure 6.10, can be trained to recognize a face by gradually adjusting the weights of connections between the (artificial) neurons.[91] Such artificial neural networks are at the basis of many recent spectacular advances in Artificial Intelligence and are enabling many different tools of automation, such as self-driving cars.

There are many examples of learning in the animal world; for instance, birds learning to sing in a particular dialect (as discussed in Chapter VII). Some learning happens by interacting with the environment; for instance, a young antelope fine-tuning its running skills by trial and error. Although such skills learned by an animal during its lifetime are not passed on to its descendants, if the individuals that are better learners are more

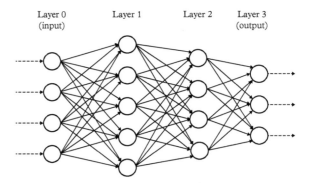

Figure 6.10 *An artificial neural network is an interconnected group of nodes (Schmidhuber, 2015), and is a simplification inspired by biological neural networks (Figure 6.9). Each node represents an artificial neuron, and an arrow represents a connections between neurons. Simplifying the concept further, we can say that each connection has a weight which can be either positive or negative. The neuron sums up all the weighted input signals, and if the sum exceeds a certain threshold then outputs it as the signal. Such an artificial neural network is trained by gradually adjusting the weights of each connection so that the neuron at the output layer gives a desired signal for the input signals. For instance, the inputs may be the pixels of a certain image, while the desired output is positive if the image represents a particular object, such as the face of a particular person.*

likely to survive, the species as a whole is 'learning'. Can evolution itself be viewed as a species learning from the environment via optimizing the fitness of its individuals?[92] If learning is viewed as recording the information relevant to performing the actions that are being learned, in evolutionary learning this information is recorded in DNA. The analogy between evolution, machine learning, and recording of information is best illustrated by the class of adaptive algorithms known as genetic algorithms, discussed in Chapter VII.

Learning can happen only if the adjustments that optimized the fitness in the past will optimize it in the future. In other words, when the relevant information is integrated, past experience tells the learner about what actions will lead to what outcomes in the future. This is true in learning individuals as well as species. Therefore, for evolutionary learning to be possible, either the fitness landscape has to stay relatively stable or its changes in the future have to be predictable from changes in the past. For instance, plant species have 'learned' that a winter is always followed by spring, summer by autumn. If the species' reproductive cycle is a year or more, the fitness can be averaged over a year, in which case we can still assume that the fitness landscape is constant. But there are dry summers and wet summers, and some winters may be cold, some may be mild. The succession pattern of these may be quite unpredictable; nevertheless, some plant species have 'learned' how to deal with this by hedging their bets.[93] If the climate changed completely, for instance if summers became systematically dryer, past experience would not help. But then again, if this is a part of a long-term pattern, for instance millennia of wet summers followed by millennia of dry ones, in principle the past experience may help, provided that memory is long enough. Evolutionary memory is relatively long, but it is not infinite; as discussed in Chapter V, genes that are not used are lost. In machine learning it is typically assumed that memory is not time-limited and what is learned is never forgotten. In evolutionary learning this cannot be assumed.

The interpretation of biological evolution as learning has its critics. The selection of the fittest works only in the present—it tries to optimize what is best at any given moment. While human learners or machine learning algorithms can plan for the future, biological evolution is 'learning' via so-called greedy algorithms—algorithms that only try to optimize each current step without looking ahead.[94] However, if there is sufficient memory, any learning algorithm can be turned into a greedy one by designing an appropriate measure that is to be optimized. The past experience has been recorded in the DNA of the species and the present is interpreted in the context of this information. We do not yet know exactly what kind of optimization functions can emerge in the process of selection of the fittest. This essentially is where the scientific debate lies. Finally, let us note that in the evolutionary context, one of the most successful approaches to machine learning, known as deep (neural network) learning, has been shown to be analogous to multi-level selection.[95]

Animal communications

Eyes and brains gave animals the ability to see and to recognize patterns, improving their ability to find food and to avoid predators. In parallel to vision, a different sensory system was evolving—the ability to detect and analyse sound. Lungfish, the fish that evolutionary is closest to land animals, had this ability some 250 million years ago.[96] Combining such hearing with the ability to produce specific sounds, animals developed a means to communicate—transmitting specific information to each other.

Clearly animals can communicate, but the use of the term information in this context seems to have caused controversy.[97] In my opinion this controversy is misplaced. The use of the concept of information does not need to imply any assumption about animal intelligence or consciousness, which is a rather different topic. We talked about information transmission when describing intra- and inter-cellular communications via signalling molecules, and we can talk about information in communications between computers. There is no reason why this concept could not be used for describing animal communication achieved via sound or vision. Information is transmitted, received, and interpreted. This is how the term information was meant when it was first introduced by Norbert Wiener and Claude Shannon in the 1940s. Comparing '*living individuals and . . . some of the newer communication machines*', Wiener wrote:

> Both of them [living individuals and machines] have sensory receptors as one stage in their cycle of operation: that is, in both of them there exists a special apparatus for collecting information from the outer world at low energy levels, and for making it available in the operation of the individual or of the machine. In both cases these external messages are not taken net, but through the internal transforming powers of the apparatus, whether it be alive or dead. The information is then turned into a new form available for the further stages of performance. In both the animal and the machine this performance is made to be effective on the outer world. In both of them, their performed action on the outer world, and not merely their intended action, is reported back to the central regulatory apparatus.[98]

Animals evolved increasingly sophisticated ways of transmitting information to each other, for instance, to warn each other of predators. Typically, animals living in close groups share many genes, therefore the ability of such signalling increases the fitness of their genes. Communications between animals have also enabled evolution of yet another level of self-organization: not only do molecules self-organize into supra-molecular structures to minimize free energy, not only do cells self-organize into complex multicellular organisms to maximize the fitness of their genomes, the organisms themselves can self-organize in cooperative structures to maximize the fitness of their genes.[99] Birds travel in formations to minimize the friction of the air (Figure 6.11) and wolfs hunt in organized packs to catch and overpower large prey.

One species, *Homo sapiens*, has evolved this ability to exchange information to a qualitatively new level by evolving the faculty of language. Thus, for the first time since the emergence of life, evolution gave life tools to communicate unlimited amounts of information by means other than DNA or other polymer molecules. This also allowed humans to develop entirely new ways of information processing. This is the topic of Chapters VII and VIII.

Figure 6.11 *Migratory birds, such as geese, swans, ducks, often self-organize in symmetric V-shaped flight formation to improve their energy efficiency.*
Image reused from © gal mashiach/Shutterstock.

VII

Evolution as a Ratchet of Information

A process which led from the amoebae to man appeared to the philosophers to be obviously a progress—though whether the amoebae would agree with this opinion is not known.

(Bertrand Russell, 1914)

If we compare the early single-cellular life that existed on Earth billions of years ago to the multicellular life forms that exist now, it seems an obvious conclusion that the complexity of life has increased. Arguably we would come to this conclusion in whatever rational way we measured complexity. But is life's complexity still growing? Has life's complexity increased, say, over the last hundred million years? Is it possible that from some point the overall complexity of life on Earth has stopped growing? Is there an evolutionary trend towards increasing complexity?

Throughout this book I have emphasized the analogy between the genome in a living cell and the tape in von Neumann's self-replicating automaton. There, the tape carries information that instructs the automaton what to do. As was already noted, the tape can carry extra information instructing the automaton to build another object in addition to its own replica. It can be argued that the complexity of the object that will be built is limited by the amount of the information on the tape. By analogy, in a multicellular organism, the genome contains information not only enabling the cell to grow and to divide but also to differentiate into particular cell types, which then self-organize into organs and body parts, together eventually developing into a fully fledged multicellular organism. Does this analogy imply that there is more information in the genome of a multicellular organism than in the genome of a microbe? If so, the growth in complexity of life would imply accumulation of information in genomes. To argue this, we need to explore the concept of information in some more depth.

The word *information* was already present in Latin; its modern meaning, however, is much younger and its origins can probably be associated with the invention of the telegraph. Before the telegraph, any transmission of information relied on transporting physical materials, such as letters. The speed at which information could travel between distant locations was limited by the speed of relocation of materials—essentially by how fast a horse could gallop, a ship could sail, or a pigeon could fly. According to a legend, in 490 BC, a Greek soldier Pheidippides ran 42 kilometres from the town of Marathon to Athens to announce the defeat of the Persians. Although the light or smoke signals

Living Computers. Alvis Brazma, Oxford University Press. © Alvis Brazma (2023). DOI: 10.1093/oso/9780192871947.003.0008

carrying one or a few *bits* of information—for instance, warnings about an approaching enemy—were used already in ancient times, sending larger amounts of information got decoupled from the transport of matter only with the invention of the electric telegraph.[1]

One of the first electric telegraphs was built in 1774, when the Swiss scientist Georges-Louis Le Sage linked two rooms of his house with 26 wires, one for each letter of the alphabet.[2] Others were experimenting with similar ideas; amongst them, the famous German mathematician Carl Friedrich Gauss, working with physicist Wilhelm Weber, built more than a mile-long telegraph line connecting the Göttingen Observatory with the Institute of Physics in 1783. They used a magnetic needle that could turn left or right, thus effectively utilizing a binary alphabet—each turn of the needle could transmit one *bit* of information. Later telegraphs used a binary alphabet of short or long electric pulses, dots, and dashes, as in Morse code. In 1838 William Cooke and Charles Wheatstone built a telegraph line running 21 kilometres along the railway from Paddington station to West Drayton in England, while in 1845 Samuel Morse and Alfred Vail opened a line running from Washington DC to Baltimore in Maryland in the United States. The first transatlantic cable connecting Europe and North America was completed in 1858, about a year before Darwin published his *Origins*. During the three weeks while the cable functioned, it had reduced the time of sending information between the two continents from 10 days to an instant. One of the about 400 messages sent during these three weeks said: 'the 62nd Regiment were not to return to England',[3] allegedly saving Britain £50,000. The information was something real; having had an impact on the real world. A lasting connection between Europe and America were finally achieved in 1866. The speed at which information could be transmitted had firmly beaten the speed of moving physical materials.[4]

The current age is so dominated by information that it is hard to imagine what the world was like before everybody instantly knew what was going on in any other part of the planet. The establishment of functioning long-distance telegraph lines had an important impact on the very formation of the concept of information. As telegrams were expensive, codebooks were introduced to send the desired information with as brief a message as possible. Morse code was designed to represent the letters that are more frequent in English with the shortest sequences of dots and dashes, thus minimizing the message length. A different codebook listed all the English words and gave each a number; the meaning of a sentence could be transmitted as a string of numbers. As telegrams did not travel in sealed envelopes, secret codes were invented to conceal their meanings to those who did not have the right code book (the right interpreter). Although none of this was entirely new, with the telegraph, for the first time in history, humanity became accustomed to the idea of coding and to information as something separate from a particular physical media and even the specific text.

For reliable long-distance information transmission to become a reality, various technological challenges had to be overcome. Although many of these were in engineering, it turned out that there were fundamental limitations too. A fundamental limitation on how much information could be transmitted in a particular time interval is imposed by thermal noise. Thermal noise in electrical currents was studied by John B. Johnson and

Harry Nyquist[5] at Bell Laboratories in 1926. Now known as Johnson–Nyquist noise, it is a consequence of the equal partitioning of energy (Chapter II) and is essentially the same phenomenon as Brownian motion. Recall that at temperatures above 0 °K, all particles jiggle. This applies not only to atoms, molecules, and larger particles but also to electrons. This is a fundamental property of matter; at temperatures above 0 °K thermal noise cannot be eliminated. And this has fundamental implications for information storage and transmission. This fundamental problem of noise was discussed by Ralph Hartley (also from Bell Laboratories) at the International Congress of Telegraphy and Telephony[6] in 1927, however the theory of information transmission was really pinned down by Claude Shannon, in his seminal papers 'A mathematical theory of communications' and 'Communication in the presence of noise',[7] published in 1948 and 1949 respectively. The scientific concept of *information* has been associated with Shannon's name ever since then.

Shannon formulated a mathematical model that included an information source generating the messages, such as strings of 0s and 1s, a transmission channel, and a receiver. The question was how to quantify what needs to be transmitted so that the receiver would know what the source is doing. To quantify this, Shannon first found a way to quantifying the unpredictability of the source. The more unpredictable the source, the more information needs to be transmitted. If the source is repeating the same word over and over again, then the receiver only needs to be told what this one word is. If a stream of unpredictable sentences is coming out of the source, then the receiver needs to be told what each and every of these words are.

The mathematical expression to quantify this unpredictability of the source turned out to be similar to that for thermodynamic entropy (Chapter II). There is a story that thinking how to name this unpredictability, Shannon consulted von Neumann. Allegedly von Neumann suggested calling it entropy, adding something to the effect that 'nobody knows what entropy really is, and therefore you will always have an advantage in any discussion'. We now call this quantity *Shannon's entropy*, or *information entropy*, or if the context is clear, simply *entropy*.

Shannon defined the *amount of information* contained in a message as the reduction in the entropy this message achieves.[8] In this way, information was not only measured in a way independent of the physical carrier but also independent of its content. As Shannon himself was keen to stress, his information measure had nothing to do with the meaning of the message. According to Shannon's information measure, a completely random (and thus unpredictable) sequence of 3 billion letters A, T, G, and C, contains at least as much or more information as the human genome. Shannon was very aware of this limitation and explicitly discouraged the use of his theory outside the intended context of telecommunications. As has been noted by several authors, Shannon's information measure quantifies the maximal amount of information that the message *could* contain, rather than what it *does* contain. One of the most important contemporary applications of Shannon's information theory is in data compression—how much an image, a movie, a telephone conversation, or a DNA sequence can be compressed for storage or transmission. This is limited by the information entropy in the data.

Shannon also introduced the concept of *channel capacity*—the amount of information that can be transmitted over a channel in a given time, independent of the specific physical medium. A telegraph, telephone, radio, or fibre-optic cable each have different fundamental limitations on their channel capacities, however to know how much information can be transmitted in a given time interval, we only need to know the channel capacity, rather than what the particular physical medium is.

Despite the mentioned limitations, as was pointed out by John Maynard Smith, Shannon's information theory does have applications to biology.[9] One of Shannon's most important contributions to information theory was the invention of various coding systems that would allow for the channel capacity to be utilized to its maximum. Shannon showed how information can be encoded efficiently and unambiguously without delimiters such as commas or spaces. There is an analogy between such delimiter-free coding and Crick's reasoning about the triplet codon in DNA: a frameshift by one or two positions completely changes the sequence of the coded amino acids, while a triple frameshift restores the code (except for one amino acid).[10] This simple observation was used to decode the triplet codon. As was noted by Maynard Smith, if scientists had studied DNA exclusively from the perspective of biochemistry, they might never have discovered the triplet code. But the application of Shannon's information to biology is not limited to such simple observations.[11] Nevertheless, it is not controversial to say that Shannon's information measure does not really correspond to what is meant by information intuitively in many situations, including in biology.

Like energy, information is one of these rather general concepts that are hard to define. In 2018 in Wikipedia, *information* was defined as '*any entity or form that resolves uncertainty or provides the answer to a question of some kind*'. Since then, the definition has changed every year and in 2022 Wikipedia defined it as follows: '*information is processed, organized and structured data*'. These definitions are arguably as good as any but they are not very concrete. The yearly changes speak for themselves.

There is a popular quotation attributed to Lord Kelvin: '*to measure is to know*'.[12] Can we quantify information in a way meaningful to genomics? Energy can be measured thanks to its conservation; information is not conserved—if I have inside information which horse to bet on and if I tell you this, we both have this information, even though its value may be marginally reduced. As a scientific concept, energy was first introduced in the nineteenth century[13] and we still do not fully understand what it really is (Chapter II); the scientific concept of information is about 100 years younger and seems to be even more elusive.

Nevertheless, most people will probably agree that information is something real in a sense that it can make an impact on the physical world. Having the right information can save time, energy, or money. Information can help us to win a bet, to navigate a labyrinth, or to assemble flat-pack furniture. Intuitively most people will agree that more information is needed to assemble an airplane than to assemble a shelf,[14] or that more information is needed to find a treasure in the shortest possible time in a larger labyrinth than in a smaller one. Thus, there is something to measure.

Information and prior knowledge

Consider a robot navigating a labyrinth, tasked to reach the exit. The robot can move forward, turn left or right, and is equipped with sensors telling it when it is at a junction. If the labyrinth has only three-way junctions and the robot does not like turning back except for dead-ends, then at each junction it has exactly two choices—go right or go left. If the robot does not know anything about the labyrinth, then by turning right or left at random, with a bit of luck, it may eventually reach the exit.

Suppose the robot has a memory device into which a sequence of instructions '*turn right*' or '*turn left*' can be recorded. Every time the robot comes to a junction, it retrieves and executes the next instruction. If we encode *turn right* as R and *turn left* as L, a sequence such as RLLLR will guide the robot for certain number of turns. Most will agree that such a sequence of instruction can be called *information*, and that having 'the right' information would help the robot to accomplish the task on average faster. In Chapter I we postulated that we could talk about information only in the context of an interpreter. Here the robot is the interpreter—it converts a sequence of symbols R and L into actions.

Can the amount of information in such a sequence of instructions be quantified? Because resolving the choice between L and R takes 1 *bit*, the sequence of n instructions seems to require n *bits*. If the labyrinth included junctions with three exits, these can be decomposed into two consecutive two-exit junctions, encoded, for instance, as L, RL, RR. A four-exit junction can be decomposed to LL, LR, RL, RR. Thus, 2 *bits* of information are sufficient for junctions with up to four exits. Similarly, 3 *bits* are sufficient to guide the robot at a junction with up to 8 exits ($8 = 2^3$), and more generally, m *bits* for 2^m-exit junctions. Given that $k = 2^m$ implies $m = \log_2 k$, we conclude that to make a choice from k options, no more than $\log_2 k$ *bits* of information are needed.[15] Consequently, to instruct our robot in a sequence of n consecutive junctions of up to k exits each, $m\log_2 k$ *bits* of information will be sufficient.

This however does not mean that $m\log_2 k$ *bits* of information will always be needed. Consider a labyrinth such that to get to the exit, at some point one has to turn right at every junction, say 20 times in a row (all the left turns are dead ends). Do we need to specify RRRRRRRRRRRRRRRRRRRR? Obviously, we can 'compress' this sequence of instructions to, for instance, 'R *20 times*' or R^{20}. Or perhaps simply R⋆, meaning '*repeat R until reaching exit*'. Similarly, instead of LRLRLRLR...LR we could write (LR)⋆. More generally, we may notice that a sequence of symbols can sometimes be represented by a shorter sequence. In other words, some messages can be *compressed*. This is also what the users of early telegraphs found—to transmit certain information, it suffices to send the shortest message that can be interpreted unambiguously by the receiver who knows how to uncompress it. A smart phone will expand some abbreviated messages as you type them. Given that the compressed message contains all the information that is in the original one, we can define the *amount of information* in the message as *the length of the maximally compressed message*. To avoid ambiguity, we can assume that the compressed message is encoded using the binary alphabet.

However, to be able to work with the compressed messages, the robot needs to be smarter. First, the binary alphabet of letters L and R, has now been extended to five letters: L, R, (,), and ⋆. These can be represented in binary alphabet using a method similar to how we decomposed *k*-exit junctions into series of two-exit ones, but this means that to encode each symbol more than 1 *bit* of information will now be needed.[16] Secondly, and more fundamentally, to understand the new 'language', the robot needs to be able to parse expressions like (LR)⋆. The robot needs to have the 'prior knowledge' that seeing the opening parathesis it needs to find the corresponding closing one, and that (*s*)⋆ means *repeat s until reaching exit*. More generally, the interpreter of the information needs specific *prior knowledge* how to decompress the messages it receives.

If the receiver of the message has the Universal Computer at its disposal, it can execute any algorithm transmitted to it alongside the actual message. Thus, the Universal Computer can be viewed as some sort of *universal prior knowledge*. We can send to the Universal Computer the compressed message together with the decompression algorithm. This may remind us about the concept of Kolmogorov complexity introduced in Chapter II. Kolmogorov complexity of an object (or a sequence of symbols) was defined as the shortest possible description of this object (of this message), and this indeed is at the basis of Kolmogorov's definition of information. According to Kolmogorov, the information content of a message equals *the shortest description of this message in the binary alphabet.*[17]

Information and complexity are two sides of the same phenomenon—the more complex the object, the more information is needed to describe it. One way to measure the complexity of an object is by how much information would need to be given to the Universal Constructor, to make this object. Recall however, that Kolmogorov complexity is not computable, it can only be approximated. Consequently, the true amount of information in a message can only be estimated, for instance by applying a good compression algorithm. The length of the binary string output by a compression algorithm cannot be less than the amount of information in the message. And the better the algorithm, the closer the output to the true information content. As with Kolmogorov complexity, his definition of information is meaningful only for long messages.

In Chapter II, Kolmogorov complexity was used to measure order—the higher the disorder, the higher its Kolmogorov complexity. A careful reader might have noticed that this appears to imply a contradiction—the most disorderly messages are the random ones, thus like in Shannon's definition, a random sequence of symbols contains most information. Are we back to square one? An important difference between Kolmogorov and Shannon's definitions of information is that Kolmogorov's measure can be applied to a single, sufficiently long message, while to use Shannon's approach we need a source generating many messages according to some distribution. However, the solution to our paradox resides somewhere else. It lies in realizing that typically we are interested in information about something particular, rather than about information 'as such'. Quoting Kolmogorov:

> *While a map yields a considerable amount of information about a region of the earth's surface, the microstructure of the paper and the ink on the paper have no relation to microstructure of the area shown on the map. In practice, we are most frequently interested in the quantity of information 'conveyed by an individual object x about an individual object y'.*[18]

Consider a labyrinth such that a particular sequence of instructions, say LLRLRRRL-RLRL, takes the robot from entrance to exit. As there is no obvious way how this sequence can be compressed, the information in this sequence can be estimated as about 12 *bits*. Consider a different sequence LLRLRRRLRLR, which is almost the same, only the last instruction L is missing. How much information does the second sequence contain about the navigation in the labyrinth to the exit? It almost does the job, but not quite. Suppose the second sequence of instructions is already stored in the robot's memory. Now instead of giving the robot the entire instruction sequence above, we can say: '*use the sequence you already have and then* L'. Now we need only 1 *bit* of information, rather than 12. Obviously, this assumes that the robot has the knowledge of how to interpret such instructions. What is already stored in the robot's memory device can be considered as specialized *prior knowledge*. The *mutual information* between the two messages above is 11 *bits* (within a precision of a constant overhead).[19] The concept of mutual information can be generalized; we can ask how much information a message contains about a given object or a given process.

In the mathematical 'world' of information *bits* and *bytes*, cellular automata, and tessellation structures discussed in Chapter I, Kolmogorov complexity is a well-defined concept. When applying this concept to the physical world of atoms and molecules, the situation becomes more complicated. Any physical system has to 'obey' the laws of physics, but physics can make some tasks easier, for instance through self-assembly or self-organization (Chapters II and III). A protein folds to minimize the system's free energy, therefore to make a protein, the robot may only need to assemble the amino acids in the right order rather than to fold the protein. We only need to describe the sequence of amino acids, rather than the three-dimensional structure of the protein. Abstractly, we can say that if the interpreter is a living cell, then physics is a part of the *prior knowledge* of the interpreter. A living cell 'uses' the minimization of free energy to read and interpret information in the DNA to build its replica. The genome does not have to describe the laws of physics. More generally, there may be many layers of prior knowledge (see Box 7.1).

In 1965, in his seminal paper 'Three approaches to the quantitative definition of information', Kolmogorov wrote: '*I believe that the approach proposed here yields, in principle, a correct definition of the "quantity of hereditary information", although it would be difficult to obtain a reliable estimate of this quantity*'.[21] The question of quantifying hereditary information was obviously on Kolmogorov's mind a decade before Fred Sanger invented DNA sequencing, which eventually gave us the means to read this information.

Box 7.1 Layers of prior knowledge

Consistent with our definition, the amount of information that a particular interpreter needs for performing a specific task can be measured as the length of the shortest binary string, which when received, the interpreter can accomplish this task. The interpreter can be a specialized robot, a general-purpose robot (a real-world implementation of the Universal Constructor), a living cell, a human person, a self-driving car, or some other apparatus. An interpreter of some sophistication will have the Universal Computer at its disposal, but in

continued

Box 7.1 *continued*

addition, it may also have some specific *prior knowledge*; for instance, an algorithm for parsing messages like (LR)*. A self-driving car will have algorithms telling it how to interpret
information it is constantly receiving from sensors. This prior knowledge—the respective
algorithm—can itself be viewed as information, quantified as the shortest description of the
code of this algorithm. There may be different layers or levels of prior knowledge which build
upon each other. An interesting discussion of layering of prior knowledge can be found in
the book by Douglas Hofstadter, *Gödel, Escher, Bach*:

> [B]*efore you can use any rule, you have to have a rule which tells you how to use that
> rule.*[20]

Consider a person observing a colleague working on a computer and, for instance, noticing that it is possible to capture the content of the screen by entering a combination of, say,
three specific keys on the keyboard. Arguably, learning this trick can be viewed as acquiring information. How much information has been acquired? The length of the binary string
describing which particular combination of the keys does the trick can be estimated as about
3 *bytes* or 24 *bits*. But is such a quantitative estimate meaningful? We made various implicit
assumptions about the prior knowledge that the observer already had; for instance, that they
already knew how to operate a computer. Clearly, in comparison to this prior knowledge, the
amount of the new information is very small. This is why the specific estimate of exactly 24
bits does not seem to make much sense.

More generally, when the amount of the new information is small compared to the information describing the prior or background knowledge, the new information can be quantified
only approximately. This is related to what was discussed in Chapter II: Kolmogorov complexity makes sense only when applied to long strings, and thus to complex objects, and
moreover, the possible shortest description can only be estimated.

Estimating the amount of new information is particularly difficult when the receiver is a
human person because the background information typically exceeds the transmitted information by orders of magnitude. Few will disagree that, for instance, a molecular biology
textbook contains more information than an instruction manual for a washing machine, or
that a large university library contains more information than any particular textbook (even
if the particular book is not in the library), but comparing the amounts of information contained in manuals for a washing machine and a dishwasher does not make much sense. If one
already knows how to use the washing machine, one does not need the manual. And even if
one does not know how to operate the machine, one may be able to figure this out using one's
general prior knowledge. In this case the manual contains 0 *bits*. This is why we rarely read
such manuals.

The analogue and the digital

The reason why we can talk about the abstract concept of information is that quite
often it does not matter how the information is represented. Whether these are holes

in a perforated tape or sequences of nucleotides in DNA, they encode information. Bioinformaticians have learned a lot about organisms by treating genomes as if they are sequences of letters. When describing computer architecture, von Neumann preferred to talk about abstract logical gates rather than their specific electronic implementations.

Some readers may have noticed that I am treating the concept of information as inherently digital—as strings of discrete symbols. Strings can be easily generalized to two-dimensional arrangements, like configurations of *Game of Life*, or three-dimensional or even four-dimensional arrangements that include the time dimension, but these would still be digital objects. Can information be analogue? Does a film photograph contain information? Of course, it does. One can argue that our visual perception itself is analogue, though this is debatable.[22] Nevertheless, we cannot *a priori* treat the concept of information as exclusively digital.

Some of the simplest examples of where the concept analogue information can be found concern regulatory feedback loops. Whether it is the centrifugal governor of James Watt regulating the pressure in a steam-engine, or an electric thermostat maintaining constant temperature in a room, or a transcription factor regulating its own expression by binding to its own promoter, all these represent regulatory feedback loops. The physical implementations are wildly different but they all share something in common: they all represent feedback circuits, or more generally, all are parts of control systems. Such commonality between control systems of physically very different nature was one of the key observations of Norbert Wiener, which in the 1940s lead to the development of a new scientific discipline—cybernetics. Information is its key concept.

Feedback loops and information are both mathematical models that can be studied independently of their physical implementations. Nevertheless, there is something in common to all physical systems, which has profound implications on how information can be represented. This is thermal noise. Thermal noise limits how much information can be recorded in limited space or transmitted in limited time (at limited energies) without resorting to digital-combinatorial representation. We already touched upon this when discussing Johnson–Nyquist noise, but to illustrate this further let us consider the following example.

Consider an analogue dimmer light-switch. Conceptually, it can be viewed as a dial with one arm, as shown in Figure 7.1. The position of the arm on the dial or the angle between the arm and a defined position can be used to record information. The more precisely we can position the arm and read its position back, the more information can be recorded. In the continuous mathematical 'world' of real numbers and infinite precision, a position or an angle will record an infinite amount of information. However, in the material world, long before we reach the limitations imposed by the discrete nature of atoms, we are limited by thermal noise and its implications on the maximal precision of a measurement.

A dial of an ordinary analogue wristwatch has 60 equally spaced marks representing minutes or seconds; thus, any particular position of the arm can reliably record $\log_2 60 \approx 6$ *bits* of information. If we divide every minute into 10 equally spaced parts and if

Figure 7.1 *An arm and dial device for recording information. The amount of information that can be recorded is limited by the precision with which the arm can be positioned and measured, which in turn is limited by thermal noise.*

the dial is large enough to have the marks ~ 0.2 mm apart, which would still make them distinguishable by the naked eye, then a position of an arm could record $\log_2 600 \approx 9$ *bits*. If we used an optical microscope to read out the position, we could increase this amount until we reach 0.2 μm resolution limit (Chapter IV), which adds another $\log_2 1000 \approx 10$ *bits*. At room temperature we would be unlikely to achieve even this limit due the thermal noise. (We could cool the system down but that would increase the size of the device and impose other limitations.)

We can increase the amount of the recorded information by making the dial larger, but that means increasing the dimensions. Or we can make repeated measurements and average the outcome, but that increases time; moreover we need not only measure but also to position the arm with required precision. If we do not want to increase time, we can use several replicate devices. But then there is no point in using just one arm—we can do much better by combining several arms on the same dial. It is not surprising therefore that analogue watches have two or three arms to represent hours, minutes, and possibly seconds. This enables one to distinguish between at least $60 \times 60 \times 60 = 216,000$ states, enough to store > 17 *bits*. In practice, in an analogue watch we use $12 \times 60 \times 60$ states, multiplied by another two given by the light or dark state of the environment.

This example is a manifestation of a general rule—*in a limited space and time, large amounts of information can be represented only by combinatorial means*; in other words, only as a combination of a small number of distinguishable objects or states. This empiric rule is a consequence of the omnipresent thermal noise. The word *digital* is essential here because to use combinatorics, we need to be able to distinguish between the different objects that we are combining. And in the real world of thermal noise, we can distinguish only between a limited, typically small number of objects. This is formalized in the so-called Shannon's second theorem, briefly discussed below. Schrödinger had come to a similar conclusion in his *What is Life* in the 1940s when discussing how it was possible for a living organism to transmit large amounts of genetic information to its descendants.

One *bit* of information can be represented by any physical system or device that has two relatively stable distinguishable states which can be inspected without changing the state, and which can be switched from one state to the other in a controlled way. The first operation is reading, the second is writing. A light switch has two states, *on* or *off*, thus it can record 1 *bit* of information. In Chapter III we discussed two-state proteins; the state of such a protein can store 1 *bit* of information. If the system has more than two stable states, it can record more information. A given position in a single DNA strand can contain a nucleotide of one of four different types—it can be in one of four different states A, T, C, or G, and therefore it can store 2 *bits* of information.

To record information, the different states have to be separated by an energy barrier (Chapter II) preventing them from switching spontaneously (Figure 7.2). One can think about a two-state information recording system as a surface with two cavities and one marble, which can be in one of the two cavities in this way recording 1 *bit* of information. The higher the barrier, the safer the information is against corruption; for instance, against erasure by thermal noise. Light switches usually do not spontaneously switch their state—the energy barrier is too high for thermal noise to play a significant role.

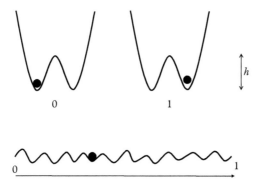

Figure 7.2 *Top. A representation of 1 bit of information in a system consisting of two cavities and a marble. The marble in the left cavity represents 0, the right cavity represents 1. At temperatures above 0 °K, the system is vibrating and the marble jumps up and down. The higher the temperature, the higher the amplitude of the vibration. Sooner or later the marble will jump over the barrier and thus the information will be corrupted. The higher the (energy) barrier* h, *or the lower the temperature, the longer the information will stay uncorrupted.* **Bottom.** *The real-world representation of analogue information. One might think that considerably more than 1* bit *of information can be recorded in this way (intermediate states representing additional information), however, at temperatures above 0 °K, the recording is unprecise and will be corrupted fast.*

Spontaneous mutations converting one nucleotide to a different one through a series of chemical reactions do occasionally occur in DNA.[23]

A device with k states can store up to $\log_2 k$ *bits* of information but in the material world of limited resources and thermal noise, the number k is fundamentally constrained. As mentioned, this observation is captured in Shannon's theorem about the transmission of information via a noisy channel. It states that if per time unit the energy of the signal equals S, and of noise is N, then the capacity of the information transmission channel is limited by $\frac{1}{2}\log_2(1+S/N)$ *bits* per transmission. The signal to noise ratio S/N intuitively corresponds to the maximum number of different energy levels that can be distinguished from each other given the noise. To distinguish between more states, we need more energy or less noise. But even then, the gain would be rather small because the logarithm grows rather slowly: $\log_2(1000) \approx 10$, $\log_2(1,000,000) \approx 20$, etc. In combination with the slowness of the logarithmic growth, thermal noise is an absolute killer of analogue information!

A single device with k states can store $\log_2 k$ *bits* of information; a combination of n such devices jointly can store $n\log_2 k$ *bits*. Thus, the number n has much higher impact than the number of states k. For instance, the number of arms on a dial has much higher impact than the number of marks on the dial. Similarly, the length of a string of symbols is much more impactful than the size of the alphabet. This is the power of combinatorics. Computers use memory elements that have just two states, 0 and 1, typically represented by an electric charge at a particular location in a silicon crystal (computer memory chip). It is much easier and more efficient to manufacture chips that contain billions of such elements than to try to increase the number of states for each element. Life has opted for the alphabet of four nucleotides. Although in part this may be a historic accident and in part dictated by the constraints of chemistry, it is also possible that any gain in increasing the alphabet would not compensate for the increase in copying errors due to noise.

What about analogue versus digital information processing? Models of analogue information processing are sometimes presented and it is claimed that these more accurately represent the way information or computations are done in real-world physical systems, such as the animal brain. Such analogue models may contain interesting theoretical ideas but unless they consider the effects of the omnipresent thermal noise and the limited precision by which things can be measured, they do not represent the real-world systems more accurately. It has been noted by some that the artificial neural networks (see Chapter VI), which are good at many computational tasks and under some assumptions can be shown to be capable of implementing the Universal Computer, can be viewed as analogue machines. However, they all are run on a digital computer and arguably for reasons more than convenience.[24] What about the human brain, arguably the most powerful computer that exists? We do not know enough about how the brain works to say if it is more digital that analogue, but it is most likely that it gets around the problems caused by thermal noise by using massive parallelism and implementing a reliable device made of unreliable components (Chapter I). Given the thermal noise, an analogue computer cannot do more than a digital stochastic computer, and by implications not more than the digital Universal Computer. See also Box 7.2.

Box 7.2 Fuzzy information or lack of information?

Let me postulate that to call a mathematical structure 'information', for every two instances of this structure, we need to be able to decide if they are identical, and if they are not, then we need to be able to tell what exactly the differences are. If we cannot do this, we are not talking about information. A road sign telling us to slow down has to be distinguishable from any other road sign. A partially obscured road sign may alert us to look out for something unspecified; this too is information telling us that we just passed an unspecified road sign. Obviously, strings of symbols satisfy this distinguishability requirement. Finite configurations in the *Game of Life* also satisfy this requirement. On the other hand, analogue photographs can satisfy this postulate only if the required resolution is specified, which effectively amounts to digitization. Once a photograph is digitized, the postulate is satisfied—we have extracted information from the photograph. Note that the digitization does not have to be based on pixels; it can be achieved in different ways, at different resolutions or by identifying specific shapes or objects in the image. For instance, is there a cloud in the sky nor not? Effectively, this postulate states that information cannot be fuzzy—if something is fuzzy, this is information missing rather than information.[25]

There are ambiguities in this postulate; for instance, I have not specified what operations are permitted when deciding if two entities are identical. We have to be careful to avoid circularity here because if the decision procedures are digital, the information becomes *a priori* digital. Nevertheless, interpreting such ambiguities in a sensible way, we will be likely to conclude that information is ultimately digital. This effectively means that although the representation of information in the physical world can be analogue (unless we go down to the scale of atoms), the information that we extract from it for processing is inherently digital. While one might argue that whether information can be analogue or not is a matter of definition, if we want to account for thermal noise then accepting that information is inherently digital becomes a necessity.

To finish this section, let me restate once again the fundamental observation that *in the physical world, in limited space and time and using limited energy, the only way to encode large amounts of information is by digital-combinatorial means.* If we accept that any information can be represented digitally, then it is quite straightforward to prove that every piece of information can be encoded as a one-dimensional string of symbols, and moreover, it can be encoded using the binary alphabet of 0s and 1s.

Information in a genome and the *prior knowledge* of a cell

A multicellular organism develops from a single cell and the information that guides this development comes from three sources—the genome, the epigenome, and the external environment. Here, by epigenetic information I mean all the information encoded in the cell, except for what is encoded in its polymer molecules, which I define as the genetic information. Typically, all genetic information is in the genomic DNA; other polymers present in the cell are encoded in the genome. The epigenetic information includes the information encoded in specific structures not specified by amino acid sequences, such as protein states or membrane structures, information encoded in the inherited abundances of various molecules, for instance transcription factors, and the information encoded as DNA methylation marks (Chapter III).

The environmental information, unlike genetic and epigenetic information, is external, nevertheless some information can be transmitted between generations via the environment. There are essentially two ways how this can happen: first, by the organism modifying its environment, which is referred to as *niche construction*; and second, through transmitting cultural information, for instance, birds learning to sing in a particular 'dialect' from their parents. The distinction between niche construction and cultural inheritance is not always clear cut: nobody is likely to view bacteria releasing digestive enzymes in the environment as cultural information, nor birds singing, as niche construction, but what about birds making nests? It turns out that the latter is neither one nor the other: it has been demonstrated experimentally that birds inherit their nest building skills genetically.[26]

How much information is transmitted via each of these means and how do these amounts compare? To answer these questions, we need to consider at least three factors. First, what is the combinatorial encoding potential of each of these channels. Second, we need to look at what the noise levels in each of these communication channels are; that is, what are the capacities of genetic, epigenetic, and environmental information transmissions channels. Third, we need to ask to what extent the capacity of each of these channels is used in practice.

Let us start with the genetic information. The combinatorial potential of genetic information encoding is enormous; it is only limited by the maximal size of the genome that the cell can carry. As each nucleotide pair can encode up to 2 *bits*, the total amount of information in the genome is the number of nucleotides or twice the number of nucleotide pairs. The genetic information is transmitted between generations extremely reliably, less than one mistake per billion letters copied (Chapter III), thus the noise level in this channel is low.

This is the upper bound, but this does not yet tell us to what extent this channel capacity is utilized. Not all the nucleotides in the genome are likely to carry information relevant to the functioning of the cell. As noted, less than 2% of the human genome codes for proteins, and though other parts of the genome may encode other functional information, it is generally believed that large parts of human genome consist of so-called junk DNA. We are interested in the mutual information between the genome and the molecular processes in the cell; we can call this the *functional genetic information*.

To estimate the lower bound of the *functional information* in the genome, that is the minimal amount, first we should realize that in the long haul, a disruption of any of the organism's proteins is likely to have an impact on the individual's fitness under some conditions (otherwise evolution would have got rid of this protein). In Chapter III we estimated that over 800 *bits* of information are needed to encode an average sized functioning protein. Thus, in small bacteria with 1000 genes, the genome needs to encode at least 800,000 *bits* of information, while a mammalian organism, such as the human, needs at least 20 times that, which is about 16 million *bits*. Measured in *bytes* and in rough approximation, this is about 100 kilobytes (KB) for a small bacterium, and 1–2 megabytes (MB) for a human. These numbers are small, but these are only the lower bounds—at least this much information is required to encode the proteins. The human genome is likely to contain a significant amount of functional information outside the protein-coding parts, so the real number can be expected to be considerably higher.

As noted, the upper bounds of genomic information can be estimated by the genome size, but we can lower it by seeing how much the particular genome can be compressed (and in this way getting rid of redundancies). In practice, the upper bounds are on the order of a couple of hundred MB for bacteria, and around a gigabyte (GB) for humans. The true amount of the genetic functional information is somewhere between the lower and the upper bounds.[27] Obviously, this window is rather wide; can it be narrowed? It is quite likely that for microbes or other species with large populations, the true amount of functional information is close to the upper bound, as the Darwinian selection is strong and most redundancy in the genome is likely to be purged. For species with smaller populations and suspected 'junk' in the genome, one might ask if it is possible to reduce their genomes while essentially preserving the phenotype? Such experiments are difficult, in part due to the need to preserve the essential epigenetic information.

Epigenetic information

The epigenetic information can be viewed as the *prior knowledge* of the cell that is needed to interpret the genetic information. Such a definition of epigenetic information is consistent with the meaning of the term *epigenetics* as introduced by the British biologist Conrad Hal Waddington in the 1940s to describe the link between an organism's genotype and phenotype. A cell may have different epigenetic states (different cell types), each of which interprets the genome somewhat differently. Moreover, a cell may be able to interpret genomes of somewhat different species too; recall Venter's experiments described in Chapter IV. After inserting a new genome into a genome-less cell, after several replication cycles, the cell got converted into a different cell which was 'described' in the new genome. A living cell is the interpreter of the genome information that tells it how to replicate and differentiate, similar to how a robot is the interpreter of the information helping it to navigate in a labyrinth.

How does the mount of epigenetic information compare to what is transmitted between generations genetically? To address this question, let us consider the following *gedankenexperiment*. Suppose we have a laboratory robot that is sophisticated enough to synthesize any specified DNA molecule, as well as assemble a living cell from molecules. If we had such a robot, we could compare the length of the shortest sequence of instructions needed to synthesize the particular genome with the length of the shortest sequence of instructions to make a working cell (assuming that the information in the genome is already given). Although the information would be quantified only in respect to this particular robot—to the particular interpreter—if our robot was general enough, these estimates would represent a certain level of objectivity.

The outlined laboratory experiment is beyond our present capabilities, nevertheless the *gedankenexperiment* demonstrates that comparing the amount of information in the genome to the amount of information in the rest of the cell is a meaningful exercise. Venter's experiments mentioned earlier suggest that not every detail of the cell mattered; a cell from a somewhat different species was good enough to interpret the information in the genome adequately. Moreover, given that eventually it was the genome that took over and converted the cell according to its own information, it is not unreasonable to speculate that the genome contained more information than the rest of the cell. This becomes even more clear in multicellular organisms (see Box 7.3).

Box 7.3 Multicellular organism as Von Neumann's replicator

Recall von Neumann's self-replicating automaton U+R+C+code(U+R+C) from Chapter I. The first part U+R+C is the actual device, while code(U+R+C) is its description written on a tape, which is read and interpreted by the device. As discussed in Chapter I, this device can be modelled as a two-dimensional cellular automaton, which in one implementation was estimated to fit into a rectangle of 32,000 cells, each with 29 possible states. About 5 *bits* of information are needed to distinguish between 29 states ($2^5 = 32$), thus about 150,000 *bits* of information are needed to describe this structure. Consistent with this, the length of the tape was estimated to contain 150,000 binary digits. (Whether the interpreter of the information on the tape is the Universal Constructor, or a human, the information that is needed to draw the two-dimensional copy of the replicator is present on the tape.) This is one of the simplest known models of an information carrying self-replicating structure.

Now let us look at the simplest living cell. The theoretical minimal living cell discussed in Chapter IV contained about 150 genes, each of them coding for a protein of some 200–300 amino acids.[28] Assuming that about half of the amino acid in the coded genes is essential for the structure of the protein, on the order of 150,000 *bits* of information are needed to describe these proteins adequately. This is about the same as information needed to describe the two-dimensional von Neumann's replicator described above. By analogy, with von Neumann's replicator, let us assume that the amount of information needed to describe the cell itself is about the same as for the genome. This is a leap of faith, but we can allow for a large error here without impacting our further reasoning.

As discussed in Chapter I, we can expand code(U+R+C) to code(U+R+C+D), where code(D) is a description of an additional device D, which would be built alongside the replica. That is, if we feed code(U+R+C+D) to automaton U+R+C, then device D is produced alongside the replica of the initial automaton. Importantly, we only need information about D on the tape, D itself is not needed, as automaton U+R+C will build U+R+C+D from the information code(U+R+C+D) given on the tape.

Using a certain level of abstraction, U+R+C+D can be interpreted as a multicellular organism, while code(U+R+C+D) is its genome. Genes in code(D) are telling the cell how to differentiate with each new division, so that the differentiated cells self-organize in organs, body parts, and the whole new multicellular organism would emerge. As argued above, the information needed to describe device U+R+C is in code(U+R+C), and thus both have the same complexity. On the other hand, code(D) can be anything, it can be much larger than code(U+R+C). By analogy, the information needed to describe a multicellular organism can be much higher than that needed to describe a single cell. If so, it follows that the amount of information in the genome of a complex multicellular organism is considerably more than the epigenetic information (prior knowledge) of the initial single fertilized egg (the initial cell).

In multicellular organisms, we need to distinguish between two types of epigenetic inheritance: the somatic epigenetic inheritance, which occurs within the organism when somatic cells divide and the organism is developing, and the *transgenerational* epigenetic inheritance, which passes information to the next generation via germline cells. Here we are interested in transgenerational inheritance.[29] Transgenerational epigenetic inheritance certainly exists and some changes in the epigenetic state from generation to generation (epimutations) produce a clear phenotype. An example of this is methylation marks in the genome, which can change the expression of specific genes, and thus have clear phenotypic effects. Transgenerational epigenetic inheritance of phenotypic effects of epimutations are well known in plants,[30] an example is shown in Figure 7.3.

In mammals, transgenerational inheritance of epigenetic marks seems to be rare and confined to rather specific cases. One reason is that there is a systematic erasure of genome-methylation marks between successive generations. In fact, there are two rounds of such reprogramming, leading to almost complete erasure of methylation marks.[31] Nonetheless, rare epigenetic marks can be transmitted to the next generation. One example is the yellow-skinned, so-called *agouti mouse*. The yellow skin colour is caused by aberrant expression of agouti gene,[32] which normally is silenced epigenetically by DNA methylation of its promoter. This epigenetic silencing is inherited across generations but an epimutation can occasionally happen.

Over 80 examples of transgenerational epigenetic inheritance in different species have been collected and systematized by the Israeli scientist Eva Jablonka and colleagues.[33] Although some of the epigenetic mechanisms can be used combinatorically and certainly different modalities can be combined with each other, none of the presented 80 examples involve significant combinatorics. Thus, it is not likely that large amounts of information are transmitted this way, certainly not in comparison to what is encoded genetically.[34]

Over the last decade, the developments and discoveries in epigenetics have generated headlines in the popular press. Transgenerational epigenetic inheritance has sometimes been claimed to enable characteristics acquired during an organism's lifetime to be inherited, thus breaking the Darwinian evolutionary paradigm. However, perhaps with few exceptions, epigeneticists themselves do not make such claims.[35] Our current understanding of transgenerational epigenetic inheritance in mammals is summarized in a recent review, 'Transgenerational Epigenetic Inheritance: Myths and mechanisms':[36]

[A]lthough the notion of adaptive epigenetic inheritance retains considerable appeal, concrete evidence from model systems is still lacking.

Figure 7.3 *Flower* Linaria vulgaris *provides one of the most cited examples of transgenerational epigenetic inheritance and an associated occasional epimutation.* Linaria vulgaris *can have radial* (**right**) *or bilateral symmetry* (**left**) *(Miska and Ferguson-Smith, 2016), as was first described by Linnaeus more than 250 years ago. The radial symmetry flower phenotype is the result of a naturally occurring epigenetic mutation—a relatively stable DNA methylation state that suppresses the transcription of the gene called Lcyc and that can be inherited transgenerationally (Cubas et al., 1999). A spontaneous switching of this methylation state—an epimutation—can occasionally happen. A non-ripening tomato phenotype provides a similar example (Manning et al., 2006). Such 'abnormal' phenotypes often are the result of a genetic mutation in a gene promoter region, typically caused by mobile genomic elements (Chapter VI), which then is suppressed by an epigenetic mechanism (Heard and Martienssen, 2014). Note that in the mentioned examples, the amount of epigenetically transmitted information is 1 bit of information.*
(Figure courtesy of Enrico Coen ©.)

Further:

> [P]*roof that transgenerational inheritance has an epigenetic basis is generally lacking in mammals. Indeed, evolution appears to have gone to great lengths to ensure the efficient undoing of any potentially deleterious bookmarking that a parent's lifetime experience may have imposed.*

Combinatorial epigenetic encoding mechanisms certainly exist and are used in somatic epigenetic information transmission.[37] In contrast, it does not appear that life uses epigenetic combinatorics in transgenerational inheritance. High noise levels in this information transmission channel may be one of the reasons. The fidelity of transmission of the nucleotide methylation state, which is thought to be one of the most robust epigenetic inheritance mechanisms, is in the range between 95% and 99%. Thus, one to five methylated nucleotides in every hundred are likely to be corrupted when the genome replicates.[38] Recall the error meltdown threshold from Chapter V. For a population of replicators to be sustainable, their genomes cannot contain more binary digits than the inverse of the transmission error per *bit*. Assuming the fidelity of epigenetic inheritance as 1% error, the limit to the amount of information that can be transmitted across generations in a sustainable way is 100 *bits*. In somatic inheritance the information needs to be sustained only for a limited number of cell divisions, and thus the error meltdown threshold does not apply. Transgenerational epigenetic inheritance, however, can create the environment on which genome mutations can work; for instance, to help a cell to acquire antibiotic resistance (Chapter V).

Niche construction

Niche construction is a process in which organisms modify their environment. '*Numerous animals manufacture nests, burrows, holes, webs and pupal cases; plants change levels of atmospheric gases and modify nutrient cycles; fungi and bacteria decompose organic matter; bacteria fix nutrients.*'[39] As noted by Dawkins, some of these examples, such as fungi decomposing organic matter, may be described better as niche alteration rather than construction. Another term used to describe these phenomena is *ecological inheritance*. Over 40 different examples of niche alteration have been described, and mathematical models show how such ecological feedback can alter evolutionary trajectory of a species.[40] From the theoretical perspective this is to be expected, as altering the environment is likely to change the species fitness landscape, which then changes evolutionary trajectories.

Assessing the amount of information transmitted from generation to generation via niche alterations is tricky. How much information does an earthworm pass to its descendants by altering the soil? How much information does a beaver transmit to other beavers by constructing a dam? Perhaps the most important niche construction/alteration event in the entire evolutionary history of life on Earth was the great oxygenation event that happened over 2 billion years ago (Figure 6.8), opening enormous new evolutionary possibilities. The outcome of the oxygenation was a provision of life forms with new materials and new ways to use energy, dramatically changing the fitness landscape. Although, this example illustrates that information is not the only factor important in evolution; such gradual changes in the environment will be eventually recorded in the genomes, as discussed below.

None of the over 40 mentioned examples of niche construction/alteration described in the article mentioned above[41] employ combinatorial information encoding. On the other hand, the space constraints for encoding information in external environment are quite permissive—the size of the habitat of the animal. Still, it does seem reasonable to assume that the alterations that happen on the timescale of a generation do not alter the fitness landscape significantly enough to transmit large amounts of information.[42]

Animal cultures

Animals can communicate information to each other, including to their descendants directly, for instance, using sound, touch, or visual signals. Animal *communications, traditions*, and *culture* are a fascinating and a broad research subject which cannot be given due credit on a few pages.[43] Here I will offer some examples from the literature to demonstrate how information can be communicated in the animal world via cultural means and to assess the capacity of this information transmission channel.

Culture can be defined as a set of *traditions*, where a tradition is a '*distinctive behaviour pattern shared by two or more individuals in a social unit, which persists over time and that new practitioners acquire in part through socially aided learning*'.[44] For instance, eating haggis and drinking whisky on Burns' night is a tradition of Scottish culture, while eating caraway cheese and drinking beer on summer solstice night is a tradition of Latvian culture. What is important here is that *traditions* are not inherited genetically, nor epigenetically. If we accept this definition,[45] then at least some animal species can be shown to have culture and thus are transmitting information between generations by means other than genetics. A combined analysis of data produced by many independent chimpanzee behavioural studies in Africa in total comprising data accumulated in 151 years of chimpanzee observations in the wild revealed 39 traditions.[46] They include different ways of tool usage, grooming and courtship behaviours, which differ between seven different chimpanzee populations. For instance, in some sites, chimps crack nuts with a wooden 'hammer' and others with a stone 'hammer', and this behaviour is consistent and persistent across the site but does not have a genetic or an obvious ecological explanation. A fascinating example of a behavioural pattern that almost certainly is inherited culturally in chimps is termite 'fishing'.

> *At many study sites chimpanzees harvest termites using a single probing tool inserted into the sides of the insects' mounds. . . . A study in the Goualougo Triangle, Republic of Congo, described chimpanzees approaching termite mounds already armed with appropriate tools, sometimes two different ones [Sanz et al., 2004]. The first is a stout stick . . . which is thrust into the ground using both hands and often a foot, puncturing a tunnel into the nest about 30 cm beneath the ground. A more delicate probe is then inserted into the tunnel to extract termites; this probe is first prepared by biting it to length, manually stripping the leaves and pulling it through the teeth to create an effective 'brush-tip'. This brush-tip method, like the use of the puncturing stick, is not known for chimpanzees harvesting termites elsewhere in Africa.*[47]

This habit is almost certainly transmitted by young chimps observing their mothers; there is evidence that the skill is acquired much earlier by females, who spend more time observing their mothers than males do.[48]

In some cases, it is possible to demonstrate that the information is inherited culturally even more directly. In one experiment, chimpanzees at Bossou were introduced to a new type of nut, cracking of which is a skill present in chimpanzees at distant locations.[49] When this introduction coincided with the immigration of a female already 'expert' in dealing with the new nuts, the 'secret of the trade' gradually spread to a majority of the community through resident chimpanzees observing the female (and presumably later observing each other). An interesting example of animals transmitting information culturally is demonstrated in the experiment involving chimpanzees that was done in a laboratory by using a fruit dispensing apparatus, called the 'Doorian Fruit apparatus', shown in Figure 7.4, which shows without doubt that animals can pass information to each other by direct communication.[50] Another example is animal pedagogy, employed

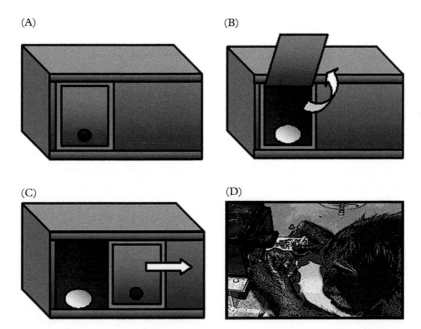

Figure 7.4 *Doorian Fruit apparatus from Horner et al., 2006. (A) The starting position with door closed. Fruit can be accessed either by lifting the door (B) or by sliding the same door sideways (C). The slide mechanism is spring loaded, thus the doors always close automatically. (D) Outlined photograph of a model animal performing the lift method. Chimpanzees were divided into three groups. One individual from the first group was trained to fetch fruit by the lift-door method and one individual from the second group was trained to use the slide-door method. Next, a second individual in each of the two groups was allowed to observe the trained individual of the group until the trained individual succeeded obtaining fruit 10 times. Then the observer was allowed to operate the apparatus. If the observer succeeded in fetching the fruit 10 times, the third individual from the respective group was brought in to observe this, and so on. The third (control) group consisted of six individuals, none of which were trained. Simplifying the description of the outcome of the experiment somewhat, in the lift-door group the skill to fetch the fruit by this method was transmitted along a chain of six chimpanzees, while in the slide-door group the skill was transmitted along five. In the control group two individuals opened the door by lifting, one by sliding, and three individuals failed to open the door. Although this experiment directly demonstrated that cultural information can be transmitted in chimpanzees amongst individuals of the same generation, it seems reasonable to speculate that similar information transfer can happen between successive generations (an experiment to test this directly would require 90 years).*
(Copyright National Academy of Sciences, USA (Horner et al., 2006).

by some predator species, such as cheetahs or meerkats, who bring partially incapacitated prey to their cubs or pups.[51]

How much information needs to be transmitted from a 'competent' chimp to the learning one, for instance, to learn the skill of cracking a nut? The general motor skills needed for this are most likely in part information inherited genetically, in part acquired from interactions with the environment. Nevertheless, additional information apparently is acquired culturally through observing a competent animal. This may be similar to a human learning to open a new type of a child-proof lid, or perhaps to learn the trick of making a screenshot on a computer (see Box 7.1). In comparison to the prior knowledge, arguably, the amount of new information is rather small.

A different set of examples of direct information transmission in the animal world is bird song, which provide an excellent experimental system. By swapping eggs between nests, or by raising birds in sound isolation or in environments where songs of different species are played, we can distinguish between the information that is transmitted culturally and what is acquired genetically or epigenetically. Are bird songs learned culturally or determined genetically? How much information does a bird song transmit? How many different messages a bird can distinguish? These questions are discussed in Box 7.4. The general conclusions are that first, bird songs are largely defined genetically and, second, the amount of information transmitted by a particular song does not seem to exceed more than a few *bits*.

Box 7.4 Animal communications

In many songbird species, males sing multiple types of song, the repertoires ranging from a few songs to over 1000, but a small to moderate size of repertoire of less than 100 songs is the most common.[52] Experiments show that differences in repertoire size between species are due to genetic inheritance; for instance, the grey catbird and the sedge warbler generate large normal song repertoires when raised in song-isolation conditions.[53] However, there are *dialects* shared by bird communities of the species living close to each other but different from the dialects of other communities of the same species. These are at least in part transmitted culturally within species but cannot cross species barriers. White-crowned sparrows have dialects, which are acquired during the first 30 to 100 days of life by learning from older individuals.[54] If however, white-crowned sparrows are exposed to songs of related but different species of a song sparrow, or to Harris's sparrow, these songs have no effect on song learning by white-crowned sparrows. Apparently, there are genetically inherited barriers, which cannot be 'overruled' culturally. Studies of other birds, for instance pied and collared flycatchers living on the Baltic island of Oland, have led to similar conclusions that 'early song discrimination has largely genetic component'.[55]

How much information is transmitted from bird to bird via a song? There are two questions—how much information can be encoded in bird songs, and how much of this information can be interpreted by the receiver. Like human language, bird songs are hierarchically organized and contain elements of a grammar, however these patterns are relatively simple.[56] How much of the information that can potentially be encoded in a bird song will be interpreted by the receiver?

One of the evolutionary explanations of why birds sing is related to mate choice in sexual reproduction. How much information a singing male transmits to attract his mates—is this simply to signal its level of fitness or does a song transmit information about more specific phenotypes—is hard to track. A more tractable phenomenon for quantitative research is alarm calls. Black-capped chickadees, a common songbird throughout North America, have elaborate system of vocalization, which includes two very different alarm signals to alert other individuals about what types of predators are endangering them. Specifically, 'when flying raptors are detected, chickadees produce a high-frequency, low-amplitude "seet" alarm call; in response to a perched or stationary predator, they produce a loud, broadband "chick-a-dee" alarm call that is composed of several types of syllables'.[57] While the first is a warning call, the second is a 'mobbing' call, which recruits other chickadees to harass or mob the predator. Importantly, the 'chick-a-dee' call encodes information about the size of the predator: there is a strong inverse relationship between the number of D notes and the body length of the predator. Thus, these chickadee warning calls encode information using combinatorics—there are two particular call types, and one of them is a composite of different syllables, a combination of which encode information about the size of the predator. Nevertheless, the combinatorics here is quite limited, to distinguish three different types of predators, each of 10 different sizes, we need to distinguish 30 possibilities, which is about 5 bits of information ($2^5 = 32$). Importantly, the 'meaning' of the combined signals at least in part can be derived from the meaning of the individual signals. Similar use of limited combinatorics has been observed in birds of some other species.[58]

Songbirds are not the only species that transmit information via alarm calls. Vervet monkeys give acoustically distinct alarm calls in response to the presence of leopards, eagles, and snakes. On seeing a leopard, a monkey gives a loud barking call; other monkeys run into the trees. On seeing an eagle, Vervets give a short double-syllable cough; Vervets on the ground look up and run into the bushes. If a Vervet encounters a python, it gives an alarm call referred to as 'chutter' and the monkeys on the ground stand bipedally and look around.[59] These calls seem to be inherited genetically rather than learned socially, though the response to them possibly is learned socially. The use of signal combinatorics has also been observed in other primates, including chimpanzees.[60]

As fascinating as these and other examples are, we still have to conclude that quantitatively the amount of information transmitted by animals other than humans can be measured in single or possibly tens of *bits*, which certainly is many orders of magnitude less information than these animals transmit to their descendants genetically. It is important to note that although animals do use combinations of signals, their use of combinatorics is not open-ended, as human language is, the number of signals in the combined sequence is restricted.

The amount of information that is transmitted from generation to generation through cultural channels by non-human animals is a hotly debated subject, however I have not been able to find any evidence to contradict the conclusion of John Maynard Smith and his co-author, David Harper, who write:

> In the wild animals acquire a number of distinct signs, or 'words', but they can be counted in tens, in contrast to the human vocabulary of thousands of words, made possible by the recognition of discrete sounds, or 'phonemes'. There is little convincing evidence that, in the wild, the meaning of animal utterances depends on the order in which signs are arranged, in contrast to the human ability to convey an indefinitely large number of meanings by arranging words in the appropriate sequence (syntax).[61]

With due respect to our animal relatives, the amount of information that non-human animals can transmit to each other culturally is minuscule in comparison to the information they pass on from generation to generation genetically. This does not mean that the amount of information that an animal acquires from the environment is not significant. For instance, when an antelope is fine-tuning its running skills, mostly likely it acquires information by interacting with the environment through trials and errors. However, this information is not passed over to the next generation; the antelopes of the next generation will have to learn these skills by themselves.

The conclusions of this chapter thus far can be summarized as follows: amongst all the above-described means of information encoding—genetic, epigenetic, and environmental—only genetics uses large-scale combinatorial means to represent information. If so, it follows that *before the emergence of human language, most of the heritable information present on Earth was stored in DNA* (or at early stages, possibly in RNA or other polymer molecules).

Measuring biological complexity

If, as argued above, most of life's inheritable information is in DNA, then there should be a positive correlation between the biological complexity of an organism and the amount of functional information in its genome. This, however, is a contentious subject in biology. To be able to test any correlation, we need a way to measure organism's phenomenological complexity. The Kolmogorov complexity-inspired approach has been used to measure the biological complexity of an organism as the shortest 'genetic program' that generates the cell types making up the organism.[62] Interestingly, this has led to a conclusion that evolution has been minimizing complexity, rather than maximizing it, as some tend to believe. This observation is consistent with our earlier conclusion that the lower bound for the amount of functional information in a genome is not very high. However, for most species we do not know how the organism develops from a single cell in sufficient detail to use such a Kolmogorov complexity-inspired approach. Moreover, one can argue that this approach is implicitly based on the genome, which is not what we want.

Phenomenological or phenotypical complexity of the organism includes at least three components. The first is the organism's morphological complexity; for instance, the sophistication of the organism's body plan, the number of different organs, tissues, and cell types. Second, there is the biochemical, or metabolic complexity; for instance, how many different biochemical reactions the organism uses, or how many different molecules it can utilize for food. Finally, there is the behavioural complexity component, such as the complexity of the social interactions between individuals of the species. These are relatively independent components; for instance behaviourally, plants are arguably simpler than most animals, but on the other hand plants have complex metabolisms (some plants have many more genes than animals).

Although the search for ways to define and quantify morphological complexity has been going on for some time, it seems that in practice little advance has been made since the US biologist John Tyler Bonner published his book, *The Evolution of Complexity*,[63] in 1988. One of the measures of an organism's complexity that he proposed was the number of cell types; a simplified version of Bonner's classification is presented in Table 7.1.

Table 7.1 *Bonner's classification of organism complexity (Bonner, 1988)*

Organism	Nr of cell types
Bacteria	1–3
The simplest algae, moulds, sea lettuce, some fungi	4
Mushrooms, kelp	7
The simplest animals, e.g. sponge	9–12
Land plants	~ 30
Insects, worms, molluscs (higher invertebrates)	> 50
Vertebrates	> 120

The number of cell types does not help us to measure the complexity of single-cellular organisms. A more general approach could be counting the different 'parts' including the number of different proteins and the number of organs, rather than only cell types. It has been suggested that not only the number of different parts but also the number of interactions between the parts and their hierarchical organization levels have to be included,[64] but it is unclear, how to combine these different aspects into one measure. When it comes to practice, even the authors of this suggestion fall back to simply counting the parts. A different way than to measure phenomenological complexity may be to quantify the complexity of the geometrical shapes of the organisms, which is what is used to characterize the complexity of fossils.

Quantifying behavioural complexity is, arguably, the most difficult part, in particular for animals with complex behaviours. Thus, not surprisingly, biologists might argue that it is impossible to characterize biological complexity by one number. However, if we want to make rigorous statements that the complexity of one organism is higher than another, or to correlate organism complexity with something else, we need at least an approximate quantitative measure. At the minimum, we need to establish a partial order, otherwise we are left waving our hands.

Given the challenges in measuring phenomenological complexity, the correlation (or lack thereof) between the amount of information in the organism's genome and its phenotypic complexity is rather hard to quantify beyond a simple observation: the genomes of complex multicellular organisms typically are larger than those of single-cellular prokaryotic organisms. Most bacteria have fewer than 10,000 genes, while complex multicellular organisms rarely have less than 13,000 genes and most have more. On the other hand, recall from Chapter VI that the two related species of algae, one of which is single cellular, the other multicellular, contain about the same number of genes.[65] An often-used example to illustrate the alleged lack of correlation between phenotypic complexity and the number of genes is the fruit fly *Drosophila melanogaster* and the worm *Caenorhabditis elegans*. The fly has less than 16,000 genes, while the worm, which has less than 1000 cells in total, has around 20,000 genes (almost as many as a human has). Few will argue that phenomenologically this worm is more complex than a fly or as complex as a human. Clearly, if there is any correlation between an organism's phenomenological complexity and its number of genes it is not strong.[66]

Part of the explanation for this apparent lack of correlation is that the number of genes is not a good indicator of the amount of functional information in the genome of a eukaryotic organism. The genes themselves have variable complexity and the organisms that we regard as more complex, typically have more complex genes (Chapter VI). Even more importantly, there are many genomic elements that regulate gene expression and many different interactions between genes, and the number of regulating elements is not necessarily proportional to the number of genes. There is a good correlation between the organism's morphological complexity and the number of different transcription factors in the species.[67] In fact, if we look closer, outlier cases aside, there certainly is a positive, even though not very strong correlation between our intuitive perception of the complexity of the organism and the amount of functional information in its genome, to the extent we can assess these. As we become better at assessing the functional information in the genome, this correlation will become more apparent. Given the limitations of epigenetic information representation, organism cannot be more complex than the amount of information in its genome. An ecosystem cannot be more complex than the total amount of information in all genomes.

Is there progress in evolution?

If we compare life before the Cambrian explosion just over half a billion years ago (Chapter VI), with the complexity of life on Earth now, there seems to be little doubt that there has been an increase in complexity. Is increasing complexity a trend in evolution? And is there any sort of progress in evolution at all?

If we take the anthropocentric view that humans are the current pinnacle of evolution then, given that the human species is less than ten million years old, certainly there has been progress. However, is there an objective reason to consider humans as more evolved than other life forms? Obviously, humans are the organisms writing books about this subject, which gives them an advantage in the argument. At the same time, bacteria are more numerous and arguably more adaptive. Can we find any specific trend in evolution that is going in a particular direction, which we could call progressive? This subject has been debated widely, with a clear majority view siding against the notion of progress in evolution.[68] But is this really so?

Darwinian selection increases fitness rather than complexity but perhaps fitness and complexity are related? In a simple and stable environment, bacteria may lose genes to increase their fitness (Chapter V), thus the selection of the fittest can lead to a decrease in complexity. But this seems to be different in diverse and changing environments. For instance, in a changing environment, where best adaptation to different condition requires a different enzyme, evolving several separate enzymes and a means for switching them on and off may be advantageous. This is an increase in complexity. And indeed, some life forms have been increasing in complexity along some evolutionary lineages.

But not all. As mentioned in Chapter VI, sponges have largely stayed the same over half a billion years, and simple microbes have not disappeared either.

It has been argued that there is a trend of selection for larger size, known as Cope's rule, named after the US palaeontologist Edward Cope.[69] Under certain circumstances larger individuals can 'bully' smaller ones, potentially gaining a fitness advantage. Larger microbes may be able to prey on smaller ones, however under some circumstances, for instance, when food is scarce, being smaller may be an advantage. Additionally, species of larger individuals reproduce more slowly. The absolute majority of the 'individuals' on Earth are still microbial forms, while dinosaurs went extinct. To be rigorous, one needs to distinguish between the trend within evolutionary lineages and between lineages.[70] As noted, along some evolutionary lineages, size is increasing, and if there is a selection for size, can selection for complexity be a side effect? None of this is proven and the majority of evolutionary biologists seem to reject any idea of either direct or even any indirect selection for complexity.

Can there be an indirect reason why the complexity of some life forms has increased since the first single cellular organisms emerged? One possible 'driver' is an increasing diversification of the forms of life on Earth, in part because of life expanding into as many ecological niches as possible. Different niches require different adaptations, which can drive diversification. Part of this diversification is speciation, where a single species gradually 'gives birth' to two new species, each adapting to a particular niche. Interesting recent observation of a speciation process is cichlid fish in Lake Massoko in Tanzania, where the formation of a new species has been observed in action and linked to the divergence of the genomes.[71] The process of microbe diversification in a heterogeneous environment can even be observed in a laboratory.[72]

The view that the increase in complexity is a side effect of an increase in diversity has been argued by Stephen Jay Gould.[73] If life is diversifying, some forms are likely to become more complex than others. Given that there is a minimum complexity below which a replicating information-carrying system, and thus life, cannot exist (Chapter I), this creates a 'wall' of minimal complexity. Consequently, diversification will drive the average complexity of life forms up, as illustrated in Figure 7.5.

But what if there is also the second wall of the maximal possible complexity on the right side in Figure 7.5? For instance, if there is a limit to the amount of information that a genome can contain. Even though noise level in the genetics channel is low, there is a limit, and with genome size growing, at some point Eigen's error threshold may be reached. If there is such an upper limit in complexity, the growth in complexity will stop once an equilibrium distribution is reached.

Is it possible that some time ago life has already reached an upper complexity limit? Can complexity be constrained by the amount of information that the total DNA on Earth could hold and transmit reliably? Williams noted such a possibility in 1960s.[74] There is no fossil evidence that the complexity of life has grown, say, during the last 100 million years.

At the same time, however, some present-day primates—humans—seem to be behaviourally much more complex, at least by some complexity measures, than any primates that lived, say, a million years ago. It is certainly unlikely that anything like

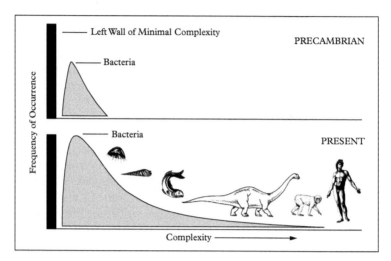

Figure 7.5 *The increase in organism complexity as a consequence of the increase in their diversity. The diversification of life leads to the diversification of complexity. Given that there is a 'wall' of minimal complexity on the left side, the diversification of complexity leads to an increase in the mean complexity. Note however, that if we have the second 'wall' representing the maximal complexity on the right side, at some point the growth of the mean complexity will stop.*
(Figure adapted from Gould, 1996.)

human language existed more than a couple of million years ago. Language adds to behavioural complexity enormously, which arguably is achieved via accumulating information in media other than DNA.

Genetic algorithms

If the growing complexity of life means accumulation of information, then what are the mechanisms of how this happens? Darwinian selection works via what we called genetic algorithms. The origins of this concept can be traced back to work of Alan Turing in the 1950s, however the idea of genetic algorithms crystallized only in the early 1970s, largely in the work of John Holland at Michigan University. Genetic algorithms are now a branch of computer science with many applications in solving optimization problems, such as finding the most aerodynamic shape of an airplane.[75]

As in Moran and Eigen's models discussed in the previous chapters, a genetic algorithm assumes an evolving set of sequences, called a *population*, and a fitness landscape, which can be viewed as a mathematical function assigning each sequence a numerical value called *fitness*. We can assume a population of binary sequences of 0s and 1s of a fixed length, say 1000 *bits* per sequence, and a population of constant size, of say 500 such sequences. Such a population then can be seen as a table (matrix) of 1000 columns and 500 rows of 0s and 1s. The population is evolving in discrete steps—generations— through recombinations and mutations in the following way. First, a chosen number of the fittest sequences, say 250, are selected for 'reproduction'. Second, these sequences are grouped in pairs and the sequences in each pair are recombined by crossover. Different versions of crossover can be defined; at one extreme, one breakpoint is chosen, and the left part of the first sequence is joined up with the right part of the second sequence. At the other extreme a new sequence can be formed by randomly picking each consecutive digit from either the first or the second sequence.[76] More than one crossover can be applied to each pair, each crossover producing one offspring. The number of reproducing pairs and the number of crossovers is chosen such that the size of the population stays constant. After the crossover, the resulting sequences are randomly 'mutated' at a specified mutation rate—with a defined small probability each digit in each sequence can be swapped over to the opposite one. Thus, the matrix of binary numbers evolves step by step; the process is repeated while the joint fitness of the population keeps growing, or stopped after a defined number of generations.

It is not hard to spot an analogy between this algorithm and the process of sexual reproduction combined with selection of the fittest. But obviously, as with any model, genetic algorithms are a simplification of reality. First, in the model above, all sequences are of the same sex. Next, in genetic algorithms, genomes have identical and unchanging lengths and mutations are confined to point mutations. In real biological populations, deletions, and amplifications of parts of genomes are common and essential to evolution—without this, large eukaryotic genomes could not have evolved. In genetic algorithms, the selection of the fittest is deterministic—the fittest sequences always win. Most fundamentally, in genetic algorithms, the fitness is a mathematical function that has been defined *a priori*; in real-world biological systems the fitness of an organism is the result of interactions between genomes, cells and the environment, and it is constrained by the laws of physics. This also means that the fitness function can change through niche alteration. Notwithstanding all these simplifications, genetic algorithms show how information can accumulate in genomes.

In each round of selection, new *bits* of information increasing the overall fitness of the population are 'learned'. We can start the process with a set of random sequences and end up with a set of highly specific sequences of higher fitness. This can be viewed as starting with 0 *bits* of information about the fitness landscape and ending with *n bits* of information, where *n* is the length of the sequences. The amount of mutual information between the sequences of the population and the fitness landscape has increased.[77]

One of the simplest possible fitness functions is addition. Specifically, every position in the genome is assigned a defined value, and the total fitness is the sum of the contributions of the individual positions. This corresponds to the absence of genetic interactions (absence of epistasis, Chapter V) and every *bit* has a fitness value independent of the rest of the genome. A particularly simple model assumes that every *bit* in the sequence contributes to the total fitness either $+1$ (good genes) or -1 (bad genes). David MacKay, from Cambridge University, showed mathematically that in this model, in the absence of crossover, 1 *bit* of information is learned in each selection round. If arbitrary multiple crossover points can be assumed, then in each step $\sqrt{n}$ *bits* are learned, where *n* is the length of the sequences.[78] We have already referred to this result in Chapter VI, when arguing that crossover can speed up evolution by a factor equal to the square root of the coding part of the genome.

In real biological evolution, populations of biological organisms are 'learning' the fitness landscape by interacting with the environment. The view on evolution as accumulation of genetic information was suggested already in 1961 by Motoo Kimura, who used non-trivial mathematical results from population genetics to conclude that the total amount of genetic information that could have accumulated in genomes since the Cambrian explosion (i.e. during the last half a billion years) is on the order of 100 million *bits*.[79] In 2000, John Maynard Smith came to similar conclusions using simpler arguments. He argued that if in every generation half the population is selected at the expense of the other half, 1 *bit* of information is learned.[80] He wrote:

> *Occasionally someone, often a mathematician, will announce that there has not been time, since the origin of the earth, for natural selection to produce the astonishing diversity and complexity we see. The odd thing about these assertions is that, although they sound quantitative, they never tell us by how much the time would have to be increased: twice as much, or a million times, or what? The only way I know to give a quantitative answer is to point out that, if one estimates, however roughly, the quantity of information in the genome, and the quantity that could have been programmed by selection in 5000 MY, there has been plenty of time. If, remembering that for most of the time our ancestors were microbes, we allow an average of 20 generations a year, there has been time for selection to program the genome ten times over. But this assumes that the genome contains enough information to specify the form of the adult. This is a reasonable assumption, because it is hard to see where else the information is coming from.*[81]

Whether any information has been added to the genomes after the Cambrian explosion has been contested, for instance by Williams, however it does seem incontestable that information in the genomes was growing at the early stages of evolution, including during the Cambrian period.

Genetic algorithms also demonstrate that life does not explore all possible genomes one by one in an exhaustive search but rather moves towards higher fitness, gradually adding useful *bits* of information to the surviving genomes. Evolution can be viewed as a (generalized) genetic algorithm working on the entire set of all genomes of all the individuals of all the species present on Earth.

The Brownian ratchet of information

As discussed above, genetic algorithms learn by acquiring information; the *bits* that improve the population's fitness are recorded, the ones that do not, are discarded. Although the mutations and recombinations are stochastic events that can go either in the direction of increased or decreased fitness, the selection of the fittest provides evolution with a direction. Therefore, some authors make a comparison between Darwinian selection of the fittest and *Maxwell's demon*.

Maxwell demon is an imaginary nanoscale creature invented by the Scottish physicist James Clerk Maxwell to study the Second Law of Thermodynamics.[82] Maxwell considered a *gedankenexperiment* involving two chambers filled with gas. The chambers are separated by a wall, which has a small hole in it. A nano-scale demon is sitting at this hole operating a door. It lets fast molecules to go from left to right, but not the other way, and slow molecules from right to left. As the result, the temperature in the right chamber is growing, while in the left one it is decreasing. This means that the thermodynamic entropy of the whole system is decreasing. If the door is operated in a reversible manner, for instance, using the 'lift door' mechanisms in Figure 7.4 (energy spent in opening the door is regained when the door is closing), the decrease in entropy would be achieved without influx of free energy. Thus, if such a demon existed, the Second Law of Thermodynamics would be broken.

Various solutions to this paradox have been proposed. Some are based on attempts to link the thermodynamic and information entropies. To operate, the demon needs information about the molecules (their velocity and location), and it is claimed that acquiring this information has to be compensated by comparable increase in thermodynamic entropy of the system. Rigorously linking information and thermodynamic entropies, and through this, the concepts of information and energy, would be a fundamental achievement. Several prominent scientists have followed this fascinating pursuit, one of the first was the famous Hungarian-American physicist Leo Szilard who, during the Second World War, alongside John von Neumann and Richard Feynman, worked on the Manhattan Project in Los Alamos, New Mexico. Szilard later played an important role in establishing the European Molecular Biology Laboratory. Nevertheless, it does not seem that the existing 'proofs' of the relationship between these two entropies have managed to avoid at least some hand-waving.[83] At the same time, a much simpler explanation was proposed by Richard Feynman. The equal partitioning of energy (Chapter II) makes the demon and the door subjects to Brownian jiggling, and consequently, the demon is unable to operate the door with the required precision. To argue this, Feynman came up with his own version of Maxwell's demon, shown in Figure 7.6, which is now known as Feynman's ratchet.

A version of Maxwell's demon can be implemented if we inject into the system free energy to compensate for the decrease in entropy (recall that free energy equals the potential energy minus entropy times temperature). Thus, if we transfer free energy into the system, the Second Law is no longer broken. A system that exploits random Brownian motion to perform mechanical work is known as the Brownian ratchet and it is the basis of molecular machines converting chemical energy into mechanical work. We discussed such a molecular motor, the burnt-bridge Brownian ratchet, in Chapter IV as a mechanism of how bacteria separate their chromosomes during replication.[84]

Coming back to Maxwell's demon, the analogy with Darwinian selection may seem quite compelling: just like Maxwell's demon sorts chaotically moving molecules, selecting the fast ones, natural selection 'sorts' mutations, selecting the fittest *bits* of information. The link between Maxwell's demon and how living cells use information is well argued by Paul Davies in his book, *The Demon in the Machine*.[85] Note though that to 'sort' mutations evolution uses energy. Thus, there is an even more compelling link

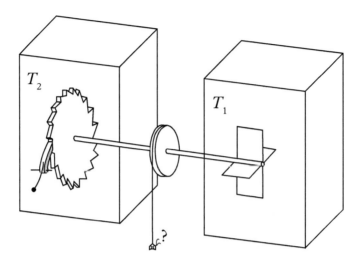

Figure 7.6 *Feynman's ratchet. The left and the right side of the device are immersed each in a different thermal reservoir, at temperatures T_1 and T_2, respectively. The device is small enough that the paddle wheel in the right reservoir is moved by collisions of the molecules from the surrounding medium against the paddles. The ratchet and pawl in the left reservoir block motion in one direction and allow it in the other. Thus, at $T_1 = T_2$ mechanical work can be extracted from thermal fluctuations to lift the small insect at the end of the thread, in violation of the Second Law of Thermodynamics. The resolution of the paradox relies in the realization that at thermal equilibrium the pawl will jiggle.*
(Figure from *The Feynman Lectures on Physics*, Vol. 1, 1963).

between evolution of life and Brownian ratchet. Just as the Brownian ratchet uses energy to convert chaotic molecular motion into directional mechanical work, evolution uses energy, ultimately provided by the Sun, to learn from the environment and concentrate the acquired information in its DNA. This essentially is the Schrödinger's observation that life uses free energy to maintain order.

All the information on Earth

We estimated the lower bound of the amount of functional information in an individual genome as the information needed to encode functioning proteins, and the upper bound as what a genome of the given size can hold. This was about individuals, but what about the amount of information in all genomes of a particular species taken together? Given that the genomes within a species are similar, there is a significant amount of mutual information between the genomes of the individuals of a species. Thus, the total information in all genomes of the species is less than the sum of that in individual genomes. (If all genomes were identical, the amount of genetic information of the species would equal the information in one individual, the only information to add would be the number of individuals.) Given the mutual information between the genomes, one way to estimate the total amount of information in the genomes of a species is take all genomes, put them together, and compress the data. The length of the compressed data in *bits* would be the amount of information in the species.[86]

How much information is held in the genomes of all life on Earth? As all species are evolutionarily related, their genomes share information. In fact, mutual information between genomes from species at different evolutionary distances are routinely used to infer the functional significance of specific genomic positions; the more similar the genome regions between species, the more likely that these regions have functional roles. Consequently, the total information in all the genomes on Earth is less than the sum of information in the individual genomes or individual species. Like in a single species, a rough estimate of the total information in the entire DNA on Earth could be achieved by taking all the genome sequences from all the species and applying a good data compression algorithm to all of them taken together. Projects to sequence genomes from tens of thousands of species representing a significant part of the biological variation on Earth, such as the Earth Genome Project,[87] will soon make such estimates possible. We do not need to sequence all species, at some point the increase in amount of new information added by additional species will start slowing down.

To see if the total information that is present in the genomes is still growing, we would also need to estimate the amount of information that was in the DNA on Earth millions of years ago. It is rather unlikely that we will ever be able to do this directly, however one can hope that indirect estimates will be possible eventually; for instance, if we establish a quantitative relationship between the amount of functional genetic information and the morphological complexity of the organisms.

As already noted, at least at early evolutionary stages of life on Earth, the information in the genomes certainly was accumulating. Whether information in the genomes has grown, say in the last 10 million years, is less clear. Maybe the growth stopped, maybe it did not. If evolution is learning, is it possible that life has reached the stage in biological evolution when as much information is lost as is learnt? Perhaps for the complexity of life on Earth to keep growing, life needed to find different carriers of information?

Since the emergence of human language in the last million years, and writing less than 10,000 years ago, information has started accumulating outside DNA. It seems quite possible that information outside genomes is now growing faster than inside them. And this new type of information too is contributing to life's complexity. Many contemporary evolutionary biologists and palaeontologists, including Stephen Jay Gould, dismiss the idea of evolutionary progress in biology as non-scientific and anthropocentric. If we measure progress (or lack of it) as the growth of information, we do not need to be anthropocentric. Even if at some point the growth of information in DNA stopped, information is now growing outside DNA as the result of cultural evolution of human species, contributing to the growing complexity of life on Earth. This is the topic of Chapter VIII.

VIII

From DNA to Language

By 40,000 years ago, populations in Europe and elsewhere were producing a wide range of novel artefacts, were burying their dead, painting on the walls of caves, and engaging in trade. It seems that something happened to make possible both the geographic spread of H. sapiens, and the burst of new technical and cultural practices. The obvious candidate is the origin of human language.

(John Maynard Smith and Eörs Szathmáry, 1995)

[C]himpanzees require thousands of trials, and often years, to acquire the integer list up to nine, with no evidence of the kind of 'aha' experience that all human children of approximately 3.5 years acquire. . . . A human child . . . grasps the idea that the integer list is constructed on the basis of the successor function.

(Hauser, Chomsky, and Fitch, 2002)

Is it good to talk? Somebody sitting on a train next to a person talking on the phone for hours is likely to say no. However, talking is what makes humans different from all other known forms of life. Animals of other species may be able to communicate using tens of different signals, some can possibly be taught to recognize and respond to hundreds; humans have tens of thousands of different words. Even more remarkably, humans have the ability to combine these words into a potentially infinite number of sentences with an unbound number of different meanings. There are no limits to the amount of information that humans can communicate to each other using a language. In this respect, there is no continuity between humans and other animals. Other traits such as bipedality, ability to use tools, having a large head at birth, or grandparenting, may each be distinguishing humans from most other animal species, but none comes even close in significance to our faculty of language.

Arguably, the faculty of language is as much the cause as the consequence of the particular way that humans evolved. Language enabled cultural evolution in the human species on a level qualitatively different from that of any other animals, enabling humans to become the dominant animal species on Earth, for better or for worse. Some will argue that it is not only language that gave humans this dominance, but that humans also have higher cognitive abilities than other animals. But these are more difficult to pinpoint,

Living Computers. Alvis Brazma, Oxford University Press. © Alvis Brazma (2023). DOI: 10.1093/oso/9780192871947.003.0009

while the faculty of language provides a clear dividing line. Just like the dividing line between *pre-life* and life.

Evolution is possible only if information is transferred from generation to generation. As discussed in Chapter VII, before the emergence of language, the only means by which large amounts of information could be transmitted between generations was via polymer molecules. In practice, for all the known forms of life, this meant via DNA, or sometimes RNA, or more abstractly, via what we call *genes*. With the emergence of language this changed: humans pass on and spread their ideas at least as much as their genes. Language provided life with means of digital-combinatorial encoding of information different than polymer molecules for the first time in the billions of years of life's evolution. Not only the information printed on paper or stored in databases is now likely to be accumulating faster than that in DNA, even more fundamentally, enabled by language, cultural evolution mostly works via the mechanisms described by Lamarck rather than the Darwinian selection of the fittest: what is learnt during ones lifetime is passed on to subsequent generations. Processing of cultural information mostly take place in neural networks of the human brain and via various algorithms rather than exclusively via genetic algorithms mutating, scrambling, and selecting bits of DNA. This makes cultural evolution much faster.

What are the unique underlying biological differences between humans and other animal species that give us this unique faculty? Can this difference be explained by genes and molecules? Humans started diverging from the lineages of our closest living relatives, chimpanzees and bonobos, somewhere between 5 and 10 million years ago.[1] It is rather unlikely that the common ancestors of these species were able to speak. At what point in the evolutionary lineage leading to humans did the faculty of language emerge, and how? Was the emergence of human language a slow process, taking a large proportion of these millions of years, or did something specific happen rapidly at a particular point? Whether there were many small steps or one big one leading to language, these changes would have to be reflected in the genome—DNA was the only capacious information carrier available to evolution before language emerged. What were these genetic changes that eventually gave us the faculty of language? Do humans have a 'language gene' or is it a combination of many genes that give us this faculty? Did the faculty of language emerge exclusively in our evolutionary lineage, or did we have talking relatives that went extinct? Do all existing languages have a common origin?

Few scientists believe that our species has a specific language gene and most believe that all languages have a common origin, but beyond that, there is little consensus. Those looking for a diversity of opinions in science need only look at studies of emergence, evolution, and nature of human language. As promised earlier, our narrative is now moving closer to speculations and scientific controversies. Very little, if anything, was controversial in the first four to five chapters of this book; some elements of controversy, such as whether evolution is progressive, or if biologically relevant information can be measured, came up in Chapters VI and VII, but when it comes to the origins and evolution of language, the lack of scientific consensus is striking.

Here is one example. In 2002, three US scientists, Marc Hauser, Noam Chomsky, and Tecumseh Fitch published a joint article, 'The faculty of language: What is it, who has it and how did it evolve?', in the Science Compass column[2] of one of the highest impact scientific journals, *Science*. A quotation from this article is given at the beginning of this chapter. The authors introduced a distinction between *the faculty of language in the broad sense* (FLB) and *the faculty of language in the narrow sense* (FLN). FLB encompasses all essential faculties that enable humans to have language but which are not unique to humans. These include the ability of *vocal learning* to imitate novel sounds, which some birds, such as parrots, and some marine mammals have, and the ability to build mental maps of their environment and to reason about it, which many non-human animals are believed to have. For instance, it has been shown that some animals have the ability to make inferences of transitivity—to infer that if animal A can beat animal B and B can beat animal C, then that animal A can beat animal C.[3] Every pet owner will confirm that animals are quite good at understanding the world around them and using this understanding to their advantage, but animals cannot tell you what they had for dinner the previous night. FLN is part of FLB unique to humans, the lack of which 'prevents' all non-human animals from communicating large amounts of information.

In their article, Hauser, Chomsky, and Fitch argued that the basis for FLN is the human ability to utilize grammar—to stream our thoughts in a syntactically organized way that can be interpreted by other human individuals. Furthermore, they hypothesized that this builds upon our ability to utilize *recursion* or *recurrence*—repeated and unlimited embedding of structures within structures in a hierarchical way, like parentheses in mathematical expressions. Recursion is a non-trivial concept (see Box 8.1), but an essential element of unlimited or unbound recursion is the notion of *and so on* or *etc.*: for instance, '*1, 2, 3, 4, and so on*'. The authors refer to this as the *successor function*. The grasp of this notion appears unique to humans. As mentioned in the quotation at the beginning of this chapter, when children learn to count, at some point they have an 'a ha!' or 'eureka!' moment—they have understood the concept of the successor: 1 is followed by 2, 2 is followed by 3, 3 is followed by 4, *and so on*. The floodgate has opened, and they can count further and further. Apparently, no other animals have this ability. With herculean efforts a chimpanzee can be taught to count up to nine,[4] but it is thought that they memorize each number as a separate entity, rather than discovering the concept of the successor function. There must be something unique in human genes that gives us the ability to comprehend '*and so on*'.

Does FLN include anything in addition to grammar? Is there anything else essential in the human ability to utilize grammar in addition to unbound recursion? These are controversial subjects and there is no scientific consensus regarding the possible answers. Pinpointing the minimum set of concepts that are needed for language is the research subject of Chomsky's *Minimalist Program*,[5] which he has been pursuing since the 1990s. The Minimalist Program is a good example of (my interpretation of) Occam's razor—try to explain as much as possible with as little as possible.

Hauser, Chomsky, and Fitch's *Science* article was soon under attack from two other prominent scientists: Steven Pinker (also known for his popular science books

Box 8.1 Elements of the Universal Grammar: The merge operator and recursion

Two main principles seem to be essential in the grammars of all known languages and therefore are likely to be an essential part of the Universal Grammar:

1) The hierarchical structure of a sentence: parts of the sentence embedded into other parts (see section 'What is language?'), which Chomsky encapsulates in the concept of the *merge operator*.[6]
2) The principle of unbound recursion or recurrence: we can apply the same construct recurrently, as many times as one likes.

Chomsky's merge operator is an operation that takes two linguistic terms—words or phrases—X and Y, and forms a composite term (X, Y), which itself can further be a subject of another merge,[7] as described earlier. The merge operator reflects the hierarchical nature of a language, however it has been argued that the ability to understand hierarchical structures is not unique to the faculty of language, nor in fact to humans.[8]

The second concept—the unbound recursion—essentially boils down to combining *merge* with the already discussed concept of '*and so on*': we can use the merge operator after the previous merge operation *and so on*. Unbound recursion generates the so-called discrete infinity, for instance the set of all integers or natural numbers 0, 1, 2, 3, 4, *and so on*. As mentioned earlier, this ability to comprehend the successor operator or discrete infinity seems to be unique to humans. If the universe if finite, then discrete infinity is an abstraction,[9] but possibly an essential one for the faculty of language.

The link between the unbound recursion as a key to the faculty of language and the discrete infinity exemplifies a possible link between the human faculties of language and computing. Parsing a sentence is computing. In theory, the ability to handle recursion is all that is needed for computing; this follows from the equivalence of the Turing Machine and Church's lambda calculus, mentioned in Chapter I. Perhaps, it is therefore not surprising that the first writing and the first numerical systems also emerged in parallel.

including *The Language Instinct*) and Ray Jackendoff, who jointly wrote a 36-page damming critique of the original article.[10] Amongst other criticisms they wrote:

> [the Hauser, Chomsky, and Fitch] *hypothesis ignores the many aspects of grammar that are not recursive, such as phonology, morphology, case agreement, and many properties of words. It is inconsistent with the anatomy and neural control of the human vocal tract. . . . The recursion-only claim, we suggest, is motivated by Chomsky's recent approach to syntax, the Minimalist Program. . . .*

Hauser and his co-authors responded with a rebuttal,[11] noting that many of the criticized aspects, such as the concept of FLN, should not be controversial at all. This generated a further 15-page criticism from Pinker and Jackendoff.[12] Later, another prominent linguist, Derek Bickerton, published an article, 'Language evolution: A brief guide for linguists'[13], largely dismissing both sides of the argument. Bickerton further extended his criticism in the popular book *Adam's Tongue*.[14] Although different opinions and discussions amongst scientists about unresolved questions are not uncommon, some of the arguments used by linguists seem to verge on personal attack and go well beyond what one would expect in scientific discourse.

There are reasons why the science about language is difficult. There is no record of language evolution until writing was invented less than 10,000 years ago. Fossilized skulls may say something about the evolution of our cognitive abilities, but this provides only limited information about the emergence of language.[15] Fossils can also shed light on the capabilities of vocal tracts and hearing organs, but rather inconclusively[16] and, moreover, producing and hearing sounds are only two elements of the faculty of language. As will be discussed below, genomic differences between humans and other animals have revealed some genes and mutations that seem to be important for language and are unique to humans; however, the genetics of human language has turned out to be a rather difficult pursuit.

Establishing truth in science proceeds through a peer review process. Scientists compare conclusions from their theories to empirical data and try to convince each other that their explanations are valid. Even when everybody is convinced, this does not yet mean that the theory is correct. However if the consensus remains unchanged for long, it becomes increasingly likely that the theory explains the evidence. Arguably, this is the definition of scientific truth.[17] Much in the theory of evolution has reached that state. However, when it comes to questions of how the human faculty of language emerged, what fundamental elements distinguish language from other animal communication, and what is the genetic basis of human language, opinions still differ wildly. Although progress has been made in the last couple of decades,[18] scientists are still nowhere near a consensus on the topic. As the spoken word does not fossilize, it is difficult to gather hard data against which theories can be directly tested, but this is not a reason to avoid the study of language evolution. A good summary of the state of the field, and a presentation of many of the different perspectives, are given in the excellent book *The Evolution of Language* by Tecumseh Fitch.[19] This chapter is too brief to include all the different views and largely presents the arguments that I personally find most convincing, and my own interpretation of these.

Not all controversies matter for our discourse. For instance, there is no consensus amongst scientists whether our Neanderthal relatives—archaic humans who lived side by side with us less than 40,000 years ago—did have the faculty of language or not (as we will discuss later). This is a fascinating subject but the answer to this questions does not impact the conclusions of this book very much. What matters is that the emergence of human language, for the first time in life's history, allowed for the transmission of large amounts of information by means different than DNA or other polymers, and that this kick-started a new and faster type of evolution: cultural evolution. It is one of the key theses of this book that *the transition from having no language to having language was as important a transition as the one from pre-life to the first information carrying self-replicating systems*. It opened a new phase in the evolution of life.

What is language?

What makes a human language special? An evident difference between language and other animal communications is that humans extensively combine signals to form new signals, moreover there is no limit to how many signals we can combine. And we already know that exploiting combinatorics is the only way large amounts of information can be recorded or communicated. We use combinatorics in our communications on two fundamentally different levels. First, we combine a small number, usually tens, of different basic sounds, known as *phonemes*, to form tens of thousands of different words.[20] The use of combinatorics to construct words is even more obvious in writing, and particularly in writing systems that use alphabets—most such systems use only 20 to 40 different characters, roughly mapping to phonemes, sequences of which represent words. Second, we combine this larger but still finite number of words to generate a potentially infinite number of sentences. There is no limit to how many different sentences exist as there is no specific limit to the maximal length of a sentence; any sentence declared as the longest possible can be extended, for instance, by prefixing it with 'Mary said that'.

The emergence of phonemes to create words has a possible explanation in biological evolution.[21] One can speculate that as our ancestors became increasingly sophisticated, their communication needs increased, and the number of different signals corresponding to objects (such as lions, antelopes, different plants), actions (such as running, hitting, hiding), or situations (such as danger, rain, cold) were also increasing. It can be argued that at some point, the number of concepts useful for communication became larger than the number of different individual sounds that our ancestors could tell apart reliably. This was like increasing the number of marks on a dial, as described in Chapter VII (Figure 7.1), where at some point we cannot tell the neighbouring positions apart. Combining discrete phonemes into larger units provided our ancestors with a solution, allowing them to communicate a much larger number of distinct signals or meanings. This was like discovering that the number of distinguishable states on a dial-based information-recording device can be increased exponentially when more than one dial hand is used. One can speculate that a gradual introduction of combinations of sounds widened the communication channels of our ancestors, possibly giving them a fitness advantage. In a sense, this 'discovery' was the first transition from analogue to digital in the human history.

Let us note here that language is not the same as *speech*, though they probably coevolved. Speech can be defined as a complex, articulated vocalization, while language, is a system of communications that maps an open-ended set of concepts onto an open-ended set of signals.[22] Language, though normally expressed through speech, does not require speech; for instance, sign languages are as expressive as spoken languages. Nevertheless, this ability to combine signals is a prerequisite for any means of using a language.

How and when the initial discovery that phonemes can be combined to increase the number of distinguishable signals came about is unknown. Although currently we can only speculate about what gives humans this ability to use phonemes, it is indisputable that combining phonemes is what allows us to distinguish tens of thousands of different words. It has been argued that non-human animals, such as dogs, that can be trained to recognize over 100 different words, recognize each word as a unit, rather than parsing them into phonemes. Perhaps this is one reason why their repertoire of signals remains limited.

There is a fundamental difference between *phonemes* and *words*—phonemes do not have meanings of their own while words do—words map to concepts in our mental maps of the physical world.[23] Thus, in language, first we combine a small number of meaningless phonemes into a much larger number of meaningful words. We then combine a still a finite number of (meaningful) words to form a potentially infinite number of different sentences, each with its own higher-level meaning derived from the meaning of the words it consists of.

This jump to infinity is what makes language special. An average high-school graduate knows over 50,000 different words, but this is still a finite number. In principle, it is possible to learn the meaning of every word by heart and probably most words are learned in this way, even though they are communicated through phonemes. But it is not possible to learn all sentences. When children learn to talk, they do not do this by memorizing every sentence they hear. Humans can invent, as well as interpret, new sequences of words that they have never heard before. Moreover, these sentences are interpreted in the same or similar way by other speakers of the language.

The existence of hierarchical levels—*phonemes, words, sentences*—is a feature common to all human languages. We further combine sequences to form stories, stories in books, and books in libraries, nevertheless the most important building block of a language seems to be a sentence, which is what gives language its open-endedness. Recall from Chapter VII that some non-human animals too have the ability to combine signals, and arguably in a way where the 'meaning' of the combined signal is derived from that of the individual ones.

However only humans have the ability to combine signals in an open-ended way, without limits on the number of signals in the combination. We combine phonemes to form words and words to create sentences in a sequential way—one phoneme after another, one word after another. This is a consequence of the serial nature of vocal communications—we have to utter phonemes or words one after another, we cannot vocalize two phonemes, nor two words simultaneously. In visual communications we can exploit two or three dimensions, oral communications are essentially one dimensional. It may seem paradoxical that the most powerful communication channel that human has—language—essentially is one dimensional. However, given that every piece of information can be encoded as a *bit-string* (Chapter VII), this should not be very surprising.

In principle, *sequencing* alone provides humans with combinatorial tools to generate an infinite number of expressions, just like sequencing nucleotides in DNA does. However, for communicating messages that have to be interpreted as they are streamed, sequencing alone is not sufficient. In addition to sequencing, language uses hierarchy that goes beyond phonemes, words, and sentences: sentences themselves have nested hierarchical structures. The simple sentence, *'a dog barked'*, already has a structure. *'A dog'* is a simple linguistic unit made of two words *a* and *dog*, which can then be combined with the verb *barked* to form a simple sentence. Consider a series of relatively simple sentences:

A dog barked.
A dog barked in the courtyard.
I told them that a dog barked in the courtyard.

Each sentence is constructed from the previous ones by joining them up in a longer sentence. The last sentence has the following hierarchical structure:

((I told) them) (that (((a dog) barked) (in (the courtyard)))).

Why do we need to use hierarchies? A possible reason for the emergence of such a hierarchical structure in language is the difficulty that most humans have in parsing even moderately long sequences of symbols or sounds unless there is an identifiable structure. For instance, when writing down a telephone number, we often introduce spaces or dashes that separate every few digits, for instance, 00-1-202-456-1111, thus introducing a structure. This in turn may be related to the limits of our short-term memory, which is thought to have the capacity of fewer than 10 *bits*.[24] One can ask why the human brain evolved this way, but that is a question for another day. The use of hierarchy is a powerful way of introducing structure in sentences, apparently necessary and sufficient to stream the mental maps in our brain by one-dimensional means.

The hierarchical structures used in sentences of a particular language are described by the grammar of this language. Grammar can be defined as a finite set of rules applied to a finite set of words to produce a potentially infinite number of syntactically correct sentences. These can then be effectively parsed by the receiver and interpreted, provided that the receiver knows the meaning of the individual words. According to Chomsky, the main common construct of the grammar of all natural languages is the so-called *merge operator* (see Box 8.1), which takes two words or grammatically correct constructs, and joins them up in a composite construct. For instance, take a construct (*A dog*) and a word, *barked,* and join them in a sentence: *((A dog) barked).* This operation can be applied recursively (or recurrently); for instance, we can construct the sentence above stating that *I told them that a dog barked . . .* by recursively applying the merge operator to increasingly larger units. According to Chomsky, the real evolutionary breakthrough that gave humans the faculty of language was the 'invention' of a potentially unbound recursive use of the merge operator. Note, that the first dividing line between language and other animal communications already occurs in the use of phonemes, though one can argue that the open-endedness of the human language begins only with a sentence.

As already noted, hierarchical grammatical structures of a language are what apparently allow humans to translate the mental maps of the external world into one-dimensional sequences of words. If one took words from a longer sentence and scrambled their order, understanding the meaning of the resulting sequence of words would not be easy. One would probably have to experiment with reordering the words until something like a meaningful sentence appeared. As the number of possible sequences grows exponentially with the number of words, such reordering might take hours even for sentences of modest length and might not be possible at all without writing the sequences down. Moreover, the meaning might not always come out unambiguously. Only grammatically correct (or almost correct) sentences can be interpreted while they are streamed.

Possibly at some point in evolutionary history, our ancestors were able to communicate by using single words or short arbitrary ordered sequences of words—using the system of communications which is referred to as *protolanguage*. Arguably, the amount of information that could be communicated this way was still limited. In his book *Adam's Tongue*, Bickerton gives an interesting example to demonstrate that a protolanguage can also pass on significant amounts of information. Here is his paragraph written in a hypothetical protolanguage:

> *Me throw dart/spear. Point hit animal. Point fall out. Wound close. Animal get away. Suppose point stay. Animal bleed. Animal get weak. Catch animal. Look this seed. Seed stick in skin. Seed get little thing. Little thing stick in. Suppose point get same-kind thing. Maybe point no fall out. Me catch animal. Me kill animal. Me eat animal.*[25]

Probably most readers will find that this describes a hypothetical 'primitive' thinking towards the invention of a spear with barbed point, like in a harpoon. The information about this thinking has been transmitted using a protolanguage and no potentially unbound hierarchy is involved. However, there is an unbound use of sequencing and a simple grammar involved in this reasoning; first, the paragraph (we could call it a super-sentence) consists of mostly two- to three-word *units*, a maximum consisting of six-words, separated by '.' (in contemporary writing we would separate these by ';'); second, no word is repeated within the unit but most words are repeated between the units, often between the adjacent ones. This is a kind of a grammar. In fact, it has a limited hierarchy: words and what we called units. An additional, more speculative conjecture is that we are able to understand the message because we can parse it into our own language, which complies with a 'proper' grammar. But it also seems that such a protolanguage lacks precision. Grammar is what binds the mental maps of the physical world to syntactic structures of a sentence. This gives language its precision, which arguably is what enabled human cultural evolution.[26]

It may be interesting to note that in genomes too, information is encoded hierarchically. In a gene, each triplet of nucleic acids maps to a particular amino acid. We can loosely relate triplets to words and genes to sentences. Together with other meaningful signals, such as genomic regulatory elements, these 'sentences' make a 'story'—the genome—which is read by a cell, instructing it how to make an organism.

Although the grammar of real human languages is complex, most linguists agree that there must be a common set of constructs at the basis of all grammars of the existing, and even the possible languages that humans can learn and understand. This set of rules can be called *Universal Grammar*. Effectively it is a grammar of grammars of all human languages. The origins of this concept can be traced back to the Roger Bacon's *Overview of Grammar* written in 1245, though in modern linguistics it is usually attributed to Chomsky.[27]

Learning by example

Ample evidence suggests that every healthy child born to parents of any ethnicity or nationality is able to learn any of the ~ 6000 languages spoken around the world. At the same time, all attempts to teach more than trivial elements of a language to our closest relatives, chimpanzees, or any other animals, have failed.[28] Therefore, there must be something unique to humans, some *prior knowledge*, that allows us to learn a language and which is missing in other animals. This is not limited to our vocal capabilities; for instance, teaching chimpanzees a sign language has not been any more successful.

Children learn languages by listening to how others talk, imitating them, trying out using what they have learned, and assessing the outcomes of such 'experiments'. Although grammar can be described as a set of rules, many of which are taught at school, children do not learn a language by learning the rules of grammar explicitly; once children get to school and receive grammar lessons, they usually are already quite fluent in speaking the language. And probably many people around the world, perhaps most, have no idea of the grammar of the language they speak, yet they can still communicate with their neighbours freely. Humans learn languages, not grammars, however the underlying grammatical structures common to the sentences of a language is what makes this possible. A finite sample of potentially infinite set of sentences is sufficient for a human to learn an infinite language. While learning a finite set of signals is not unique to humans, learning a potentially infinite language is.

The process of learning an generalizing from examples can be studied mathematically[29]; this is the goal of the so-called algorithmic learning theory, or more broadly, of what historically was known as the theory of indicative inference—how general rules can be inferred from a finite set of examples (see Box 8.2). One of the most important conclusions of this mathematical theory is captured in Gold's Theorem,[30] named after one of the founders of inductive inference, the American mathematician Mark Gold. In 1967 Gold proved that there is no algorithm that could learn every possible infinite language from a finite set of examples. For such a learning algorithm to exist, the class of the languages that we want to learn needs to be restricted in advance.

The Universal Grammar can be viewed as the restriction placed on the class of languages that humans are able to learn. As Martin Nowak puts it '*the existence of Universal Grammar is a mathematical necessity*'.[31] Unless we accept a version of neo-vitalism, there must be a genetically inherited brain structure that gives humans this *prior knowledge* providing the ability to learn languages compliant with the Universal Grammar, which distinguishes us from other animals. As noted by Hofstadter in *Gödel, Escher, Bach*:

> It seems that brains come equipped with 'hardware' for recognizing that certain things are messages, and for decoding those messages. The minimal inborn ability to extract inner meaning is what allows the highly recursive, snowballing process of language acquisition to take place ....

Other animals apparently do not have this *prior knowledge*—they do not have the genetically inherited brain structure for Universal Grammar, or whatever we call it. We can

Box 8.2 Learning by examples and inductive inference

In the mathematical theory of languages, a *language* is defined as a set of sequences (sentences) over a certain finite set of characters, called the *alphabet*. For instance, we can have an alphabet α, β, γ and define a particular language as the set of all sequences that begin with one or more characters α, followed by the same number of characters β, and ending with exactly one character γ. Using '*and so on*' we can express this as $\{\alpha\beta\gamma, \alpha\alpha\beta\beta\gamma, \alpha\alpha\alpha\beta\beta\beta, \ldots \text{ and so on}\}$. Here, we do not distinguish between words and letters; both are of a finite number and thus they are merged into one concept—*alphabet*. We can view such an *alphabet* as the set of Chinese characters—each character representing a word.

Grammar is defined as a finite set of rules that defines which sequences (sentences) belong to the particular language, that is, which sequences are *correct*. A language can be finite or infinite. If a language is finite, it can be described simply by listing all its sentences. For instance, the set of three sequences $\{\alpha\beta\gamma, \alpha\alpha\beta\beta\gamma, \alpha\alpha\alpha\beta\beta\beta\gamma\}$ is a finite language. Obviously, an infinite language cannot be defined by listing all its sentences. Nevertheless, if we have the concept of '*and so on*', a finite alphabet and a finite set of rules can define an infinite language. For instance, if a letter x repeated n times is denoted by x^n, then $\{\alpha^n\beta^n\gamma$, where $n = 1, 2, \ldots$ *and so on*$\}$ defines an infinite language. Note that in order to define this language, the use of *and so on*, either explicitly or implicitly, is inevitable. Perhaps the absence of this concept in the minds of non-human animals is what prevents them from understanding infinite languages.

The language introduced above is rather trivial. A slightly more interesting language could be obtained if we extended alphabet to introduce the parenthesis symbols '(' and ')', and defined an expression to be valid if all the parentheses were closed properly. For instance, expressions $(\alpha\beta\gamma)$ and $((\alpha\beta\gamma)\beta\gamma)$ are valid, but expression $(\alpha\beta\gamma(\alpha\beta\gamma)$ is not. To define this language a recurrent use of the merge operator or something equivalent will be needed. Obviously, natural human languages are much more complex than the ones given above. Nevertheless, it is quite possibly that the evolutionary discovery of the concept of *and so on* is what made humans unique.

The mathematical theory of algorithmic learning or, as it was called historically, the theory of inductive inference, which studies how general rules can be inferred from specific examples, was developed in 1960s and 1970s. Given expressions $\alpha\beta\gamma$, $\alpha\alpha\beta\beta\gamma$, $\alpha\alpha\alpha\beta\beta\beta\gamma$, $\alpha\alpha\alpha\alpha\beta\beta\beta\beta\gamma$, one could infer that the language is $\{\alpha^n\beta^n\gamma$, where $n = 1, 2, \ldots$ *and so on*$\}$. This defines an infinite language. However, it is also possible that $n = 1, 2, 3, 4$ rather than any integer, and that the language actually is finite. How can one decide which one is it? We can argue by Occam's razor's principle that if n is large, it is likely that the language is infinite (because the description of the infinite language will be shorter than the finite list of all sequences). But how large exactly should n be to draw such a conclusion? This demonstrates the difficulty. Mark Gold used this difficulty to prove that not all grammars can be learned from examples and that for a learning algorithm to exist, the class of grammars has to be restricted *a priori*.

Universal Grammar of the human languages is this factor that restricts what languages a human can learn. It describes the rules that must be true in any language that a human child can learn. But because the Universal Grammar itself does not have to be taught to a child, it seems safe to assume that it is inherited genetically.

be quite certain that these unique human capabilities are encoded genetically because, as argued in the previous chapter, before language itself emerged, DNA was the only means by which large amounts of information could be passed on to subsequent generations. And it is fair to assume that the amount of information needed to encode this *prior knowledge* is quite substantial—if it could be described in just a few *bits*, some other animals could be expected to have acquired this faculty too, just accidentally.

Are all languages that we can learn equally expressive? It has been argued that all real-world natural languages are universal in a sense that any thought that one might have can be expressed in any language. But how do we know what thoughts we have until we express them? Linguists have a concept of an internal language, or *i*-language, which is claimed in some way to be at the basis of all spoken languages. When using this concept, however, one has to be careful to avoid circular arguments. But one can ask a more tractable question—can anything that can be expressed in one language be expressed in a different language? Probably most readers will agree that at least in principle, we can take any book written in any language and translate it into any other language. If so, we can indeed argue that all languages are equally expressive, and thus, in some sense, universal.

The last decade has brought impressive successes in teaching computers to communicate with humans using the human language or to translate between different languages. These successes are mostly based on 'training' large artificial neural networks (Chapter VI) on human texts. Such artificial neural network machine learning approaches can be characterized as *black box* approaches as not necessarily it becomes clear what the computer has learned—the algorithms that allows the computer to use the language are encoded in billions or trillions of *weights* of the connections linking the artificial neurons. To understand what is it that allows the computer to use a human language, we need to investigate the underlying black box, just like we are investigating the human or animal brain. These machine learning successes also takes us back to one of the questions asked in the Introduction—are humans closer to other animals or to computers?

From animal communications to language

How did humans evolve their unique faculty of language? Was this a gradual process consisting of many small steps or few large ones? Can the roots of language be found in animal communications or did human language evolve as a new trait? One of the essential features of human language, distinguishing it from animal signals, is its controlled use. Most scientists believe that legacy of animal communication in human behaviour is in the uncontrolled expressions of emotions, such as laugher or cries of pain, while language evolved independently, possibly for reasons other than communications.[32] As bird feathers initially evolved for purposes other than flying, it is possible that the prerequisites to the faculty of language evolved for something else, for instance as a part of the evolution of our cognitive abilities. Like the emergence of life itself, the origin of language is one of the greatest puzzles of modern science.

What were the evolutionary pressures and mechanisms that led to the emergence of language in humans but not in other animals? Darwinian evolution compels individuals to compete with each other, so why would one share one's information with others? As in evolution of any cooperative behaviour, discussed in Chapter VI, the answer may be in the genes. If genes are the main objects of Darwinian selection, it may be to their advantage that relatives share information with their kin–with individuals with whom they share genes. Thus, possibly language started as communication between relatives. But it is also possible that as our ancestors learned to recognize specific (friendly) individuals, a reciprocal altruistic behaviour emerged, facilitating and facilitated by wider exchange of information. Primates are social animals; possibly they had more pressure to evolve a means of communication than most other species.

The closest living human relatives are chimpanzees and bonobos, with whom we shared a common ancestor around 7–8 million years ago.[33] The human and pre-human species that have a common ancestor with great apes are traditionally referred to as *hominins*. Some decades ago, the story of human evolution was assumed to be largely linear. To simplify the story dramatically, it was thought that the ancestor of humans and other great apes lived on the edge of tropical forests; the human ancestors left the forest and moved into the savanna, gradually evolving to become bipedal, which allowed them to free up their hands for making and using increasingly sophisticated tools. This resulted in the need to develop higher cognitive abilities. Hominin brains were evolving to become increasingly larger. A few hundred thousand years ago this lineage evolved into the Neanderthals, which then evolved into modern humans—*Homo sapiens* (*H. sapiens*). Although this is a gross simplification of the story as it was told by paleoanthropologists, a linear sequence of events was its important part. Somewhere in this sequence of events, the faculty of language must have emerged.

This linear story was replaced by the *out of Africa hypothesis*:[34] the cradle of human evolution was Africa, and migratory waves of different pre-human species poured out of there periodically. The last wave of modern humans came less than 50,000 years ago, colonized the world squeezing out all earlier hominin species. However, this out of Africa hypothesis too is now undergoing a revision.

Now we know that for millions of years, the Earth was co-inhabited by more than one hominin species, as shown in Figure 8.1. Individuals of different hominin species occasionally interbred, thus strictly speaking hominin evolution cannot be described as a phylogenetic tree. New discoveries keep emerging and the hypothetical evolutionary tree of hominins shown in Figure 8.1 itself keeps evolving.[35] For instance, a new addition to this picture is the hominin species named *H. floresiensis*, a small-skulled hominin that lived on the island of Flores in Indonesia as recently as 50,000 years ago. The history of divergence and mixing of different pre-human species or sub-species is looking increasingly complex. The details of this partial phylogenetic tree are not essential for our discussion, the reason I present Figure 8.1 is mostly to show that for over a million years, more than one hominin species coexisted. Was it *H. sapiens* acquiring language that ended this cohabitation?

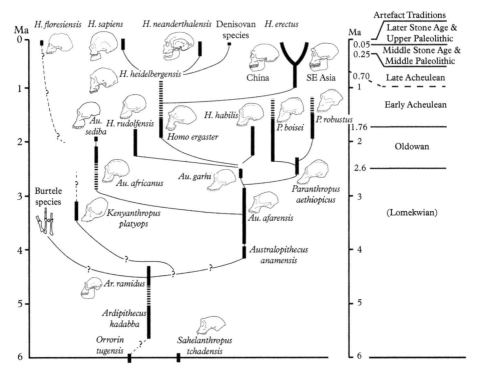

Figure 8.1 *Left. A hypothetical phylogeny of species most closely related to humans.* **Right.**
Approximate timing of the events in million years ago (MYA) and the respective archaeological periods.
By far the best-characterized early hominin is Ardipithecus kadabba, *which lived an estimated 5.77 to*
5.54 MYA. Australopithecus, *which were terrestrial bipeds but continued to use trees for food and*
protection, appeared around 4 MYA; their brain size was similar to chimpanzees and gorillas. Some of
the earliest fossils of the genus Homo *were found in East Africa and are dated to 2.3 MYA. In brain*
and body size, these early specimens are similar to Australopithecus *but has different molar teeth,*
which suggests a change in diet. By 1.8 MYA, they were using primitive stone tools to butcher animal
carcasses, which added energy-rich meat and bone marrow to their diet. The oldest member of genus
H. habilis (2.3–1.4 MYA) is found in East Africa and is associated with butchered animal bones and
simple stone tools. H. ergaster *is an extinct species of the genus* Homo *that lived in eastern and*
southern Africa about 1.9 million to 1.4 million years ago. It probably gave birth to the lineage that led
to H. heidelbergensis, *H.* Neanderthalensis, *and* H. sapiens, *but also to another species,* H. erectus,
which has been found throughout Africa and Eurasia, persisting from 1.9 MYA up to as recently
100,000 years ago (and perhaps even more recently). This means that H. erectus *possibly cohabited*
with Neanderthals and even with H. sapiens. *As indicated in the figure,* H. heidelbergensis *was*
thought to be the last common ancestor of H. Neanderthalensis *and* H. sapiens, *but this is now*
contested, possibly all three lineages coexisted.
(From Klein, 2017).

The last split in the hominin tree probably was the split between *H. sapiens* and *H.*
neanderthalensis, which happened no later than half a million years ago,[36] most likely in
Africa. Today, two closely related archaic human species, Neanderthals and Denisovans,

are distinguished. It is possible that they both first split from modern humans and then from each other. They migrated to Europe and Asia and continued evolving there, while the ancestors of the modern humans, *H. sapiens*, continued evolving in Africa.

Genome analysis reveals that the last common ancestor of all present-day humans lived about 300,000 years ago,[37] possibly slightly earlier. This is when the Khoe-San population in the Kalahari Desert in southern Africa started diverging from the rest of the African population, which is the earliest known genetic split within *H. sapiens* populations. The fossil record, on the other hand, suggests that the anatomical features specific to modern humans emerged much later.[38] To reflect this continuous evolution of *H. sapiens* in Africa, the concept of 'archaic *H. sapiens*' has been introduced.

H. sapiens did not firmly establish themselves outside Africa and the Middle East until about 50,000 years ago, even though periodic migratory waves of *H. sapiens* out of Africa probably began much earlier.[39] At around 40,000–45,000 years ago in Eurasia, the migrating *H. sapiens* came into a sustained contact with Neanderthals and Denisovans, though episodic contacts might have already happened before.[40] The species cohabitated and interbred until the archaic humans went extinct by less than 40,000 years ago. From comparing DNA extracted from bones of Neanderthals or Denisovans with genomes of modern humans we now know that these species were not reproductively isolated: in total about 8% of the contemporary *H. sapiens* gene pool come from the archaic humans and 1–2% of the genome of any human individual of European or Asian origin comes from the archaic humans.[41] Speciation is a continuous process and at early stages there is no full reproductive isolation. It has been questioned by some if *H. neanderthalensis* and *H. sapiens* can really be viewed as different species, however the majority view is that these groups are different species. Ever since the first archaic human DNA was uncovered and sequenced by the Swedish scientist Svante Pääbo in the late 1990s, the possibility that *H. sapiens* and *H. neanderthalensis* had cohabited and interacted with each other has captured the human imagination. In 2022, Pääbo was awarded the Nobel Prize in Physiology and Medicine 'for his discoveries concerning the genomes of extinct hominins and human evolution'. There is no doubt that the pioneering work of Pääbo has revolutionized our understanding of human evolution.

When did our ancestors start to talk?

When and where in the hominin 'tree' did the faculty of language first emerge? What evidence to speculate about this exists? An important element in hominin evolution was the manufacturing and the use of tools. The oldest well-documented stone tools that we are aware of have been dated as being over 3 million years old.[42] This was the beginning of the Stone Age, which lasted until about 10,000 years ago, when humans first mastered the use of metal. From what we know, the design of the early stone tools did not change much until less than 2 million years ago, when new types of tools characterized by distinctive oval and pear-shaped handaxes began to appear.[43] It has been speculated that this might have been the time when a protolanguage was emerging, as suggested by Bickerton in *Language and Species*.[44]

Despite modest technological advances, the period of up to about 300,000 years ago has been characterized as 'two million years of boredom'.[45] On the other hand, there is evidence that 300,000 years ago, hominins began using the first 'composite tools', for instance axes mounted on wooden (or bone) handles.[46] Were these changes related to the emergence of language? Could the skills to make such more sophisticated tools evolve via humans observing each other, which apparently is the basis of chimpanzee cultures (Chapter VII), or was language necessary for this to occur? It is not an unreasonable hypothesis that for the art of more complex tool-making to develop, significant amounts of information had to be passed on from individual to individual and to subsequent generations culturally. Given that 300,000 years ago is also the current estimate of the age of the last common ancestor of modern humans, it is tempting to assign special importance to this time, even though this is somewhat speculative.

Paleoanthropologists attribute special importance to artefacts that have no apparent practical utility and that can be viewed as *symbolic*, such as ornaments. These are associated with the emergence of symbolic thinking and behaviour, which has some commonality with the use language. Symbols work only if they can transmit messages, such as showing the social status of the individual. But for this to happen, the symbols need to have a similar meaning to all who use them—to the transmitters and the receivers—which is what language achieves. One particular evidence of symbolic behaviour comes from burial sites, when various artefacts have been buried next to the deceased, possibly to show their social status. The oldest known evidence of an intentional burial comes from Skhul Cave in Israel, dated at 115,000 years old; more impressive evidence of symbolism is found in Blombos Cave in South Africa, just over 70,000 years ago. Alongside carefully crafted tools such as spearheads, it contains symbolic objects such as beads made from seashells. Was the period between about 300,000 and 70,000 years ago when the faculty of language was evolving and spreading amongst *H. sapiens?*

Did Neanderthals possess language? The level of sophistication of Neanderthals and to what extent they exhibited symbolic behaviour is hotly debated. There is evidence that occasionally Neanderthals buried their dead and that later Neanderthals had some ritual behaviour. Nevertheless, there seems to be a qualitative difference between any possible use of symbolism by Neanderthals and the modern humans (for discussion, see the captivating book by Chris Stringer, *The Origin of Our Species*).[47] There is also an ongoing debate about whether the Neanderthal vocal tract would have allowed them to articulate the range of sound frequencies needed for speech, though whether such inferences can be made from fossils at all is disputed.[48] If Neanderthals did have a language, then the faculty of language must have been present more than half a million years ago, unless it evolved independently in two lineages, which does not seem likely.

How much time did the evolutionary process that led to language take? Some linguists, including Ray Jackendoff, argue that language started evolving as early as 2 million years ago and that it was a gradual multistep process.[49] On the opposite side, Richard Klein argues that the emergence of language happened as recently as 50,000 years ago, leading to the colonization of the entire world soon after, a view also taken by

Chomsky.[50] Bickerton takes a similar view, but he disagrees with Chomsky's 'camp' on the mechanisms by which this rapid evolution of language occurred. As will be discussed later, Chomsky argues that this must have been the result of a specific genetic mutation that gave humans the faculty to comprehend the successor operator (or the concept of '*and so on*'); Bickerton denies that the emergence of language was due to genetics at all.

Possibly, in a stepwise process, first hominins mastered combining phonemes to form words and combined a small number of words to describe more complex concepts or situations, but initially without using grammatical constructs.[51] This is what is referred to as a *protolanguage* stage, pre-dating grammar, which certainly was a major evolutionary invention—the ability to combine signals to form new signals. Perhaps the protolanguage stage was also when the faculty of speech was evolving. No contemporary primates other than humans have the ability for vocal learning. Thus, the ability of vocal learning must have evolved somewhere in the path from our common ancestors with chimpanzees to modern humans, before or simultaneously with other faculties needed for language.

The second major step, or possibly a parallel process, was the evolution of our ability to utilize grammar to form longer interpretable linguistic structures. Whether it was one large-effect mutation, as suggested by Chomsky, or a number of gene variants each with a small effect coming together, it is quite likely that this transition took tens of thousands of years. The right genes, or their combinations, had to spread through the entire population of *H. sapiens*. The described two-step evolutionary process is speculative; some authors, including Pinker and Jackendoff, argue that the process was more gradual, consisting of many smaller steps.

Given that the last common ancestor of all modern humans lived as early as 300,000 years ago, the theory that language emerged 50,000 years ago may need rethinking. Is it possible that positive selection for the 'language gene' was so strong that after the 'language mutation' happened just 50,000 years ago, this gene swept through the entire, already genetically diverse *H. sapiens* population? The genetic evidence does not seem to favour this.[52] That said, there is evidence that during the last 120,000 years, the world's human population went through one or more genetic 'bottlenecks', at times possibly shrinking to fewer than 15,000 individuals.[53] One may speculate that this bottleneck could have purged *H. sapiens* from those genotypes that did not fully support language, thus perhaps rescuing the late origin hypothesis. Also, in a small population, mutations are more likely to spread, even if they do not initially contribute positively to fitness, or even if they are somewhat negative (Chapter V). Consequently, some 'cryptic' mutations could have occurred during these bottlenecks, later enabling the evolution of language faculty.

Nevertheless, most genetic evidence now seems to point to the origin of the language earlier than 50,000 years ago. The analysis of phonemic diversity of existing languages points to the window of the emergence of the human language between about 150,000 and 350,000 years ago in Africa.[54] The beginning of the colonization of the world by modern humans too has been pushed back to earlier times. Finally, few scientists believe that the faculty of language was a result of one mutation. Nevertheless, on the timescale of natural biological evolution (as opposed to artificial selection), ~ 300,000 years is not very long, and it does not seem likely that many relevant mutations could have happened and spread in such a short time. Thus, given that there are no consequential intermediate stages between communications observed in contemporary animals and contemporary humans, the hypothesis of one or a small number of crucial genetic changes leading to language remains attractive.

Why did all hominin species except for *H. sapiens* die out? Why did Neanderthals die out? Was this anything to do with the emerging faculty of language in *H. sapiens*? It seems quite certain that by the time the anatomically modern humans came into persistent contact with Neanderthals, *H. sapiens* already had a well-developed faculty of language.[55] Although different factors, such as climate change, may have contributed to the demise of Neanderthals, given the persistence of many coexisting hominin species for millions of years, it seems likely that something changed with the emergence of *H. sapiens*. It is tempting to speculate that language was the key factor.

On a broad evolutionary timescale, it makes little difference whether the faculty of language began emerging 2 million or just 50,000 years ago. If life on Earth is over 2 billion years old, then 2 million years represent no more than 0.1% of life's existence, while 50,000 years represent 0.0025%. By any measure, language appeared only at the last moment. This conclusion stands even if we begin the comparison with the emergence of the first animals, say over half a billion years ago, or even with the mammalian lineage, about 200 million years ago. And even if we begin counting the time from the split of the primate lineage some 50 million years ago, still language has existed for no more than 5% of that timescale. Thus, even in the primate lineage, for 95% of its history or more, most information was transmitted from generation to generation via DNA. Regardless of how many discernible steps occurred in the path to language, regardless of whether this process took thousands, tens, or hundreds of thousands years, or even over a million years, on the timescale of evolution of life on Earth the emergence of language took only a moment. This was a 'phase' transition from finite to infinite—from the ability to represent and transmit small amounts of information, to unlimited amounts. With some imagination we can see an analogy between the emergence of life and the emergence of language. The first was the emergence of the ability to record and transmit unlimited amounts of information via polymers, the second was the emergence of the ability to do the same by other means.

We can say almost certainly that when *H. sapiens* were firmly established around most of the globe no later than 15,000 years ago, our species had a strong command of language. At that time, most likely different languages were already spoken around the world, but they all had the ability to communicate unbound amounts of information. And as more recent history shows, even when speakers of very different languages meet, they quickly learn how to communicate with each other. But what exactly is it that gives humans this unique faculty of language, a faculty missing in all other animals?

From genes to language

If the faculty of language has a genetic basis, this should be reflected in the difference between the human genome and the genomes of our closest relatives, the great apes.

The chimpanzee (chimp) genome was sequenced in 2005, four years after the first human genome, and genomes of other great apes, including gorillas, orangutans, and bonobos, were sequenced more recently.[56] Comparing these genomes should shed light on the genetics of language: what is common and unique to all known human genomes has a potential to be related to the unique human faculty of language.

Any possible hopes that the genome sequences of great apes would lead to a rapid understanding of what gives humans language have been dashed. On the one hand, the human and chimp genomes are rather similar—both contain about 20,000 protein-coding genes and most of these genes are in a one-to-one relationship between the two species. Moreover, almost a third of the 20,000 genes are identical between humans and chimps, and on average the difference between the respective human and chimp proteins is only two to three amino acids from a total of ~ 300–400, which is less than 1%. Still, given the number of genes, this adds up to ~ 50,000 different amino acids. If we look at the genomes overall, one position in every 150 is different. And in addition, there are millions of deletions, insertions, and genome rearrangements, in total adding up to 20 million genomic differences.[57] How can we identify which of these 20 million are relevant to language? Although these genome changes occurred after humans and chimps diverged, most of them result from neutral drift (Chapter V). And even of these changes that do contribute to phenotype, how many are likely to be related to our cognitive abilities, and how many to the faculty of language? Is trying to find them akin to looking for a needle in a haystack?

We can reduce the 'search space' if we identify which genomic differences are linked to human features related to language. One such anatomical difference between humans and all other primates is the brain volume relative to the size of the body, which amongst large animals, is unusually high in humans.[58] Given how brain:body ratios change with body size amongst primates, the human brain would be expected to be 50% larger than that of a chimp, but in reality it is 200% larger. The fossil evidence shows that along the evolutionary path leading to modern humans, starting from about 2.5 million years ago, brain size has been gradually increasing[59] (Figure 8.2). Overall, evolving a larger brain clearly was a consistent trend in human evolution, so in this sense there is linearity in human evolution. This almost certainly lead to an increase in the number of neurons and therefore probably boosting the brain's information processing capacity. The relative size of the part of the brain linked to language abilities, the frontal cortex, is also larger in humans.

There is a series of genetic mutations in the primate lineage leading to humans that can be linked to the evolution of a larger brain. For instance, the gene called *NOTCH2NL* has four copies in humans; chimps have one.[60] Deletions of genes of this family are linked to the human genetic disease called microcephaly, associated with smaller than normal brain size.[61] In contrast, an abnormally high copy number of this gene in humans is related to macrocephaly—an abnormally large brain. Two other genes linked to brain size, and possibly even directly to the faculty of language, *ASPM* and *MCPH1*, also seem to have evolved rapidly in the human lineage.[62] Another gene that is potentially important in human brain development and which has undergone three consecutive amplification events in the hominin lineages, approximately 3 million, 2 million, and

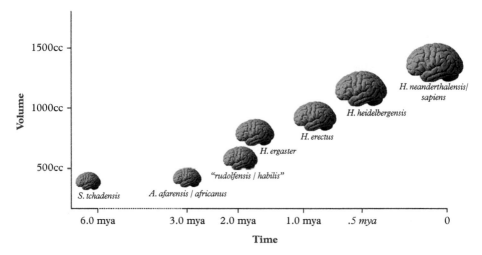

Figure 8.2 *A crude plot of average hominid brain size over time. There is a consistent enlargement of hominid brains over the past 2 million years. These brain volumes are the averages across several independent lineages within the genus* Homo *and they are likely to represent the preferential success of larger-brained individuals. For instance,* H. habilis *had a cranium that was 1.5 times larger than modern chimps,* H. erectus *had twice the size of a chimp, while the* H. heidelbergensis *brain was three times that of a chimp.*
(From Bolhuis et al., 2014; Image credit: Gisselle Garcia.)

then 1 million years ago, is *SRGAP2*.[63] When the human version of this gene is inserted into mice, some human-specific neuronal features develop. Still, none of this explains the faculty of language. Although there is a correlation between brain size and cognitive abilities in animals, neither whales nor elephants, as intelligent as they are, possess the faculty of language.

Those who have been reading recent literature about human evolution will almost certainly have come across the gene called *FOXP2*. This gene is present in humans and other great apes as well as in mice.[64] It encodes a transcription factor—a protein that regulates expression of other genes by binding to specific locations of the DNA (Chapter III). Importantly, this transcription factor regulates genes guiding the brain development. In humans, a mutation in *FOXP2* can lead to a language disorder affecting the individual's ability to break words up into phonemes and the ability to comprehend syntax, while at the same time non-verbal mental abilities are relatively intact. Thus, the *FOXP2* gene seems to be specifically related to language.

If we compare the *FOXP2*-encoded protein in humans and chimps, only two amino acids differ. These minor differences however are sufficient to change where in the genome this transcription factor binds to DNA. Thus, this protein regulates different sets of genes in humans and different in chimps. Three amino acids are different between the *FOXP2* proteins in human and in mice. Intriguingly, when the human version of the

FOXP2 gene is inserted in mice, those animals learn faster and produce sounds different from those of ordinary mice.[65]

If our faculty of language emerged after the *H. sapiens* lineage separated from the archaic humans, the language-defining genetic differences should show up in the comparison of their genomes.[66] Amongst the 3 billion nucleotides of the human genome, there are about 30,000 that differ between modern and archaic humans and also between humans and chimps, but at the same time are not known to vary in healthy modern humans.[67] These almost certainly represent mutations that have happened after the *H. sapiens* lineage separated from the archaic humans. These changes affect only 87 different proteins, many of which are found in the parts of the brain known to be associated with language. This small number of genes can be explored for function one by one. Additionally, about 3000 modern human-specific genomic variants are in genome locations likely to affect gene regulation, which too is not an enormous number for systematic analysis. This avenue of exploration makes most sense if we accept the hypothesis that the faculty of language was not present in archaic humans.

When it comes to *FOXP2*, Neanderthals and Denisovans have the same protein coded by this gene as modern humans, however, it is differently regulated. There is a mutation in an intron (Chapter VI) of the *H. sapiens* version of the *FOXP2* gene that alters a binding site, where another transcription factor, named POU3F2, binds. As POU3F2 is thought to regulate the *FOXP2* gene, apparently *FOXP2* is regulated differently in the modern and the archaic humans. So, maybe *FOXP2* is the 'language gene' after all.

How many different genetic changes were involved in the final stages of evolution resulting in humans acquiring language? Is there one particular genetic change that can be considered as the most important? Is this change related to our ability to comprehend the concept of '*and so on*'? Was this anything to do with the *FOXP2* gene? Or were the crucial mutations already present in the genome of our common ancestor 300,000 years ago, and the faculty of language emerged due to the right combination of genes, which already existed in humans, coming together? Over how long a period did these changes happen? These are all open questions.

As was already noted, some linguists, most notably Bickerton, argue that genetics was not the driver of the initial language evolution at all. According to Bickerton, even if there were mutations related to language, these came after language had already emerged, and that any brain 'rewiring', if it happened at all, happened later. In *Adams Tongue*, Bickerton attributes the emergence of language to 'niche construction' without elaborating how niche constructions, which are external, could lead to humans acquiring the faculty of language, which is internal. Because he also mentions gene expression in the same context,[68] it is possible that he includes epigenetic mechanisms in what he refers to as niche construction. As we discussed in Chapters V, VI, and VII, neither niche construction nor epigenetic encoding are likely to transmit significant amounts of information robustly across many generations. It is therefore hard to see how Universal Grammar can been inherited by these means. That said, if indeed it was just one mutation that mattered, just a flip from one state to another, it could in principle have been an epimutation. Acquiring a particular epigenetic state indeed can provide the organisms with means to exploit this state and solidify it by a genetic mutation later.

None of the described genetic differences answer the question of what the physical manifestation of Universal Grammar is and how this prior knowledge needed for the faculty of language is encoded in the genome. Can we find these differences in brain structures? Subtle human and chimp brain differences, such as gene activity in specific neurons, number of neuron synapses per unit volume of brain, or changes in the patterns of brain development have been described, and human-specific differences in brain organization and neuronal connectivity patterns have been suggested.[69] Nevertheless, none of these are conclusive, and it is still not clear if humans have any brain structure, cell type, or specific neural circuit unique to our species. It is easier to compare genomes than to compare brain structures—genomes are, in a sense, one dimensional. Perhaps an alternative is to train two sets of artificial neural networks, one by communicating with humans, the other by communicating with chimps, and then try to understand if there are any principal differences between them.

Genotypes and mnemotypes

Language gave humans the ability to communicate large amounts of information and pass it on from generation to generation. However, the sound of a spoken word disappear fractions of seconds after that word is spoken. For tens of thousands of years, until writing was invented, information transmitted via language was recorded only as modification of synaptic wiring or strength of connections between neurons in the human brain. These can be abstractly referred to as *brain images* or *brain impressions*. We still do not know exactly how the human brain encodes this information and it is most likely that each human brain records the same information somewhat differently, nevertheless the information itself, when transferred from individual to individual, remains essentially the same. To reflect this, in *Gödel, Escher, Bach*, Hofstadter talks about 'mappings between brains' and about an abstract, symbolic representation of information in the brain: '*Let us from now on refer to these hypothetical neural complexes, neural modules, neural packets, neural networks, multineural units—call them what you will, whether they come in the form of pancakes, garden rakes, rattlesnakes, snowflakes, or even ripples on lakes—as symbols.*' These *brain images* or *impressions* or *symbols* form by acquiring and recording information during one's lifetime, either through language or otherwise, but language is what allows humans to transfer them from brain to brain efficiently.

The brain dies when a person or animal dies, and the information in it is lost. Nevertheless, at least some of the information may continue living because it has been transferred to another brain before the individual died. This is similar to how genes keep living longer than the lifespan of any individual. To describe the phenomenon of such *brain images* and their ability to live beyond one generation, the US physiologist Harold Blum introduced the term *mnemotype*,[70] writing in the popular science magazine *American Scientist* in 1963. He defined *mnemotype* as the personal collection of memory images in a single brain, analogous to how genotype is the collection of genetic features. Like genes, parts of this information are transferred from generation to generation, but here it is neither stored in nor transmitted via DNA. For instance, the ability to speak a particular language is reliably transmitted from generation to generation, even though this is not encoded in DNA. Similarly, the skills to make or use a wheel can be passed on from generation to generation via a *mnemotype*. This new way of transferring information is what enables cultural evolution.

As already discussed in Chapter VII, in other animals too not all synaptic wiring is inherited; different skills develop when animals learn from observing other animals or from interactions with the environment. These skills are most probably reflected in the brain via altered synaptic wiring. But there is a principal difference between humans and other animals. Unless there is language, only relatively small amounts of non-genetic information can be transmitted between individuals. By using language, humans can pass on parts of their *mnemotypes* from one individual to another, similarly as genes are passed on from parents to offspring via DNA. A good illustration of how this can happen is in so-called oral traditions—ancient sagas and epics that have been passed on from generation to generation without being written down. It is believed that this is how the ancient Greek epics, the *Iliad* and the *Odyssey*, were originally passed on through generations (even though today we know about them because eventually they were written down).

By enabling the transmission of unlimited amounts of information from generation to generation, humans raised *cultural evolution* to a qualitatively new level. Language enabled humans to accumulate enormous amounts of cultural information much faster than information can accumulate in DNA. As noted, biological evolution works through genetic algorithms (Chapter VII) where new information is acquired through the selection of the fittest genomes—the gene variants and their combinations that give an individual higher fitness are kept, the others are dismissed. The information that individuals learn during their lifetime is not recorded in their genomes and thus is not passed on. In cultural evolution, information learned during one's lifetime may be passed on to future generations as in the evolutionary mechanism described by Lamarck (Chapter V). Cultural information accumulates in the form of *mnemotypes*, whether we want to use this terminology or not, much faster than genetic information in DNA.

There is another term—*meme*—that has been used in the literature to denote pieces of information in the context of cultural evolution. This term, however, has often been used, in my opinion, to over-emphasize the analogy between biological and cultural evolution. Although Darwinian selection may be a part of the cultural evolution too, in my view the most important features that makes cultural evolution much faster is that it operates via the Lamarckian mechanism.

The emergence of language was either the cause or the consequence, or both, to the emergence of so-called symbolic thinking—a type of thinking or information processing, which appears to be unique to humans. Understanding what is going on inside the human brain and how it differs from the possible 'thought' processes in an animal brain is a fascinating subject, but it is beyond the discussion in this book. My view is that information transmission enabled by language alone could have accelerated cultural evolution significantly and qualitatively. This does not mean that the increasing human cognitive abilities did not play a major role too; humans develop new ideas by algorithms more sophisticated than the Darwinian trial and error. And finally, the differences between the cognitive abilities of humans and other animals too may be rooted in the human ability to comprehend the concept of *and so on*.

From language to genes

We can learn a lot about human biological evolution by comparing genomes of individuals of different ethnicities and in different geographical locations, and by comparing genomes of contemporary humans to those extracted from human bones thousands of years old.[71] An interesting account of what has been learned about human history from such comparisons can be found in the excellent book by David Reich, *Who We Are and How We Got Here*.[72] Comparative genomics confirms that the emergence of language did not slow down but possibly accelerated human biological evolution, revealing interesting interactions between biological and cultural evolution.[73] For instance, presumably due to the human cultural invention of the ability to make primitive clothes, *H. sapiens* was able to move northward into higher latitudes of Eurasia, where the sunlight was weaker. With this, genes for lighter skin pigmentation were spreading to cope with vitamin D deficiency.[74] The colder climate also affected human salt intake, which is regulated by the salt-sensitive hypertension gene *CYP3A5*, a gene that has undergone recent evolution.[75]

A major event in human cultural evolution was a transition from foraging to farming around 10,000–12,000 years ago. This was a product of cultural evolution—it is not likely that we can find a 'farming gene' or even a combination of genes that enabled farming. However, arguably there are genes that changed because of cultural evolution. The most archetypal of such genes is *LPH*, which codes for an enzyme that gives humans the ability to digest lactose, which is present in milk. The abundance of this protein decreases in the human body before adulthood in most human populations outside northern Europe, and with it the individual's ability to digest lactose also decreases. However, individuals in northern Europe, where cattle farming has been particularly important, often have the so-called lactase persistence trait, caused by mutations linked to this gene. In Swedish and Danish populations, 90% of individuals have a particular mutation enabling them to digest lactose in adulthood. It is known that the ancestral *H. sapiens* genotypes did not provide for lactase persistence and this change in the northern Europeans happened only in the past ~ 7000 years.[76] Mnemotypes have fed back to genotypes.

It is also quite likely that human brains continued evolving after the human faculty of language had already emerged, moreover it is quite possible that this faculty itself has continued to evolve biologically. Indeed, mutations in the human genes related to brain development might have happened as recently as 5800 years ago.[77] There is evidence for a strong genetic component related to so-called educational attainment, which is defined as the highest grade completed in the educational system of the country where the education was received.[78] Interestingly, there is also evidence for recent selection *against* genome variants linked to higher educational attainment, possibly due to the link with human fertility.[79] Thus, not only biological evolution and genes enabled language and cultural evolution, cultural evolution feeds back to evolution of genes. Recently this feedback has become even more direct due to the development and use of genetic engineering and gene editing technologies.

From speaking to writing

Without lessening the importance of analogue information artefacts, such as cave paintings, it is fair to say that before writing was invented, that is, in prehistoric times, most cultural information was 'recorded' as *mnemotypes*. This situation began changing sometime between 5000 and 10,000 years ago, with the invention of record keeping and then writing, probably soon after humans started practising farming. The importance of the invention of digital ways of permanent recording of cultural information is reflected in the difference in how little and how much we know about human societies before and after writing—between prehistoric and historic times.

It seems a reasonable hypothesis that as the productivity of farming increased and humans were able to stock up on farming products for later use or for trade, the need to keep records emerged. The earliest tools used for this may have been clay tokens, often found amongst ~ 10,000-year-old archaeological artefacts (~ 8000 BC) in the Middle East.[80] For a set of tokens to represent a number, they need to be kept together. Indeed, perforated clay tokens, which could have been kept together on a string, appeared around 6000 years ago, for instance, in the ancient cities of Uruk and Susa in the Middle East. Tokens could also be 'archived' in sealed containers, and indeed, 'envelopes' with a diameter of around 5–7 centimetres, dated between 5500–5700 years old, shown in Figure 8.3, have been found in Uruk, as well as in Susa.[81] These are also the sites of the oldest artefacts that have been interpreted as representing the earliest writing.

Whether these clay envelopes really represent the earliest accounting systems is a speculation, however there is a general consensus that the world's oldest known writing system, Sumerian cuneiform, which developed in Mesopotamia more than 5000 years ago, was largely predicated by the needs of public economy and administration.[82] The earliest, so-called Proto-Cuneiform tablets and the related Proto-Elamite clay tablets, shown in Figure 8.4, resemble spreadsheets more than freestyle documents. Even though this resemblance is superficial, according to historians, these tablets indeed were used for tasks for which we would nowadays use spreadsheets.[83] What matters most is that these early writing systems created means of storing and communicating information digitally. According to one of the authorities on early writing systems, John Baines, the earliest scripts represent

> *a system for recording information indirectly through the combination of 'signs' belonging to a specific repertory.*[84]

This is the very definition of digital information.

Figure 8.3 *Clay envelope that contained tokens and is bearing impressed markings corresponding to these tokens, found in Susa Iran, currently in the Louvre (Schmandt-Besserat, 1992). Some tokens and their imprints have been interpreted as referring to objects, such as sheep or goats. If the envelope carries imprints of the tokens found inside it, the tokens are redundant—the impressions on the envelope itself carry all the information. Is this how the first writing began? This is the view of Denise Schmandt-Besserat (University of Austin, Texas), who proposed her theory in the article in* Scientific American *in 1978 (Schmandt-Besserat, 1978). Although this hypothesis does not represent the full scientific consensus (see book review Zimansky, 1993, which clearly shows the absence of such a consensus), it does seem to make sense.*
(Photograph with permission from Paris, Musée du Louvre.)

Around the same time or slightly later, a different writing system emerged in Egypt, possibly not entirely independently from the Mesopotamian one.[85] The Egyptian writing may have had a different purpose and was more 'freestyle', but given that Egyptians used papyrus, which is much more perishable than clay, less is known about the earliest Egyptian writing.

One has to distinguish between a *universal* or *full writing system*—a system that permits recording everything that can be expressed in a language—and a *partial writing system*, which does not.[86] Arguably, many accounting tasks do not require a full writing system. All known writing systems seem to have begun as partial systems and many evolved into full writing systems. The first Mesopotamian writing system was not a full system, but within half a millennium to a millennium, it gradually evolved into one. The same probably happened to the Egyptian writing. Full writing systems need to support grammar in some way and indeed, in both Mesopotamian and Egyptian writing systems, grammatical symbols were added over centuries. A similar development independently happened in Mayan writing in the New World. It is noteworthy that the first writing

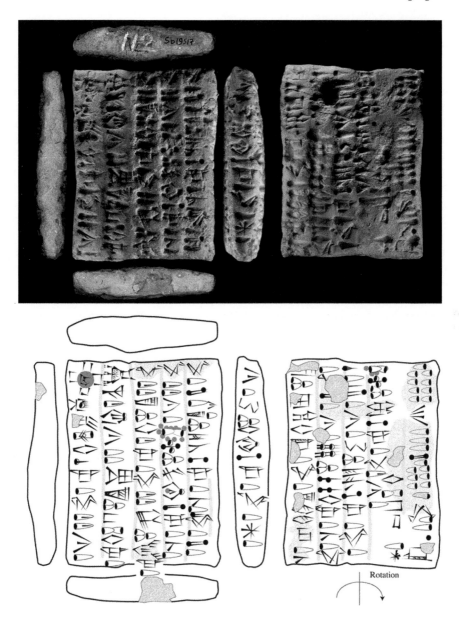

Figure 8.4 *Top. An about 5,000 years old clay tablet from Susa in the territory of modern Iran. Bottom. It is most likely that this proto-Elamite tablet represents an administrative document. In total about 1,900 different signs, which cluster in approximately 500 different forms mapping to discernible characters, have been identified in the known proto-Elamite tablets (Englund, 2004). These tablets are some of the earliest examples of relatively permanent combinatorial recordings of information in media other than polymer molecules such as DNA.*

(Top: Photograph courtesy of Musée du Louvre, Paris. Bottom: Transcript of the tablet by Cuneiform Digital Library Initiative project (MDP 17, 112, Cuneiform Digital Library Journal, 2015:001, ISSN 1540-8779, CC-BY licence).)

systems emerged at about the same time as the first numerical systems enabling arithmetical operations.[87]

The evolution of writing is a good example of cultural evolution.[88] One of the authorities on early writing systems, Ignace Gelb, in his influential book, *A Study of Writing*, suggests that all writing systems undergo similar evolutionary stages.[89] They start as pictography or iconography, the use of symbols depicting physical objects or actions, and then move to semiography, symbols that still depict physical objects or actions directly but no longer pictographically. The next step is use of logographic symbols, representing the spoken words or morphemes—the smallest meaningful semantic units. This is followed by syllabaries, written signs to represent spoken syllables, introducing more abstraction and reducing the number of symbols. The final stage is the invention of an alphabet—a system of symbols representing phonemes. Although today his model is considered an oversimplification, some of Gelb's basic observations seem to be true.[90] For instance, there are no known examples where a writing system has gone back from phonography to semiography, or from alphabet to syllabary. Note that the progression from logographs through syllabaries to alphabets goes towards the minimization of the number of different symbols and maximization of the usage of combinatorics. Such a maximization of the use of combinatorics is a progression towards a more efficient way of representing information (Chapter VII). Once a more efficient system is invented, it cannot be uninvented, even though it can be lost if it is not in use. This is true about cultural as well as biological evolution.

Does a full writing system necessarily have to represent a spoken language, or can there be other ways of representing the same information directly, bypassing the spoken word?[91] Numbers can be represented in two ways—by writing down how they are pronounced, or by using symbols, such as 1, 2, 3, 4, which are independent of any particular pronunciation. In English, the number 11 will be pronounced out loud as *eleven*, rather than *one-one*. Road signs are another example. Thus, alternatives to relying on a spoken word are possible. Nevertheless, in most, if not all currently existing writing systems, written words represent the spoken language. Linguists say that this also applies to languages such as Chinese, where most symbols denote entire words: if the spoken word for two different objects is the same, the symbol is the same. Regardless of how writing systems started, they all evolved to represent the spoken language. For Mesopotamian writing it took about a millennium to evolve to represent sound, for Egyptian it took 600 years.[92] If writing was independent of speaking, to become literate, effectively one would have to learn another language, though there are other reasons for the move towards phonography, for instance, the need to represent the proper nouns.

The invention of accounting, partial writing, and then full writing systems, like the invention of farming but unlike the 'invention' of language, was a product of cultural rather than biological evolution. Nevertheless, some human disorders, such as dyslexia, have a genetic component[93] and it cannot be ruled out that Darwinian selection for genes associated with the ability to acquire literacy has been in action since the invention of writing. As noted above, mutations in genes involved in brain development may have occurred as recently as 5,800 years ago which, almost certainly by coincidence, is about the same time when writing was invented. As the emergence of language was the result of biological evolution, and the invention of writing was a result of cultural evolution, the similarities and differences between how each evolved are noteworthy. As noted earlier, all languages are thought to have a common origin. By contrast, writing has been invented independently at least twice, in the Old and the New World, or possibly more times.[94]

From clay tablets to silicon chips

When talking about writing, we often abstract from the particular physical media in which the information is recorded. Mesopotamians mostly used clay tablets, Egyptians papyrus; we refer to both as writing. Carving texts into a stone is a method that has been used throughout history and is still used on tombstones. We refer to the abstract concept of writing as any somewhat durable combinatorial recording of information. Today, typing a text on a computer keyboard is often referred to as 'writing'; the entered information can be stored in various ways, for instance, as changes in electron distribution in a silicon chip. The physical aspect of writing and each particular physical substrate has, however, always affected how the particular writing system evolved. For example, the process of writing in clay is quite different from writing on papyrus, and there is little doubt that this had an impact on what Mesopotamian and Egyptian scripts were like.

Information written in different media have different durability—overall the higher the energy barrier that has to be overcome to record each *bit* of information, the more durable the record. The Sumerian clay tables are relatively well preserved, Egyptian papyrus less so, while engravings on tombstones are some of the most enduring. But this also affects the speed with which information can be recorded, copied, and disseminated, and more generally, how the recorded information can be processed and used.

In about the second century of the present era, parchment was invented in Asia Minor, soon becoming the most important media for recording information. By the ninth and tenth centuries, each abbey in Europe had its own scriptorium, where monks were copying manuscripts, scribing from parchment to parchment. A monk could copy about four 'leaves' of a size 35–50 cm × 25–30 cm per day.[95] I am not aware of any estimates of how many characters such a sheet contained, but looking at a small sample of digitized manuscripts from that time, it seems that about 20 lines of 50 characters each might be a reasonable estimate; thus, on the order of 1000 characters per sheet. Assuming that each such character can encode up to 2 *bytes* (B) of information, it is probably quite safe to assume that a scribe was copying not much more than 10 kilobytes (KB) of information per day, and consequently, a thousand scribes could copy up to 10 megabytes (MB) a day. Today a personal computer can copy this amount of information from one silicon chip to another in fractions of a second.

Two inventions, both of which happened in China, had a major impact on the human ability to disseminate large amounts of information: the invention of paper and the invention of the printing press. It is not known exactly when paper was invented, possibly some 400 years after parchment. The Chinese guarded their secret closely, but eventually the knowledge of how to make paper was passed on through the Mongolians to the Persians in Samarkand, and then brought to Sicily and Spain. By the thirteenth century, paper manufacturing centres were set up in Europe.

One can trace the origins of the printing press to the invention of the seal, screw-press, and the realization that different seals can be combined. A seal—a carved object, which can be used to impress a relief on soft material, such as clay—is an effective tool for wide dissemination of small amounts of durable information. The earliest seals have been found in Asia Minor and Mesopotamia, dated to the fifth or sixth millennia before the present era.[96] Seals were probably used mostly as a proof of ownership or authenticity, but this too is information. Seals can be viewed as predecessors of the printing press the same way as smoke signals are predecessors of the telegraph.

The screw-press was first used for pressing grapes and later for impressing patterns on textiles. By the fifteenth century, the press was used to print words carved on wooden blocks. However, arguably, the key invention leading to the printing press was movable characters, which could be reassembled in different configurations and used combinatorically to represent and then replicate any written text. Apparently, the Chinese used such movable letters from the eleventh century; there is evidence that the first book printed using movable letters was produced in China in around 1390. In Europe, in 1447, Johannes Gutenberg of Mainz and his friend Peter Schoeffer were the first to use similar techniques. In 1450, Gutenberg used this technique to print the *Bible*. Just over 10 years later, the printing press was spreading through Europe and information was spreading with it fast, further accelerating cultural evolution. The so-called hand-press could produce a maximum of 300 sheets per day, thus not even a hundred times more than a medieval monk could handle, though probably cheaper. For over three centuries this did not change much, however by the second decade of the nineteen century, the production rate had become 1100 sheets per day; by 1828, *The Times* newspaper had reached 4000 sheets per day, and in 1939 about 40,000 per hour.

With the invention of the telegraph and radio, other means of information dissemination were developing in parallel. These developments created a need to communicate information not only between human individuals but also between humans and machines, and later between machines. Different ways of encoding information and different materials were needed for this. In 1725 in France, a weaver from Lyon, Basile Bouchon, invented a loom that accepted commands from a punched tape (Chapter I), later replaced by metallic punch-cards. The machine read information encoded as the absence or presence of a hole in a defined location on the card and translated this information into respective weaving patterns.

Babbage planned to use punch-cards to enter information into his Analytical Engine in 1830s, although the machine was never built (Chapter I). More than 50 years later, in the United States, a young technician, Hermann Hollerith, developed a machine able to read information from punch-cards, which was used in the 1890 and 1900 US censuses and for keeping the New York Central Railroad accounts. Hollerith later founded his own company, and through series of mergers, in 1911 he became one of the co-founders of the Computing Tabulating Recording Company, which in 1924 was renamed the International Business Machines Corporation, or IBM. Later IBM played an important role in developing punch-card and other human-machine communication technologies widely used in electro-mechanical and then electronic computers. Punch-cards gave way to magnetic tapes and magnetic drums, and then to semiconductor-based transistor devices as media for machine-readable information storage. In the 1960s and 1970s, IBM was the dominant manufacturer of electronic computers.

The history of how the transistor was invented is complex, with many claims and counterclaims, but what we can say for certain is that in 1956, John Bardeen, Walter Houser Brattain, and William Bradford Shockley were awarded the Nobel Prize in Physics 'for their research on semiconductors and their discovery of the transistor effect'. This was the time when the first programmable electronic computers were developed (Chapter I). Initially computers used vacuum bulbs, but in the 1950s solid-state germanium- and then silicon-based transistor devices replaced the bulbs in most devices, reducing their size and increasing reliability. A simple two transistor circuit, called a *flip-flop*, has two stable states and therefore can be used for storing 1 *bit* of information. By implication, a device of, say, a thousand transistors can store 500 *bits* of information.

It soon became clear that it is possible to put more than one transistor on a single 'chip' and that silicon can also be used to connect transistors in a circuit. Advances in photo-lithography—one of the principal steps in manufacturing integrated circuits—resulted in Moore's law, the empirical observation made in 1965 by Gordon Moore, that the number of transistors on a single chip approximately doubles every two years.[97] This has turned out to be roughly true from the 1970s to the present day. As we can associate a transistor with recording half a *bit* of information, the amount of information that can be recorded on one silicon chip also double every two years. From 2022, Samsung Electronics began commercial production of a chip with transistor features of 3 nanometres (nm) in size. Given that on a silicon crystal, atoms are about 0.2 nm apart, 3 nm dimension means about 15 atoms. At this size, some 5000 atoms may be sufficient to make a transistor. We also need to allow for space to separate the transistors from each other, which increases the number of atoms needed, thus lowering the information storage density possibly to 1 *bit* per a few hundred thousand or million atoms. Future technologies will be able to reduce this,[98] though fluctuations caused by thermal noise will make it necessary to have redundancy and to make a reliable device from unreliable components. This will again increasing the necessary number of atoms. At some point quantum effects will become important, though quantum computing also opens new possibilities.

Recall from Chapter III that fewer than 50 atoms are sufficient to encode 1 *bit* of information in DNA. These 50 atoms per *bit* do not include what is needed in the machinery to read or write this information, nor for storing the DNA itself. Moreover, in a silicon chip, these are electrons, rather than atoms, that record information, atoms only providing the 'infrastructure'. As a result, reading and writing information is much faster in silicon than in DNA. Recall that the speed of DNA polymerase—the speed of copying information from one DNA to another—is about a thousand letters per second; the clock rate of a modern microprocessor can be more than a billion cycles per second. On the other hand, arguably, with regard to DNA information, its processing can be parallelized easier than in silicon. Additionally, DNA provides higher storage durability.

When life and the recording of information in polymer molecules first emerged, the size of the physical object representing one or a few *bits* of information was on the order of a nanometre. A few billion years later, when life (humans) first learned to record digital information by pressing characters into clay, the size of a *bit* increased millions of times. Then, in less than 10,000 years, this has again been reduced to roughly the original size of a nanometre.

As noted earlier, the use of DNA as a medium for preserving cultural information has been demonstrated.[99] It may well be that in the future, DNA will store not only genetic information but also information that has originated in cultural evolution, such as music or books. With the emergence of human language, for the first time, large amounts of information broke out of the DNA; some 100,000 years later, we are now putting some of it back into DNA again.

IX

Epilogue

Beyond Language

For some time, there was a widely held notion (zealously fostered by the daily press) to the effect that the 'thinking ocean' of Solaris was a gigantic brain, prodigiously well developed and several million years in advance of our own civilization, a sort of 'cosmic yogi', a sage, and a symbol of omniscience, which had long ago understood the vanity of all action and for this reason had retreated into an unbreakable silence. The notion was incorrect, for the living ocean was active. Not, it is true, according to human ideas—it did not build cities or bridges, nor did it manufacture flying machines. It did not try to reduce distances, nor was it concerned with the conquest of Space (the ultimate criterion, some people thought, of man's superiority). But it was engaged in a never-ending process of transformation, an 'ontological autometamorphosis'.

(Stanisław Lem, Solaris, 1961)

An animal brain is incessantly processing information incoming from sensors—eyes, ears, and nose—helping the animal to survive and reproduce. With cultural evolution advancing, new types of information processing tasks were becoming important. For instance, with the growth of global shipping in the eighteenth and nineteenth centuries, ocean navigation was growing in importance and with it, the need to know one's position in open seas. Nowadays, this is achieved via the satellite-based global positioning system, but in the nineteenth century the only way to tell a ship's position was to measure the locations of celestial bodies and then perform cumbersome calculations. To facilitate this, astronomical tables were produced and constantly refined by *computers*. These *computers* were trained humans, who were scrupulously carrying out routine calculations by hand, following fixed arithmetical procedures, which today we would call algorithms. In a way, *computers* were just manipulating symbols. But humans are not infallible in such mechanical procedures and as the result the tables were rife with errors, which was a major concern for shipping. In 1842 the British astronomer, John Herschel, wrote to the Chancellor of the Exchequer, Henry Goulburn, that '*an undetected error in a logarithmic table is like a sunken rock at sea yet undiscovered*'. Herschel was a friend of Charles Babbage, who not only realized that it must be possible to mechanise such routine calculations but was also prepared to do something about this.

The history of the mechanization of computing is at least as long as that of writing, possibly longer. The most basic computational tools are pebbles. If I have seven goats and

Living Computers. Alvis Brazma, Oxford University Press. © Alvis Brazma (2023). DOI: 10.1093/oso/9780192871947.003.0010

I want to know how many more I need to purchase to have 15, I can take 15 pebbles, subtract seven, and I will see that I need eight to make up the number I want. This is trivial though; I could find this out by counting goats directly. But if I become rich and have thousands of goats and sheep, life becomes more complicated. Do I need to count thousands of objects one by one every time I want to trade? Not really. I can use different, maybe larger, or darker pebbles to represent, say, tens of goats. I can then represent, say 35 goats as three large and five small pebbles. This is the decimal counting system. I can do computations with such pebbles, by adding or subtracting the large and the small pebbles separately, and converting every 10 smaller ones into a larger one when necessary. Babylonians had this figured out. They mostly used a sexagesimal counting system. We still use sexagesimal counting system in timekeeping: 60 seconds is a minute; 60 minutes is an hour. But the real breakthrough in counting came in the fourth or the fifth century, when the Indians invented the positional system—the order of the magnitude of a digit defined by its position in a string of digits jointly representing the number. This is the counting system that we use nowadays. Arguably this was one of the most important inventions in the history of computing, possibly the most important. It makes implementing the basic arithmetical operations such as adding, subtracting, multiplying, and dividing straightforward, by writing sequences of numbers below one another and then following well-defined manipulations.

Those who have lived in the Soviet Union may recall that in every grocery store at the checkout, to add up the costs of the purchased items, a person, usually a woman, would use a wooden *abacus*, similar to the one shown on the left panel of Figure 9.1. The invention of abacus can be traced back some 4700–4300 years to the Sumerian culture in Mesopotamia. It was a kind of table of successive columns, which delimited the successive orders of magnitude of their sexagesimal number system. By the thirteenth century the Chinese had invented an abacus that resembled the recent Russian one and that could be used for carrying out all arithmetical operations.[1]

People operating the abacus at grocery checkouts were as creative individuals as any, but the task of using an abacus to sum up the costs did not involve creativity: 'cashiers', as they were called, were just mechanically moving their fingers following a routine. The speed at which some of them did this was amazing. A device such as the Russian abacus is a pseudo-mechanical calculator. Although the manipulations are mechanical, the operators are utilizing their brain to guide their fingers to execute the algorithm. But if no real creativity is involved, it should be possible to mechanize such calculations completely. In the seventeenth century, several mathematicians, including Wilhelm Schickard, Blaise Pascal, and Gottfried Leibniz figured this out, and built the first fully mechanical calculators. Leibniz's invention of the *stepped drum* in the 1670s dominated mechanical calculator design for two centuries.

The first mechanical calculators were rather specialized for specific information processing tasks, such as multiplying two given integers. But later, these devices became increasingly more general and flexible. As discussed in Chapter I, in the nineteenth century Babbage designed his *Difference Engine* 2, which was able to compute values of any polynomial function. I will not describe the complete history of computing here; for this see the comprehensive book by Georges Ifrah, *The Universal History of Computing*,[2]

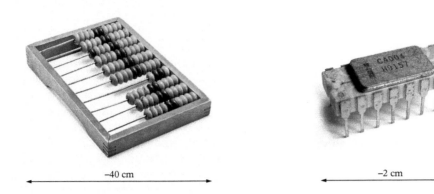

Figure 9.1. *Left. A Russian abacus, size is approximately 25 cm × 40 cm.* **Right.** *Intel 4004—the main information processing unit of the Universal Computer on one silicon chip. Including the case this is about 2 cm long, the silicon chip itself is only about 3 mm × 4 mm in size.*
(Left: Image reused from © Petr Malyshev/Shutterstock. Right: Figure adapted from Wikimedia).

but let us recall from Chapter I that Babbage also designed a mechanical computer that in principle could be programmed to implement any information processing task. In the twentieth century, Alan Turing was the first to conceptualize the idea of the Universal (mechanical) Computer.

As discussed in Chapter I, while moving mechanical parts is slow, moving electrons is much faster. The first electronic computers had to be programmed by rewiring, but in the late 1940s, von Neumann was one of the first to realized that a program could be stored in the computer memory alongside the data to be processed, and the general-purpose electronic computer became a reality. The vacuum tubes used to implement the logical gates were replaced by transistors in the 1960s, and by semiconductor integrated circuits in the 1970s. In 1971, a company called Intel began commercial production of the first complete microprocessor unit—the main information processing unit of the Universal Computer—on a single silicon chip. This chip, called Intel 4004, shown in Figure 9.1, was 12 mm^2 in size, contained about 2300 transistors, and was able to process information with a clock rate of 740,000 operations per second. Modern microprocessors may contain over 100 billion transistors and have a clock rate of more than 1 billion cycles per second.

Initially, humans and computers exchanged information via perforated tapes, punch-cards, and typewriters, then keyboards and display screens. Computers got connected to each other via the Internet and started exchanging information between themselves. Soon computers were bypassing human involvement in many operations. It is quite likely that today, when one goes to a bank to ask for a loan, the clerk is only an interface between the client and a computer. The computer 'knows' the client's credit history, income, and lifestyle, and makes the decision about whether or not to grant the loan. But one can also apply for a loan directly to the computer network from home. Might it happen that not only the clerk but also the client become redundant in some sense? Can computers start 'life' of their own?

The sophistication of real life as we know it is hard to beat. The molecular machines driving life are as miniature as possible given the atomic nature of the physical world. Their complexity per unit of space is as high as possible in principle; life is implemented via nanotechnology to the extreme. The degree of parallelism that life employs in information processing and interpretation is as colossal. In every little bacterial cell, there are thousands of RNA polymerases reading information from DNA, and ribosomes interpreting it to make proteins. There are some 10^{30} microbial cells functioning and evolving in the ocean, and 3×10^{13} complex cells in every single of the several billion human individuals existing on Earth, all processing information non-stop. Biological evolution has invented the use of electrons for information processing in animal brain, and even though silicon chips work at faster rate than the animal brain, life more than compensates for this by employing an enormous degree of parallelism. Can any other 'implementation' of information processing compete with the sophistication of life as it exists now? This is hard to imagine.

Nevertheless, on the timescale of astronomical events, there is a problem—the solar system is approaching its middle age and life on Earth may be approaching its old age. Life evolved all this complexity for the conditions existing on Earth and utilizing energy provided by the Sun. Our universe is believed to have been around 14 billion years ago, the solar system formed 4.5 billion years ago, and the first life on Earth possibly soon thereafter. At the current rate, the Sun fuses about 600 million tonnes of hydrogen every second, releasing more energy in a second than what can be obtained from burning down 10,000 trillion tonnes of coal.[3] Thermo-nuclear reactions will continue to fuel the Sun for about five more billion years, but before the Sun burns out, it will start expanding, slowly becoming a so-called red giant, big enough to engulf the closest planets Mercury and Venus, and possibly even Earth. And long before that, the Sun will be getting brighter, and about a billion years from now, the Earth will become so hot that all water will evaporate. Life on Earth will cease to exist. Assuming that life emerged 4 billion years ago, on the timescale of a human generation, the life on Earth is over 70 years old already.

The situation looks very different though from the perspective of cultural evolution and civilization. The human species is just a few hundred thousand years old and civilization less than 10,000, which is 0.001% of a billion. Equating the age of the Solar system of 4.5 billion years to the human lifespan of some 90 years, the current civilization can be viewed as being less than two hours old. Civilization is at the stage when it is most vulnerable. It has already acquired the ability to destroy itself via extensive niche alterations or use of energy at levels not compatible with preservation of information, while at the same time it has not quite yet acquired the ability to defend itself from threats reliably, such as meteorites or viruses. Will human civilization survive a few days? If it survives the first year of the equivalent of human lifespan, which is about 50 million years in reality, its chances of survival until the time when the Solar system becomes uninhabitable will increase.

But does civilization have to end in a billion years as the Earth becomes uninhabitable? Some will rightly argue that given our current fragility, there are more immediate problems to worry about than what happens in a billion years' time. Nevertheless, a billion years is the upper limit for an Earth-based life and Earth-based civilization. At that point, the temperature will become so high that no continuation of any information processing, not to mention biochemical reactions, will be possible on Earth. To keep going, our civilization will have to find a way to relocate to a different part of the Universe and to find a different source of energy. Others will say that the Universe will end anyway, but that prediction is not based on the same degree of established physics and indisputable observations of how stars evolve.

Does the end of the Solar system mean that in order to preserve our civilization, we will have to transport DNA and proteins to some other place in the Universe, or can we continue in some other way? In a billion years, what will be more important—the information in the DNA or the information in other media? Can DNA be replaced with something as physically different from it as a transistor from a vacuum tube?

For a civilization to exist for any meaningful timespan, it needs to be able to adapt to a changing physical environment. All known life achieves this via individuals competing with each other in Darwinian selection of the fittest. Nevertheless, the current physical implementation of life has its limits and in its present form life will not outlive the solar system. Can greater adaptability be achieved by other means? Can there be a civilization without life as we know it?

This book began with a list of questions, and it is finishing it with a list of related questions. The difficulty in answering them is that they largely boil down to just one fundamental question—*what is life?* If we define life as what exists on Earth now or existed in the past, then we can try to answer this question by describing how living organisms work and extracting what is common between them. I hope this book has given the readers knowledge they did not have before reading it and helped them to think about the possible answers to this question. If we accept that life can potentially be very different, then answering this fundamental question becomes even harder. But we can still try to generalize from the common features of life as we know it. In my opinion, recording and processing information to learn from the environment and to adapt when it is changing is the most fundamental element of any life and any civilization that can be imagined. Regardless of whether one agrees with this or not, I hope this book will be a help the readers to think about this more general question as well.

Notes

Introduction

1. In fact, the test Turing proposed was more sophisticated. There are three players: an interrogator and two others, say a man and a woman. The task of the interrogator is to find out which of the remaining players, is which. The Turing test was as follows: if one of the two players behind the curtain is replaced by a computer, will the interrogator be able to tell it apart from the human player? (Turing and Haugeland, 1950).
2. Williams, 1966; Dawkins, 1976.
3. Maynard Smith and Szathmáry, 1995.
4. Davies, 2019.
5. Nurse, 2020.

Chapter I

1. Von Neumann and Burks, 1966.
2. Aspray, 1990.
3. Von Neumann, 1993. Later von Neumann got involved in a bitter patent dispute with two leaders of the EDVAC project, J. Presper Eckert and John Mauchly, who had founded the Eckert–Mauchly Company to commercialize the invention. Their patents got rejected, partly because similar developments were also underway in Britain and at least one of the elements used in EDVAC was patented there, while some others were shown to be in the public domain. Von Neumann became a strong advocate of putting the fundamental inventions concerning computing in the public domain. As we will see Chapter III, there is an analogy between these events and how the first human genome project developed.
4. The largest computer manufacturer at the time, International Business Machines (IBM), announced their vacuum tube data processing machine IBM 709 in 1957. According to folklore, to meet the order from the US Air Force, this machine was 'transistorized' without changes to its design and renamed to IBM 7090. It was in operation by 1960, while IBM 709 was discontinued (Ceruzzi, 2003).
5. To assess this complexity, von Neumann noted that the minimum number of 'elementary' parts might be in the millions (Von Neumann and Burks, 1966, p. 80).
6. Von Neumann, 1951.
7. Von Neumann and Burks, 1966, p. 75.
8. This list was reconstructed by von Neumann's colleague Arthur Burks at Princeton from notes of von Neumann's lectures (Von Neumann and Burks, 1966, p. 81).

9. The AND gate has two inputs and gives a signal in the output when both inputs get a signal at the same time and not otherwise. OR gate has two inputs and gives a signal in the output, if any of the inputs receive a signal. Finally, NOT has one input and gives signal in the output if and only if it does not get a signal in the input.

10. Brenner, 2012b.

11. The word *quine* comes from the so-called Quine paradox named after philosopher Willard Quine: "'Yields falsehood when preceded by this quotation' yields falsehood when preceded by this quotation.' Is this statement true or false?

12. Hofstadter, 1979.

13. Gardner, 1970.

14. This is a consequence of the existence of configurations in *Life* that behave like the Universal Turing Machine. A different mathematical problem in *Life* is to find the smallest possible initial configuration that cannot be produced from any other configuration following the rules of *Life*. Such a configuration is called the *Garden of Eden* as it is a configuration that does not have a predecessor in *Life*. It is a 'legitimate' configuration that can be drawn on squared paper and will develop further according to *Life's* rules, but it cannot be produced by *Life's* rules out of any preceding configuration. Possibly, for a biologically minded person, a better name for such a configuration might be *the First Life in the Universe* (or, to be precise, one of them). Such configurations indeed exist, one of the latest ones, which was discovered in 2016, fits in a rectangle of 8 × 12 cells and in 2017 a configuration fitting on 9 × 11 cells were found (https://conwaylife. com/wiki/Garden_of_Eden). An even more interesting problem is finding a *Garden of Eden* that develops in an interesting way that neither stops nor becomes periodic for a long time.

15. Burks, 1970.

16. In Conway's *Life*, each cell has eight *neighbours*, which (together with the cell itself) determines the cell's future state. But the neighbourhood can be defined in a more general way. For instance, in a honeycomb pattern, a cell has six immediate neighbours. Or on squared paper, if we exclude the diagonal cells from the 'official' neighbourhood, then each cell has four neighbours. Or we can also extend the 'official' neighbourhood to the second 'ring' of neighbours, in total to 24 neighbours. Von Neumann's self-replicating automata used a four-cell neighbourhood. In *Life* each cell is a two-state automaton—the states are *alive* or *dead*. This too can be generalized—the cells can be defined to have any finite number of states; for instance, 10 states can be represented by numbers 0 to 9. The rules that determine the next state of the cell, can be defined in various ways too; for instance, assuming that cells have 10 states, the next state of the cell can be defined as equal to the sum of the states of its neighbours, unless this sum greater than 9, in which case the state is set to 0. Any rules are allowed as long as they are the same everywhere on the board and do not change in time.

17. In additional to solving technical difficulties, such an approach was essential for avoiding the problem arising from the Garden of Eden. It can be shown that Garden of Eden configurations for the von Neumann's 29 state automata exist (Moore, 1970). The Universal Constructor cannot construct the Garden of Eden configuration in principle, as otherwise this configuration would not be the Garden of Eden. Thus, strictly speaking for the von Neumann's 29-state automaton the Universal Constructor does not exist. The way this contradiction is solved, is that in von Neumann's setting, the Universal Constructor is only required to build configurations of unexcitable states.

18. Von Neumann and Burks, 1966.

19. Burks, 1970.

20. Kemeny, 1955.

21. Sipper, 1998.

22. A related question is how many steps would be needed for the replicating configuration to make a new copy. Statements in this respect have been made, for example see Pesavento (1995), however it seems that no quantitative estimates have been published in peer-reviewed scientific literature.

23. Ulam, 1970.

24. Maynard Smith and Szathmáry, 1995.

25. As will be discussed in Chapter VII, the amount of information that biological organisms pass on to their descendants can be estimated from their genomes, which typically are millions of letters for the simplest bacteria, to billions for plants and animals. Thus, it does make sense to talk about 'potentially unlimited heritability'.

26. Hofstadter, 1979. One could argue that in the context of interpreter, it would be more appropriate to talk about *knowledge* rather than *information*. However, we reserve this term for *prior knowledge* (Chapter VII). The amount of *information* received by the interpreter is the difference in the 'knowledge' that the interpreter has after receiving this *information* and prior to this. Those who are familiar with Shannon's definition of information will notice that this postulate takes our definition outside the realms of how Shannon defined this term. We will discuss Shannon's definition alongside other means of defining and measuring information in more detail in Chapter VII.

27. Turing, 1936.

28. It can be shown that if we are allowed to use two different symbols on the tape, for instance 0 and 1, the minimum number of states needed to build the Universal Turning Machine is 15. If we are allowed to use four different symbols, as in DNA, then six states are sufficient (Woods and Neary, 2009).

29. A classic example of such a non-computable problem is finding if a computer, when executing a particular algorithm (coded as a computer program) with particular input information, will stop. This is a typical information processing problem: the input is the program and the data given to the computer as a string of symbols, the output is 'will stop' or 'will not stop'. How do we know how long to wait before we can confidently say that the program will not stop? It can be proven mathematically that there is no algorithm that, given an arbitrary computer program and data, will be able to tell if the computer when executing this program will eventually stop. For some particular programs it may be possible to discover this, but not for all. For the proof of this, and a more in-depth discussion of computability, see *Gödel, Escher, Bach* by Hofstadter or any other popular book about computability. As anything that can be solved on one computer can be solved on any other computer (with sufficient memory), information processing tasks come in two flavours—the ones that can be solved on a computer and the ones that cannot: computable and non-computable information processing problems.

30. Church, 1932.

31. For those who are familiar with computer programming, let me note that the programming languages LISP and R have its roots in lambda calculus.

32. The original definition by Church stated: *Suppose there is a method which a sentient being follows in order to sort numbers into two classes. Suppose further that this method always yields an answer within a finite amount of time, and that it always gives the same answer for a given number. Then: Some general*

recursive function [a function that can be computed on Turing Machine in finite time] exists which gives exactly the same answer as the sentient being's method.

The history of this fundamental discovery is non-trivial, as is discussed in some detail by Martin Davis (Davis, 1982). The first scientist who noted that the principle of computability should be treated as a natural law was Emil Post as early as the 1920s, though his work though remained unpublished until 1943 (Post, 1943). Church's thesis, without being called as such, was first 'announced' to the mathematical world in 1935 at the meeting of the American Mathematical Society in New York City. This was before Turing's paper and his proof of equivalence were published; Church's paper describing his thesis were published in 1936 (Church, 1936). Nevertheless, Turing's role in developing this thesis was paramount; Turing was the first to introduce the idea of computations as mechanical operations explicitly. As described in Martin Davis' article, Kurt Gödel (famous for his proof of incompleteness of mathematics) was initially critical of Church's approach and sceptic of the possibility to formalize computability. After Turing presented his machine, Gödel finally accepted that this was possible. Turing became aware of Church's work only after he had finished the draft of his 1936 paper and subsequently added the proof of equivalence to lambda calculus in a revision prior to the final publication.

33. Pinpointing exactly which was the first truly programmable computer is difficult: for instance, Colossus and ENIAC were programmed by physical rewiring rather than by loading the code in the computer's memory. Arguably, it was von Neumann who first came up with the idea that the program itself can be stored in a computer's memory, though conceptually this idea can be attributed to Turing.

34. Swade, 2000.

35. The Church thesis captures yet another important observation: whatever device we can think of adding to logical switches, this will not enable the computer to solve an information processing problem that could not be solved using a Turing Machine. Going outside the binary alphabet of zeros and ones to more characters and more sophisticated operations does not add anything. Information processing using Chinese characters cannot achieve more than information processing using zeros and ones. Nothing anybody has ever thought of has helped: a task that does not have an algorithmic solution still does not have one. Analogous computers too cannot do anything that binary digital ones cannot (due to the inevitable thermal noise, see Chapter VII). And assuming that in the end we want one definite answer rather than a quantum superposition of answers, quantum computing does not change this either (Aaronson, 2008).

36. The Universal Turing Machine can be encoded as a configuration of the *Game of Life*, thus the *Game of Life* has a potential to satisfy von Neumann's requirements. Here, by 'encoding' we mean the following. There is an initial configuration in *Life* (representing the Universal Turing machine) such that adding to it (in a defined location on the plane) a configuration that encodes the information to be processed, this joint configuration (developing according to the rules of *Life*), produces a new configuration encoding, the answer to the information processing problem (encoded as a *Life's* configuration). The requirement of universality can be somewhat relaxed in the mathematical models of information carrying self-replication; for instance, see Langton, 1984.

37. Berlekamp et al., 1982.

38. Poundstone and Wainwright, 1985.

39. Aron, 2010.

40. Von Neumann, 1956.

41. To talk about reproducibility we assume that there is a correct solution (or a set of the correct solutions) to the problem, which is then reproduced with a probability larger than ½. By running the algorithm repeatedly we can find out which solution is it with any required probability.

42. One example of a stochastic algorithm working faster than any possible deterministic one is the so-called Miller–Rabin algorithm for finding whether a given number is a prime number (a number that does not divide by any other number different from itself or from one) (Rabin, 1980).

43. Gould, 1989.

44. Swade, 2000.

45. For those who have not studied biology, let me note that the system of two-part names originates in the classification of living species first introduced in the eighteenth century by the Swedish botanist, physician, and zoologist Carl Linnaeus, the father of the modern classification of species. The first part of the two-part name—*Escherichia* or *E.* refers to the so-called *genus*, while the second part *coli* to *species*. For comparison, the Latin name of the African lion is *Panthera leo*, where *Panthera* stands for the genus and *leo* for the species. The common name 'black panther' refers to two different species: in Asia and Africa they are leopards, *Panthera pardus*, and those in the Americas are black jaguars, *Panthera onca*. In a popular science book like this, for the African lion we would normally use the common name *lion*, however for bacteria we have no other choice than to use the binomial names. The *E. coli* genus was named after Theodor Escherich, a German paediatrician who first described it 1886, while *coli* is the Latin word for colon, the lower intestine. Thus, in fact, *E. coli* is not really a Latin name. For most biologists *E. coli* is simply a way to refer to a particular bacterial species, simply an identifier.

46. Tenaillon et al., 2010.

Chapter II

1. Schrödinger, 1944.

2. Brenner, 2012a. Max Perutz, who won the Nobel Prize in Chemistry with John Kendrew, wrote in 1962: '*Sadly, a close study of his book and of the related literature has shown me that what was true in his book was not original, and most of what was original was known not to be true even when the book was written.*' Perutz, 1989, pp. 234–51)

3. Kolmogorov, 1965, 1968; Chaitin, 1969.

4. This may appear counterintuitive—why not just enumerate all the possible descriptions and choose the shortest? Obviously, the exact length of the shortest description depends on the language that is used, but this is not the main difficulty—it can be proven that if the number of objects is large, then for the absolute majority of the arrangements, the length of the shortest description will be about the same, regardless of the language used. Thus, the particular language does not really matter. The difficulty is more fundamental, it is related to the problem that there is no algorithm that could tell whether an arbitrary computer program will stop (Chapter I). The shortest description can be viewed as the length of algorithm that generates the original arrangement. Thus, we not only need to enumerate all the algorithms that potentially could generate the arrangement but we also need to check if they indeed generate the description. But for this we need to know how long to wait until the algorithms finishes, which we cannot tell. One example that may help to reconcile

our intuition with the non-computability of Kolmogorov complexity is a sequence of integers: 97932384626433832795028841971693993751O. Can we describe this arrangement of digits in a shorter way? It looks quite random; no obvious pattern is there, and it is not clear if this sequence can be compressed. Those who think of searching for this string on the Internet will find that it is a part of the decimal expansion of the number π—the ratio of the circumference to the diameter. Concretely, it is the part of this sequence between the 12th to 50th digits. If we took, say, the 12th to 50,000th digits, it would take a little booklet to print this sequence, but we do not need more than a page to describe the algorithm that computes π with any required precision, thus generating this sequence of digits. If the sequence were, for instance, the binary representation of the even numbers of some 'remote' part of the expansion of π, then it would be rather unlikely that one would ever discover this. And there are many other irrational numbers we could use instead of π. A sequence that may look completely random may in fact be an output of quite a simple algorithm and thus has a short description. Given that we cannot tell if an arbitrary algorithm will stop, we cannot enumerate all possible ones and choose the shortest. What if an algorithm starts well and generates a good initial part of the sequence but then keeps going and going and going, without printing anything additional in the sequence? How long should we wait until giving up? We cannot tell.

5. To be more precise, in classical thermodynamics (as opposed to quantum thermodynamics), only the change in entropy can be quantified; the entropy's absolute value is not defined. Temperature is a quantity that links the amount of energy added to or subtracted from the system to the change in entropy. It would be difficult to apply these concept to a deck of cards in a meaningful way. A different concept—*information entropy*, introduced by Claude Shannon (discussed in more detail in Chapter VII), can be used to measure the 'orderliness' of a deck of cards, however this does not provide a direct link to the Second Law. Attempts to link thermodynamic and information entropies rigorously have been made; however, most of them include significant hand-waving. A notable exception is the work of Rolf Landauer (Landauer, 1961); the so-called Landauer's principle states that energy needs to be dissipated to erase information. However, Landauer's principle cannot be considered proven; see, for instance Norton, 2005.

6. Sometimes a distinction is made between the concepts of *self-organization* and *self-assembly*, however this distinction is not well defined and these terms are used in different ways by biologists, physicists, and engineers. For instance, what some call self-organization in biology, others call dynamic self-assembly in engineering.

7. Rosato et al., 1987.

8. If we do want to record the number too, the total description length will be proportional to $m \times \log(n)$, but for large n, we have $m \times \log(n) \ll n$.

9. Feynman et al., 1963, vol. 1.

10. Griffith et al., 2005.

11. More specifically, the kinetic energy of a molecule equals 3/2 kT, where k = 1.38×10^{23} is the so-called Boltzmann's constant and T is the temperature in Kelvins. The kinetic energy of an object equals $mv^2/2$, where m is its mass and v velocity. If we make calculations, than given that the mass of a water molecule is about 3×10^{-23} grams, we find that at room temperature the mean velocity of a water molecule (or any other molecule with a similar mass) is around 500–600 metres per second.

12. Feynman et al., 1963, vol. 1.

13. The *degree of freedom* is a non-trivial concept of physics: each particle can count for several *degrees of freedom* (for instance, moving or rotating along each of the three spatial dimensions account for a degree of freedom; additionally, composite objects can also have internal degrees of freedom, such as vibration). Nevertheless, the more particles, the more degrees of freedom.

14. The free energy is minimized if the temperature of the entire system remains constant. In an isolated system the temperature may vary and the thermodynamic entropy will tend to the maximum.

15. Consider the case shown in Figure 2.9. Suppose there is a large number of A and B. The potential energy will be minimized if all A and B are free (that is, there are no molecules AB). However, at temperatures above 0° K, to minimize the free energy there will always be a mixture of all A, B, and AB. The ratio of molecules in different states will be proportional to $e^{-(E_2-E_1)/kT}$, where e is Euler's number, E_1 *and* E_2 are the potential energy of different states, k is the Boltzmann constant, and T is the temperature in Kelvin. This is known as *Boltzmann's distribution*. If we introduce another atom C, which can combine with A or B, the atoms will recombine to reach a new state of equilibrium. Thus, free energy minimum depends on the concentration of different molecules in the environment.

16. Pradzynski et al., 2012.

17. Nakaya, 1954.

18. Libbrecht, 2006.

19. Libbrecht, 2005, 2007; Gravner, 2008.

20. Gravner, 2008.

21. Barrett et al., 2012.

22. Libbrecht, 2013.

23. Cairns-Smith, 1966.

24. Brenner, 2012a.

25. This layer will be highly dynamic, the participating molecules in it will change all the time.

26. Lipowsky, 1991.

27. Jun and Mulder, 2006; Jun and Wright, 2010.

Chapter III

1. Johannsen, 1909.

2. Williams, 1992.

3. Roller, 1974.

4. A couple of years earlier, the term *information* was used in the context of biology with an ironic connotation, as described in the excellent book by Matthew Cobb (Cobb, 2015, p. 88).

5. The branch of mathematics known as *recursion theory* was developed in the works of David Hilbert, Kurt Gödel, Emil Post, Alonzo Church, and Alan Turing, amongst others, during the first half of the twentieth century, later forming the basis of what is now known as theoretical computer science. The term information was not at the center of this discipline initially, however, once this concept was introduced by Claude Shannon in 1948, it was gradually becoming apparent that computing and processing of information is the same thing.

6. Keller, 1995.

7. The sequence taken from UniProtKB database (accession member of the protein P00560). In a mature protein the first M is chopped off.

8. Gonzalez-Pastor et al., 1994.

9. UniProt, 2015.

10. The term *protein family* usually it refers to proteins that are evolutionarily related.

11. Berman et al., 2003.

12. Orengo et al., 1997.

13. Anfinsen, 1973.

14. Using the term *self-assembly* in the context of protein folding would sound somewhat strange—amino acids do not assemble in polymer chains by themselves. However, both terms—self-organization and self-assembly—have been used in the scientific literature in this context (Jaenicke, 1998; Baker, 2000; Dobson, 2003).

15. Some proteins, or parts of a protein, may not adopt a well-defined structure at all—these are the so-called *unstructured* or *disordered* proteins. Nevertheless, it is correct to say that a protein can assume only a small number of structures distinguishable by a living cell, not having a defined structure being one of the possibilities.

16. In late 1950s, Lionel S. Penrose, an English psychiatrist, medical geneticist, paediatrician, mathematician, and chess theorist, demonstrated similar seeding on a macroscopic scale using plywood building blocks (Penrose, 1957, 1959). These papers are sometimes cited as demonstrating self-replication on a macroscopic scale, but in fact they demonstrate a mechanism of how an initial condition of the system can serve as a seed, spreading one of the possible structures to the entire system. This may happen when different crystal lattices form depending on the seed. Such 'replication' transmits only one or a few *bits* of information, like a prion protein.

17. Westergard et al., 2007; Caiati et al., 2013.

18. We will discuss how to quantify information in Chapter VII, but for now we can think about the amount of information needed to describe the environment in the following way. We make a 'dictionary' of the possible structures of the protein linking each structure to the description of the respective environmental conditions. Then, assuming that this dictionary is provided, we only need to select the respective dictionary entry. If the number of different structures of the given protein is small, then the number of entries in the dictionary will also be small. The amount of information that is needed to specify the entry measured in *bits* equals the logarithm of the number of entries. Similarly, if we knew all the possible amino acid sequences that are found in nature, we could enumerate these and make a dictionary. However, the number of these entries would be measured in many billions or trillions. Thus, taking the dictionary approach to quantifying information in the context of protein structures, we can still conclude that much more information is needed to describe their polymer sequences than to describe the environment.

19. Ellis and Hartl, 1999.

20. Sawasaki et al., 2002.

21. Fraenkel, 1993; Unger and Moult, 1993.

22. Jumper et al., 2021.

23. Anderson, 1972.

24. Ahnert et al., 2015; Marsh and Teichmann, 2015.

25. Here I refer to bacterial ribosomes; ribosomes in more complex organisms can be about one-third larger.

26. Traub and Nomura, 1968.

27. Nierhaus and Dohme, 1974.

28. Nierhaus, 1991.

29. Rong et al.,2011.

30. The attraction force of a single hydrogen bond creates a potential energy barrier about 10 times higher than typical thermal energy of a molecule at a room temperature. These barriers can hold protein molecules together sufficiently long to maintain cellular structures.

31. Adding to or removing a phosphoryl group from a molecule is a biologically important chemical reaction, used by living cells for many different purposes. There are two different classes of enzymes, kinases and phosphatases, which perform this task: kinases normally add phosphoryl groups to specific classes of molecules, phosphatases remove them. The protein shown in Figure 3.1 is one such enzyme.

32. Fang et al., 2015.

33. Any protein that catalyses the change in a chemical bond between two atoms can be considered an enzyme. The numbers above relate to the so-called metabolic enzymes—the enzymes that primarily work on small molecules (Ouzounis and Karp, 2000; Romero et al., 2005).

34. Rees et al., 2009.

35. Li et al., 2013.

36. It is worth noting that in appropriate conditions, the FtsZ proteins can form a ring on an artificial lipid membrane vehicle. Moreover, when energised, the artificial Z-ring constricts, similarly as in a living cell Osawa et al., 2008.

37. ATP stands for adenosine triphosphate, GTP for guanosine triphosphate. Conceptually, ATP (or GTP) can be in one of three states: fully charged—ATP, partly charged—ADP, and completely relaxed—AMP. Strictly speaking ATP, ADP, and AMP are three different molecules defined by the number of phosphoryl groups. The ATP molecule ($C_{10}H_{16}N_5O_{13}P_3$) has three phosphoryl groups PO_3. In ADP one is chopped off and two left, while in AMP two are chopped off and only one left. The ATP or ADP can be used to energize a protein or some other molecule, as by switching from ATP to ADP or to AMP, it releases the potential energy difference between the states, which can be transferred to the protein.

38. Rothfield et al., 2005. The principles of the mechanism behind the Min system were proposed by Alan Turing decades ago to explain how zebras get their stripes (Turing, 1952).

39. Moreover, the negative entropy term in the free energy calculations is absolutely nothing to do with the information entropy in the DNA.

40. Blattner et al., 1997.

41. Church et al., 2012; Goldman et al., 2013; Callaway, 2015; Zhirnov et al., 2016.

42. Crick, 1958.

43. Crick, 1970.

44. Crick, 1970.

45. In some textbooks, discussing the central dogma, only the part DNA → RNA → proteins is mentioned. This indeed is how Watson presented it in his famous textbook *Molecular Biology of the Gene*. However, this is not how Crick formulated this concept originally.

46. Watson and Crick, 1953.

47. Pulling apart two wires wound around each other does not work for long: the wires get tangled (one can confirm this in a simple experiment where two electric wires wound each around the other are pulled apart). To solve this problem, yet another molecular complex, topoisomerase, moves along the DNA ahead of the helicase, periodically cutting one of the strands, enabling the tangled strands to unwind, then 'soldering' the ends back together.

48. Cairns, 1963.

49. In fact, there are three DNA polymerases at work. On the so-called *lagging strand*, the polymerase is working in the backwards direction. To make this more efficient, there are two polymerases being loaded on the lagging strand periodically by yet another molecular sub-complex called a *clamp loader*.

50. Johnson and O'Donnell, 2005.

51. Baker and Bell, 1998.

52. Fowler and Schaaper, 1997.

53. Here we define *gene expression* in a rather narrow sense as transcription plus translation. Strictly speaking, *gene expression* includes other processes such as modifications inflicted on proteins after translation, or the regulation of the protein's abundance in the cell (as discussed later).

54. Sequence taken from the Saccharomyces Genome Database (SGD), PGK1 / YCR012W. By convention, the triplet codons refer to the strand complementary to the one the RNA polymerase is transcribing. This strand corresponds to the transcribed mRNA molecule and thus code for the amino acids. Also note that the triplet ATG not only signals a beginning of a gene, but also codes for the amino acid methionine. Therefore, one could expect that all proteins must begin with methionine, however there is special cellular machinery that chops the first methionine off.

55. Crick, 1958.

56. Hoagland et al., 1958.

57. To be more precise, lactose gets converted into a different molecule, allolactose, which then binds to the repressor protein, changing the shape of the repressor, and making it leave the DNA.

58. Macia and Sole, 2014.

59. Ikeda et al., 1989; Aldridge and Hughes, 2002; Macnab, 2003; Minamino et al., 2008.

60. Fiers et al., 1976.

61. Sanger et al., 1977.

62. Fleischmann et al., 1995.

63. To be precise—the result of the Human Genome Project was a 'draft' human genome, which did not represent a genome of a particular individual, but rather was a mixture of several individuals. As will be discussed in Chapter VI, individual human genomes are highly similar to each other, but not identical.

64. The 3 billion letters of a complete human genome have been printed in 24 volumes of books, on display at the Wellcome Collection at 183 Euston Road, London, UK.

65. Every protein-coding stretch in the genome begins with ATG and ends with either one of TAA, TAG, or TGA. However, if we simply slide a three-letter window through the genome, not every ATG will be a true beginning of a gene. How can one tell which ATGs are the real start signals? In bacteria, this is relatively simple: if no in-frame stop-codon appears amongst the next few hundred letters, then this is most likely to be the correct reading frame.

66. In bacteria, on average, a gene is about 1,000 nucleotides long. In multicellular species genes typically consist of pieces that code for amino acids—the so-called *exons*, interspersed with pieces that do not—*introns* (Chapter VI) and computational gene prediction becomes much more complicated that in bacteria.

67. Brenner, 2012a.

Chapter IV

1. Porter et al., 1945.

2. Kurner et al., 2005.

3. Conrad et al., 2011; Gahlmann and Moerner, 2014.

4. Kukulski et al., 2011.

5. Woese and Fox, 1977; Woese, 1987. To be precise, Woese compared sequences coding for ribosomal RNA, which can be found in virtually all known organisms.

6. For instance, *Thermotoga maritima* is a bacterial species that resides in hot springs as well as hydrothermal vents. Eukaryotic unicellular red algae *Cyanidioschyzon merolae*, live in hot springs. Archaea have been detected in soil, fresh water, and ocean (Wang et al., 2020), though less is known about the particular species.

7. Achtman and Wagner, 2008; Bochner, 2009.

8. Chisholm et al., 1988; Morris et al., 2002; Rocap et al., 2003; Giovannoni et al., 2005; Biller et al., 2015.

9. Nierman et al., 2001; Matroule et al., 2004.

10. To be precise, the strand of *M. genitalium* that was first sequenced had 482 protein-coding and 43 RNA genes (Fraser et al., 1995).

11. The definition of *free-living* varies in the literature, sometimes this term is associated with the ability to utilize nitrogen N_2 from the atmosphere (so-called *nitrogen fixation*). *P. ubique* are unable to fix nitrogen, thus by this definition it is not truly *free-living*. Here by *free-living* we mean the ability to grow and divide on simple, mostly inorganic molecules.

12. Schneiker et al., 2007.

13. Counting genes in complex multicellular eukaryotic organisms is not straight-forward due to their complex structures (Chapter VI), the number 50,000 is a debatable estimate.

14. Wang et al., 2013.

15. The dimensions of protein complexes vary widely, in *E. coli* a protein complex on average size occupies space of about 10 nm × 10 nm × 10 nm.

16. Elowitz et al., 1999.
17. Gan et al., 2008; Vollmer and Bertsche, 2008.
18. Yao et al., 1999.
19. Gan et al., 2008.
20. In the so-called *gram-positive* bacteria, the cell wall is several times thicker, but these bacteria have only one inner membrane. *E. coli* belongs to gram-negative bacteria, according to traditional microbiology classification.
21. Jones, 2004.
22. Dandekar et al., 1998.
23. Garner et al., 2011; Dominguez-Escobar et al., 2011.
24. Furchtgott et al., 2011; Amir and Nelson, 2012.
25. Typas et al., 2012; Kashyap et al., 2013.
26. For instance, in *C. crescentus* this genome location is defined by a sequence of about 250 nucleotides, containing multiple repeats of the 'motif' TTATCCAC (Mott and Berger, 2007; Haeusser and Levin, 2008).
27. In a fast-growing *E. coli*, a new round of DNA replication starts before its previous replication has finished, and thus there may be up to six replisomes working in parallel. In *C. crescentus* the genome replication starts only after the cell has divided, and thus there are only two replisomes.
28. Jun and Mulder, 2006.
29. Bier, 1997; Ptacin et al., 2010; Ptacin et al., 2014; Wang et al., 2013.
30. Erickson et al., 2010; Goley et al., 2011; Li et al., 2013.
31. Kiekebusch et al., 2012; Monahan and Harry, 2013.
32. Cho et al., 2011.
33. Marshall et al., 2012.
34. Matroule et al., 2004; Biondi et al., 2006; Goley et al., 2007; Laub et al., 2007; Shen et al., 2008.
35. Laub et al., 2000; Holtzendorff et al., 2004.
36. Laub et al., 2000.
37. Christen et al., 2011.
38. Giovannoni et al., 2005.
39. Rappe et al., 2002; Carini et al., 2013; Biller et al., 2015.
40. Morris et al., 2011.
41. Elowitz et al., 2002; Kaern et al., 2005; Rosenfeld et al., 2005; Sigal et al., 2006; Adam et al., 2008; Raj and van Oudenaarden, 2008. Proteins will not only differ in abundance, but also in sequence. The 'fidelity' of transcription is about one error in ten to hundred thousand letters, while that of translation is about one in a thousand to ten thousand letters. Given that a protein is 300–400 amino acid long, it is possible that in a cell, one protein in three may be somewhat different from what is coded in the genome.
42. Monod and Wainhouse, 1972.

43. As we will see later, once an error is passed to future generations, the only realistic way to correct it is through exchange of parts of genomes between individuals (through DNA recombination).

44. Drake, 1991; Kibota and Lynch, 1996; Lee et al., 2012; Wielgoss et al., 2011.

45. Lynch, 2010.

46. Koonin, 2000.

47. The Minimal Cell Modelling (MCM) project to design an artificial cell containing only genes needed for the basic cellular processes was proposed by Luigi Luisi from the Swiss Federal Institute of Technology in Zürich in 2002, and then developed by George Church group in Harvard (Luisi, 2002; Forster and Church, 2006).

48. Hutchison et al., 1999; Gerdes et al., 2003; Christen et al., 2011.

49. Hutchison et al., 2016. Venter had already demonstrated that such genome transplantation was possible (Lartigue et al., 2007; Gibson et al., 2010). The reason to use *M. mycoides*, rather than the smaller *M. genitalium* was that the former grows and divides much faster, making the experiments easier.

50. Bi and Lutkenhaus, 1991. The only known bacteria without this gene are two *Mycoplasma* species, which apparently divide using a different mechanism. One is *Mycoplasma pneumonae* (Lluch-Senar et al., 2010). Although many archaea species have FtsZ or a similar protein, others apparently have different division mechanisms.

51. Henderson and Jensen, 2006.

52. Krause and Balish, 2001.

53. Mercier et al., 2013.

54. Koonin, 2012a.

55. Blain and Szostak, 2014.

56. Samson and Bell, 2011.

57. Maynard Smith and Szathmáry, 1995.

58. Leigh, 1971.

59. Guerrier-Takada et al., 1983; Zaug et al., 1986.

60. Przybilski and Hammann, 2006.

61. Gilbert, 1986.

62. Blain and Szostak, 2014.

63. Szostak et al., 2001.

64. Johnston et al., 2001.

65. Wochner et al., 2011.

66. Lartigue et al., 2007; Gibson et al., 2010. Not every piece of 'foreign' DNA introduced into a cell will survive because a cell has various defence mechanisms that recognize 'foreign' DNA, chop it up, and destroy it (Marraffini and Sontheimer, 2010. But one could introduce similar 'defence mechanisms' in von Neumann's replicator too.

67. Note that another requirement was that the replicator had to contain the Universal Computer. It has been shown that enzymes existing in nature are sufficient for performing universal

computations by cutting and pasting DNA molecules, which is known as DNA-computing (Adleman, 1994; Csuhaj-Varju et al., 1995). This, however, is not what a cell normally does: cells do not process information via manipulating DNA but by means of multi-state proteins and transcription-regulation networks. Nevertheless, the need to have the Universal Computer is clearly satisfied.

Chapter V

1. Lenski et al., 1991; Good et al., 2017.

2. Edith Saunders was a prominent British geneticist working at Cambridge University in the first half of the twentieth century.

3. Any real-world replicator includes an element of stochasticity and, thus, it is not likely that all replication cycles happen in synchrony: the replication times are distributed according to the so-called exponential distribution, with an average replication time Δt. As the result, the population growth is e^n, where $n = t/\Delta t$, t is the time since the start of the replication, and e is Euler's number, which approximately equals 2.718. Since $e > 2$, the growth is faster than in the case of synchronous replication. This acceleration has a relatively simple explanation: the replicators that happened to replicate faster than average, contribute to the population growth more than the ones that are slower.

4. Lenski and Travisano, 1994.

5. Elena et al., 1996; Cooper and Lenski, 2000; Wiser et al., 2013.

6. Travisano et al., 1995.

7. Pennisi, 2013.

8. Blount et al., 2008. Speciation is a gradual process and pointing to an exact moment when a new species has emerged is difficult.

9. First, they sequenced the genome of one lineage at generations 2000, 5000, 10,000, 15,000, 20,000, and 40,000 (Barrick et al., 2009; Jeong et al., 2009).

10. Tenaillon et al., 2016; Good et al., 2017.

11. Moran, 1958.

12. Nowak, 2006.

13. The *fitness gain* is defined as $f-1$, where f is the relative fitness of the mutant, assuming the ancestral fitness equal to 1.

14. Fagundes et al., 2007; Li and Durbin, 2011; Henn et al., 2012.

15. Kimura, 1968; Kimura, 1983.

16. Wielgoss et al., 2013; Tenaillon et al., 2016.

17. Toprak et al., 2012.

18. Viklund et al., 2012.

19. Blount et al., 2012.

20. Jones et al., 1944; Frieden et al., 1993; van Soolingen et al., 1995; de Steenwinkel et al., 2012; Borgdorff and van Soolingen, 2013.

21. Galardini et al., 2017.

22. As discussed later, the assumption that roughly half of the mutations are neutral is supported by experimental evidence.

23. Ochman et al., 1999.

24. The favourite natural environment of *E. coli* is the human gut. It is estimated that every gram of faeces contains 10^7–10^9 *E. coli* cells (Tenaillon et al., 2010), and in total, there are around 10^{20} *E. coli* cells on Earth, roughly half of which live inside the gut of host organisms, the other half outside. The replication rate inside the human gut is once every couple of hours and probably slower in most other natural environments.

25. McClelland et al., 2000; McClelland et al., 2000.

26. Ochman and Wilson, 1987; Lawrence and Ochman, 1998.)

27. Henz et al., 2005; Ciccarelli et al., 2006; Letunic and Bork, 2011.

28. When *E. coli* genomes are aligned, sometimes the alignment algorithm has to 'skip' over a missing gene. For instance, it is possible, that one strain has genes ABCDFGH . . . and the other one ABDE-FGH . . . . Bioinformatics algorithms can find such 'gapped' alignments efficiently. To determine how similar two genes need to be to be considered variants of the same gene is not a simple problem. In the calculation reported here, genes sharing 90% of the DNA sequence were considered as variants of the same gene. Clearly such genes have a common evolutionary origin. It is less clear how to define 'a variant of the same gene' based on the function of this gene, particularly because proteins often have more than one function.

29. Medini et al., 2005.

30. Some theoretical predictions show that it can be expected that the *Prochlorococcus* pan-genome is 10 times as large and possibly consists of up to 60,000 genes (Baumdicker et al., 2012). The core genomes are small, pan-genomes are large. Apparently, bacteria 'do not like' to keep more genes than necessary under any particular circumstances. Having a gene comes at a cost to the organism. First, more DNA means more mutations, more of which are deleterious than beneficial. Second, expressing a gene costs energy, and for bacterial populations these costs often are high enough to be visible to the selection (Kettler et al., 2007 and Lynch and Marinov, 2015). To a larger or smaller extent, the existance of pan-genomes are a feature of all species, including humans.

31. Assessing the frequency of horizontal gene transfer events between bacteria of the same species is difficult, nevertheless, it is thought that the closer the two species, the more frequently gene transfer between them. See Barlow, 2009; Zhaxybayeva et al., 2009; Andam and Gogarten, 2011; Williams et al., 2012; Soucy et al., 2015.

32. Ochman et al., 2000; Frost et al., 2005; Lang et al., 2012.

33. For instance, over 700 genes have been introduced into *E. coli* from *S. enterica* after these species diverged 100 million years ago (Lawrence and Ochman, 1998; Ochman et al., 2000). Overall, around 16,000 DNA letters, have been introduced into *E. coli* every million years. Lateral gene transfer events have also been found within *Procholorococcus* genus species and between *Procholorococcus* and other close species (Coleman et al., 2006; Zhaxybayeva et al., 2009). At least three genes have been transferred between evolutionarily and biochemically very distant marine bacteria *P. marinus* and *P. ubique* (Viklund et al., 2012).

34. Dagan and Martin, 2007; Vos and Didelot, 2009.

35. Ohno, 1970.

36. Lynch, 2002; Conant and Wolfe, 2008; Kondrashov, 2012.

37. Romero and Palacios, 1997; Sandegren and Andersson, 2009.

38. Achaz et al., 2002.

39. The process of gradual accumulation of deleterious mutations is known as *Muller's ratchet* (named after Hermann Joseph Muller). (Muller, 1932, 1964; Lynch and Conery, 2000; Lynch, 2002).

40. Sandegren and Andersson, 2009; Kondrashov, 2012.

41. Typically, gene duplications give genes the ability to evolve towards more specialized functions. After amplification of the original gene that code for a 'generalist' protein, each of the amplified proteins evolves more specialization. For instance, enzymes often catalyse some secondary reactions in addition to the 'main' one; a duplication may allow one of the gene copies to be optimized for such a secondary reaction. Different gene families have expanded in numbers through such sub-functionalization in the *E. coli* and other species (Yamanaka et al., 1998; Serres et al., 2009; Furnham et al., 2012; Martinez Cuesta et al., 2014). About three-quarters of the ~ 300 known transcription factors in *E. coli* have arisen as a consequence of gene duplication (Babu and Teichmann, 2003). If not all, then almost all evolutionary innovations in gene function do seem to be associated with gene duplication in some way (Kondrashov et al., 2002; Conant and Wolfe, 2008).

42. In Lenski's long term evolutionary experiment, 1.4% of the *E. coli* genome was lost in 50,000 generations. The minimal genome of about 1300 genes in the minimalistic *P. ubique* is thought to have originated from bacteria with a larger genome by shedding as many as 800 genes (Viklund et al., 2012). Similarly, the minimalistic *P. marinus* strain MED4 may have lost around 40% of genes from its ancestral genome (Batut et al., 2014). Such gene loss is known as the *Black Queen* hypothesis (Morris et al., 2012; Batut et al., 2014). Bacterial cells 'want' to shed genes for proteins that they can 'steal' from other organisms in the environment.

43. Khan et al., 2011.

44. Maynard Smith, 1970.

45. Draghi and Wagner, 2008; Pigliucci, 2008.

46. Ohtsuki et al., 2006.

47. One of the largest intragenic fitness studies looking at the entire gene was done on the green fluorescent protein of jellyfish *Aequorea Victoria*, called avGFP (Sarkisyan et al., 2016). This protein emits light when excited—this is how jellyfish produce flashes of light. The experiment is based on mutating amino acids in this protein and recording how different mutations change its florescence. It turned out that about a quarter of single amino-acid mutations had no detectable effect on the fitness (they did not change the fluorescence), three-quarters of mutations showed reduced fitness, and around one-tenth lead to a completely non-fluorescent protein—0 fitness. It also turned out that accumulation of mutations makes the protein 'fall off a cliff'—any combination of 10 mutations made the protein to lose its fluorescence completely.

48. Kauffman, 1993.

49. Phillips, 2008.

50. Tong et al., 2004; Kvitek and Sherlock, 2011; Nichols et al., 2011; Woods et al., 2011; Ryan et al., 2012; Covert et al., 2013; Babu et al., 2014; Fischer et al., 2015.

51. Imhof and Schlotterer, 2001; Perfeito et al., 2007; Sniegowski and Gerrish, 2010.

52. Stevens and Sebert, 2011.

53. Eigen and Schuster, 1979.

54. Maynard Smith, 1983.

55. We can assume without loss of generality that the death rate c is the same in both populations x and x'. This is because we can adjust the birth rates f and f' to fit this assumption.

56. Given that $\Delta t > 0$ and $x > 0$, from the first expression we have $fQ > c$. Now given that $c = (fx + f'x')/(x + x')$, we obtain $Qx + Qx' \geq x + (f'/f)x'$ and given that $Qx < x$, obtain get $Q > f'/f$.

57. Here we interpret $\log(f)$ as the natural logarithm of f, sometimes denoted by $\ln(f)$, however the particular base of the logarithm does not matter much, as changing the base only introduce a multiplicative constant.

58. For instance, we can assume $f = e$, where f is the relative fitness of the best replicator in comparison to the mean fitness of all other replicators and e is Euler's number (~ 2.7). We do not explicitly use the fact that there are only a small number of positive mutations, instead, we assume that the fitness of the best replicator is approximately e times the mean of the rest of the population.

59. As will be discussed in Chapter VII, strictly speaking we should say '*one mutation per functional part of the genome*'.

60. Jee et al., 2016.

61. As we will see in Chapter VI, in multicellular organisms we can also talk about so-called somatic inheritance, which happens as the cells divide within an organism. In somatic inheritance the number of generations is limited, and thus Eigen's threshold does not apply.

62. Laland, 2018.

63. Kawecki et al., 2012.

Chapter VI

1. Wootton, 2015.

2. Rudwick, p, 35.

3. Hutton's work was subsequently published in a book, *Theory of the Earth; or an Investigation of the Laws observable in the Composition, Dissolution, and Restoration of Land upon the Globe* (see Baxter, 2003). Hutton suggested that the processes of formation of geological structures such as rocks must have been slow and hypothesized that the same processes that were shaping the early Earth were still in operation to date. He called this the principle of *uniformitarianism*—a fundamental principle of geology that explains the features of the Earth's crust via natural processes over the geologic timescale. He concluded that Earth must be millions of years old or even older. Even though Hutton did not manage to convince his peers to pay his theory the attention he believed it deserved, the years of believing in a young Earth were numbered. After the UK lawyer and geologist Sir Charles Lyell published *Principles of Geology* popularizing Hutton's work in 1830–1833, Hutton's uniformitarianism became widely accepted.

4. The idea that the Earth formed from gaseous clouds slowly rotating and gradually collapsing due to gravity was first proposed by Immanuel Kant in 1755, and later independently by the French scholar Pierre-Simon Laplace in 1976, who worked this hypothesis out in more detail.

5. Burchfield, 1975.

6. Dalrymple, 2001; Wilde et al., 2001.

7. See Oehler and Schopf, 1971; Schopf and Packer, 1987; Schopf, 1993; Knoll, 2003; Brasier et al., 2015. Others argue that the earliest really undisputable fossils of life are only 2 billion years old, e.g. Barghoorn and Tyler, 1965; Moorbath, 2005.

8. Kamminga, 1988.

9. Luisi, 2006.

10. Koonin, 2012b.

11. Budin and Szostak, 2010.

12. Nowak and Ohtsuki, 2008.

13. Nowak and Ohtsuki, 2008.

14. Similarly, as in Eigen's model, Nowak assumed that the efficiency of each reaction depends on the particular polymer p and that they happen at rates a_{p0} and a_{p1}, respectively. All polymers decay at the same rate d. Thus, if we denote the number of polymers p in the media by x_p and the number of predecessors of sequence p' by $x_{p'}$, then we can express the rate of change Δx_p in time period Δt via a system of equations, not dissimilar to the ones in Box 5.2. It can be demonstrated mathematically that there is a selection of the fittest polymers going on, even though there is no replication.

15. Derr et al., 2012.

16. Obermayer et al., 2011.

17. The RNA World, 2006; Schrum et al., 2010.

18. Benner et al., 2006.

19. Kauffman, 1993; Dyson, 1999.

20. Maynard Smith and Szathmáry, 1995, pages 67–72.

21. Vasas et al., 2012.

22. Joyce and Orgel, 2006.

23. For instance, if we have a complex consisting of four RNA molecules of 10 letters each, then 4×4^{10} different letter combinations possible, describing which requires only 24 *bits* of information rather than 80.

24. Kun et al., 2005.

25. Schrum et al., 2010.

26. Maynard Smith and Szathmáry, 1995.

27. Joyce and Orgel, 2006.

28. Cairns-Smith, 1966.

29. Martin et al., 2008; Lane, 2015.

30. Lane (2015).

31. Proskurowski et al., 2008; Dodd et al., 2017.

32. $10^7 \times 10^{18} \times 365 > 10^{27}$.

33. $10^{27} = (10^3)^9 \approx (2^{10})^9 = (4^5)^9 = 4^{45}$.

34. For estimates, see Eigen, 2013, page 545.

35. Charlebois and Doolittle, 2004; Koonin, 2012b.

36. O'Malley, 2010.

37. Brocks et al., 1999; Cavalier-Smith, 2009; Eme et al., 2014.

38. Cavalier-Smith, 2009.

39. Arendt et al., 2015.

40. To be precise, in addition to the 23 chromosome pairs, humans also have so-called sex chromosomes X and Y.

41. In some species mitochondria have been degraded to smaller organelles, known as mitochondria-like small organelles.

42. It has been suggested that mitochondria enabled significant intensification of the ATP production and thus made the subsequent evolution of complex multicellular organisms possible (Lane and Martin, 2010). This however has been questioned. The membrane synthesis is one of the most energy-consuming process in a cell, and given that mitochondria have large internal membrane surfaces to run their ATP 'chargers', the total energy gain seems to be small (Lynch and Marinov, 2016). A 'mitochondria-late' hypothesis states that mitochondria joined an already well-formed eukaryotic cell (Pittis and Gabaldon, 2016). Indeed, eukaryotes harbour bacterial genes from many different bacteria, not exclusively alpha-proto bacteria. This though can also be explained by horizontal gene transfer, which could have happened later.

43. Williams et al., 2013; Poole and Gribaldo, 2014.

44. Spang et al., 2015.

45. Jorgensen et al., 2012.

46. Koonin et al., 2004; Koonin, 2012b.

47. Jablonka and Raz, 2009.

48. There are even more cell types if we consider differences between cells at the molecular level. Defining a cell type in a complex organism is difficult. In principle, ultimately there is no other way to define cell types than by comparing their epigenomes. A large-scale project, known as the Human Cell Atlas, to define all human cell types was started in 2017. Until the project is completed, we will not know how many different cell types a human body has.

49. Bell and Mooers, 1997; Knoll, 2011.

50. Plaga and Ulrich, 1999; Velicer, 2003; Kolter and Greenberg, 2006; Berleman and Kirby, 2009; Flores and Herrero, 2010.

51. Grosberg and Strathmann, 2007.

52. These are *Agaricomycotina*, which include the vast majority of edible mushrooms, and *Pezizomycotina*, which includes penicillin-producing fungi (Stajich et al., 2009).

53. Flores and Herrero, 2010; Knoll, 2011.

54. Knoll, 2011; Sharpe et al., 2015.

55. Herron et al., 2009.

56. West et al., 2015; Shik et al., 2012.

57. Williams, 1966; Dawkins, 1976; Dawkins, 1982.

58. Ohtsuki et al., 2006.

59. Cooperation works in a straightforward way only if there are no cheaters. If one cell 'decided' not to synthesize that enzyme, it could save energy, which it can redirect to faster replication, while still benefiting from the partially digested molecules. To become a cheater, a genetic mutation would

need to happen, but after such a mutation the genomes are no longer identical. It is not impossible that cheating first emerges as an epigenetic trait, however it would have to be stably inherited for a significant number of generations, and arguably, it would have to show up in the genome eventually. Such cheating indeed can happen; for instance, this is what causes cancer. Here we are back to the problem of biochemical communism. Animals have evolved mechanisms to recognize such cheaters and to try to kill them.

60. During the formation of the fruiting body, depending on the particular conditions of development, up to 90% of the cells commit 'suicide' (Dworkin, 1996; Velicer, 2003; Bayles, 2014).

61. Mechanisms by which a few bits of information can be encoded epigenetically and passed on from a mother to the daughter cell are discussed in Chapter IV; for instance, this can be achieved via the abundance of a particular transcription factor. If the transcription factor positively regulates its own transcription, the information encoded in its abundance in a cell may be quite persistent and stable across many cell divisions. However, the amount of information that can be encoded in this way reliably will only be one *bit*—is the concentration of this transcription factor high enough to switch on its own transcription? More than one bit can be encoded and passed on epigenetically by combinations of transcription factors. Two transcription factors can be combined in four different ways, three in eight, and so on. Such combinatorial epigenetic memory is one of the necessary requirements for the maintenance of different cell types of complex multicellular organisms.

62. Prochnik et al., 2010.

63. Grosberg and Strathmann, 2007.

64. King, 2004; Srivastava et al., 2010.

65. Sebe-Pedros et al., 2016.

66. Rohland et al., 2007; Miller et al., 2008.

67. Scally and Durbin, 2012.

68. One could also ask why there is a need for two sexes; why not allow any two individuals to exchange genetic information? This would create a conflict between individuals. If two different genomes compete for resources, why would they voluntarily exchange information that helps the other to perform better? Having two separate sexes, which have to cooperate to reproduce, resolves this conflict. See Maynard Smith, 1971 and 1978.

69. See MacKay, 2003, page 270. Given that the human genome has ~ 20,000 protein coding genes, on average each with more than 1000 nucleotide-long coding part (exons), we can estimate that the 'effective' coding part of the human genome is at least a million nucleotides. The square root of 1 million is a thousand, so under these assumptions, sex speeds up human evolution about a thousand times.

70. Kondrashov, 1988.

71. McDonald et al., 2016; Goddard, 2016.

72. Dawkins, 1982, page 131.

73. For discussion see Ågren, 2021.

74. Attributed to Lewontin, in Dawkins, 1982, page 131.

75. See Ågren, 2021.

76. de Silva et al., 2004.

77. Traulsen and Nowak, 2006.

78. Lander et al., 2001.

79. Koonin and Wolf, 2009.

80. Michael Lynch emphasizes the importance of neutral drift in the explanation of genome expansion (see Lynch, 2007). This, however, is in no contradiction with the 'selfish' genomic elements being parasitic. Small population sizes of species with larger individuals keep the potentially negative effects on individual's fitness under the selection horizon. Note that the population of human species is not small anymore, but the growth in its size is a phenomenon of the last few thousand years associated with rapid cultural evolution.

81. Williams, 1992, page 10.

82. In reality these are only atoms that interact. Information is an abstract concept that models an emerging property of complex arrangements of atoms, where it is not the individual atoms that matter most but how different atoms relate to each other. Often the atoms can be replaced by different ones in a systematic way, but the result of the interactions remain the same. Then we say that the information encoded is the same.

83. Han and Runnegar, 1992; Knoll and Carroll, 1999; King, 2004; King et al., 2008; Nielsen, 2008; El Albani et al., 2010; El Albani et al., 2010; Donoghue and Antcliffe, 2010; Knoll, 2011.

84. King et al., 2008; Dayel et al., 2011; Arendt et al., 2015.

85. Arendt et al., 2015.

86. Gilbert et al., 2019.

87. See Knoll and Carroll, 1999. Scientists are still debating what caused this diversification (Morris, 2000; Peters and Gaines, 2012). One hypothesis is that the driver of the Cambrian explosion was increasing levels of oxygen dissolved in the ocean. According to geological records, increased oxygenation of the ocean indeed happened just before this period. Oxygen is needed for efficient utilization of chemical energy and at the time, most life on Earth was in the oceans. Animals need oxygen, for instance to be able to 'burn' energy fast enough to move around. The driving role of oxygen in the Cambrian diversification has been questioned lately though, and a range of other, more subtle factors have been suggested, without denying that the rising oxygen levels could have played a major role (Fox, 2016). Another reason for this sudden appearance of fossil diversity may be the ascertainment bias. One important Cambrian innovation was hard skeleton, which opened enormous evolutionary possibilities; 80 different types of skeletons can be found in the Cambrian fossils. The emergence of hard skeleton may cause an ascertainment bias in the fossil record, as such skeletons are preserved much better than soft tissues.

88. Arendt, 2008; Srivastava et al., 2010; Galliot and Quiquand, 2011; Sebe-Pedros et al., 2016.

89. Knoll and Carroll, 1999.

90. Herrou and Crosson, 2011. Animal photoreceptors evolved more than 600 million years ago, the eye probably about 540 million years ago (Lamb et al., 2007); the early bilaterians already had photoreception organs ((Arendt and Wittbrodt, 2001). New genes can be associated with increasing complexity of eye in subsequent evolutionary steps (Arendt, 2003) and many successive evolutionary innovations over half a billion years gradually led to the sophisticated vertebrate eye (Arendt, 2003; Lamb et al., 2007). Lampreys, which have eyes quite like human ones, evolved about 500 million years ago. See also Nielsen, 2008; Galliot and Quiquand, 2011; Arendt et al., 2015, 2016.

91. Note that artificial 'neurons' typically are modelled using a general-purpose computer (or to be precise, graphical processing units or GPUs) rather than implemented as specialized hardware.

92. Watson and Szathmary, 2016.

93. Gremer et al., 2012.

94. Zliobaite and Stenseth, 2016.

95. Watson and Szathmary, 2016.

96. Christensen et al., 2015.

97. Seyfarth et al., 2010.

98. Geyer and Van der Zouwen, 1994.

99. Camazine, 2003.

Chapter VII

1. A different type of a 'telegraph', an optical semaphore system, was built in France towards the end of the eighteenth century by the French engineer Claude Chappe, which eventually included 556 semaphore stations. In 1794 a semaphore line informed Parisians of the capture of Condé-sur-l'Escaut from the Austrians, less than an hour after the event. The information capacity of this system was relatively limited; the age of fibre-optic cables was still 200 years away.

2. International Telecommunication Union and Michaelis, 1965).

3. International Telecommunication Union and Michaelis, 1965.

4. Transmitting of information, like storing it, has to have a material manifestation. For fast information transmission this is provided by electromagnetic field, or in its quantum mechanical description—by photons. Electromagnetic field can transmit information with the speed 3×10^9 m/s, which for the distances on our planet in practice is virtually in an instant. For satellite communications however the limitations are notable.

5. Nyquist, 1928.

6. Hartley, 1928.

7. Shannon, 1948, 1949.

8. Shannon considered a stochastic process generating messages according to some probability. distribution. Message i occurs with a probability p_i. Shannon defined the concept of (*information*) *entropy* as $H(i) = -p_i\log(p_i)$ summed over all possible messages i. This can be shown to characterize the uncertainty of the process generating the messages; the entropy is the highest if all messages have the same probability $p_i = 1/n$, where n is the number of different messages. Now suppose there is a linked stochastic process generating messages j and that knowing the message j may tell us something about the message i. That is, if we know the message j, we know that the probability of message i is p_{ij}. If $p_{ij} > p_i$, then the message j has told us that the message i is more probable than if we did not know about j, in other words, j contains information about i. We can introduce the *conditional entropy* $H(i|j)$ of i given message j. It can be shown that if both stochastic processes generating i and j are independent, then $H(i|j) = H(i)$, while if they are not independent, then $H(i|j) < H(i)$. Shannon measured the *information* $I(j;i)$ contained in the messages j about the messages i, as the reduction in entropy $H(i)$ provided by knowing the j, concretely $I(j;i) = H(i)—H(i|j)$. Thus, if the stochastic processes generating i and j are independent, there is zero information in j about i, otherwise j contains some information about i.

9. Maynard Smith, 2000.

10. Crick et al., 1961.

11. Donaldson-Matasci et al., 2010; Rivoire and Leibler, 2011.

12. Apparently, this is not exactly what Kelvin said; those who have studied his biography claim that the accurate quotation is: '*I often say that when you can measure what you are speaking about, and express it in numbers, you know something about it; but when you cannot measure it, when you cannot express it in numbers, your knowledge is of a meagre and unsatisfactory kind; it may be the beginning of knowledge, but you have scarcely, in your thoughts, advanced to the stage of science, whatever the matter may be.*' (Lecture on electrical units of measurement, 3 May 1883, published in *Popular Lectures* Vol. I, page 73). This can be interpreted in different ways. One is that in hard sciences to know means to know how to measure. Whether one agrees with this or not, most will agree that '*when you can measure what you are speaking about . . . you know something about it*'. Importantly for later discourse, to compare the values of a certain entity (such as the complexity of an organism), we do need to know how to measure this entity.

13. The ideas about mechanical energy and its conservation can be traced back to the beginning of the nineteenth century, and in particular, to the works of the English physician Thomas Young. The law of energy conservation was first clearly formulated by the German physicist Hermann von Helmholtz in 1849, almost exactly 100 years before Claude Shannon's proposed how to quantify information.

14. In the flat-pack furniture example, a straightforward application of Shannon's approach would effectively mean assuming that instruction manuals for assembling all possible furniture are available (e.g. in one's drawer), and that one only needs to know which manual to use. The amount of information needed to choose the correct manual is what Shannon's information would give us. If there were only one piece of furniture possible, one would need no information at all. In fact, we used similar reasoning in Chapter III when discussing how to quantify information in the environment about a protein's three-dimensional structure. For instance, a prion protein has two possible structures, and which particular one it folds into depends on the environment. We concluded that in this case, 1 *bit* of information was needed to specify the environment. The justification of this approach was that a protein folds in a particular way consistent with the laws of physics and we were not trying to quantify information in the laws of physics.

15. The presented reasoning only shows the relationship $m = \log_2 k$ in case when k is a power of 2, however Shannon showed that this is true for any k greater or equal to 2.

16. We could, for instance, use 00 to denote 'L', 01 to denote 'R', 100 to denote "(", 101 to denote ')', and 11 to denote '*'. Thus, the sequence (LR)* would be written as 100000110111 and thus will take no more than $3 + 2 + 2 + 3 + 2 = 12$ *bits* to encode it, which is less than the 14 *bits* of sequence of LRLRLRLRLRLRLR. Note the use a so-called *prefix code*—none of the strings used is a prefix of any other one. This allows a unique 'decoding' of these *bit* strings to the original five symbols when reading from left to right.

17. To be more precise, Kolmogorov defines complexity $K(x)$ of a string of symbols x, as the shortest binary string that given as an input to the Universal Computer produces x. Next, the conditional complexity $K(x|y)$ of x given another string y, is defined as the shortest binary string restoring the string x, assuming that the Universal Computer has y at its disposal. Then, the information that y contains about x is defined as $I(y; x) = K(x) - K(x|y)$ (see Kolmogorov, 1965). Although there is a similarity with Shannon's definition, Kolmogorov's definition does not depend on probability distributions and is applicable to an individual sufficiently long string. Note that in some contexts y can be viewed as prior knowledge. As already discussed in Chapter II, the Kolmogorov complexity of x can be seen as the length of the most compressed form of x. Information in the string y about

the string x can be viewed as the gain in compression that can be achieved given that y is known. It has been proven that Kolmogorov's definition of information, which came about 15 years after Shannon's work, is more general (see Li and Vitányi, 2008). Shannon assumes that the messages are generated randomly in a so-called Markov process (a particular type of a stochastic process); Kolmogorov allows for an arbitrary algorithm.

18. Kolmogorov, 1965, 1968.

19. More precisely, mutual information between two messages is defined as the gain in compression that can be achieved on one of the sequences, given that we know the other. The instruction '*use the sequence of instructions you have except for the last junction*' is information, however this piece of information is of a constant length, independent on the labyrinth. The information to encode '*use the sequence . . .*' will stay the same even if the sequences s and t were much longer; in other words, this only adds a constant 'overhead'. The longer the sequence s, the smaller the overhead in comparison to s. When s is very long (when the length of s tends to infinity), the overhead becomes negligibly small. Ignoring this constant overhead, we can say that there are about 11 *bits* of mutual information between s and t: given t we can restore s with just 1 additional *bit*.

20. Hofstadter, 1979, page 170

21. Kolmogorov, 1965, 1968.

22. VanRullen and Koch, 2003.

23. Zhirnov et al., 2016.

24. One could think that there is only a practical rather than fundamental reason why artificial neural nets are not implemented physically as analogue machines, but this is rather debatable. Given the omnipresent thermal noise, for such an analogue implementation to achieve the necessary accuracy, the computational processes would need be run in replicates and averaged (see Boybat et al., 2018; Garg et al., 2022). But then we are back to the arguments around the analogue clock in Figure 7.1.

25. This does not mean that there cannot be information about probabilities. For instance, statement such as 'there is a 40% possibility of rain' is information. Note that the statement itself is not fuzzy, even though its interpretation may be.

26. Gould and Gould, 2007.

27. Note that the genome may not necessarily be the shortest description of these instructions—genome sequences can often be compressed. As the cell is not likely to have the 'prior knowledge' necessary to decompress any compressed genome sequence, the cell needs the sequence of nucleotides as they are arranged in the genome.

28. Forster and Church, 2006.

29. Transgenerational inheritance is not the same as parental or grandparental inheritance, which refers to features that disappear after one or two generations (Grossniklaus et al., 2013). For instance, in mammal pregnancy, the mother and foetus are both exposed to the same environmental influences, and so are the foetus' primordial germ cells, which will eventually produce the grandchildren. Thus, we should not necessarily be surprised if grandchildren are affected by the environmental exposures of their grandparents; for instance, in grandparents affected by famine. One example sometimes mentioned in the context of epigenetic transgenerational inheritance in humans is related to the Second World War 1944/1945 winter famine in the Netherlands—the grandchildren of this famine's sufferers exhibit a range of specific phenotypes. However, there is no consensus

amongst scientists regarding the interpretation of these observations and a close look seems to suggest that the reported phenomenon is likely to be a grandparental effect (Heard and Martienssen, 2014).

30. Cubas et al., 1999; Manning et al., 2006; Heard and Martienssen, 2014.

31. Reik et al., 2001; Morgan et al., 2005.

32. Morgan et al., 1999.

33. Jablonka and Raz, 2009. The examples are grouped in four categories: self-sustaining loops, structural inheritance based on spatial templating, chromatin marking, and RNA-mediated epigenetic inheritance. Self-sustaining loops are inherited concentrations of transcription factors that positively regulate their own transcription, thus providing a stable inheritable state. Structural inheritance includes the inheritance of prion protein states (Chapter III), as well as inheritance of membrane-based structures. Chromatin-marking-based inheritance includes trans-generationally inherited DNA-methylation patterns, as in *agouti mice*. Finally, RNA-mediated epigenetic inheritance is based on the presence, absence, or abundance levels of inherited RNA, typically short RNA molecules. These are known to play role in some animals, particularly in *C. elegans*. The most diverse epigenetic inheritance patterns seem to be present in the nematode worm *C. elegans*, but they are likely to be examples of environmentally triggered epigenetic states, the mechanisms of which are determined genetically (Remy, 2010; Greer et al., 2011).

34. Jablonka, 2002.

35. In 2013, the high impact scientific journal *Nature Reviews Genetics* asked five prominent epigeneticists to answer questions about how much transgenerational epigenetic inheritance takes place in nature and what impact it has (Grossniklaus et al., 2013). The answers were somewhat varied, though when analysed carefully, none of them really expressed support for anything that would break the paradigm of Darwinian model. As was noted by one of the respondents, *'many studies [claiming inheritance of acquired traits] failed to exclude parental effects, were not sufficiently replicated to distinguish stochastic effects from treatment effects or did not integrate phenotypic characterizations with more detailed molecular and genetic analyses'*. Not all respondents were this sceptical; the opposite views were also expressed (even though some of the examples provided seemed to result from not distinguishing between grandparental and transgenerational inheritance). It was also noted that if an existing adaptive mechanism, encoded genetically, is induced by the environment, and then maintained for generations epigenetically, which indeed happens (so-called environmentally triggered epigenetic states), such phenomena do not break the Darwinian paradigm. On the other hand, some of the examples that are given as potentially breaking the Darwinian paradigm in a rather limited sense, such as stress-induced hypermutability, do not necessarily require epigenetic mechanisms.

36. Heard and Martienssen, 2014.

37. In theory, significant amounts of information could be passed on from generation to generation via the genome methylation patterns. For instance, the promoter (the genomic regulatory region) of a gene can be in either a methylated or demethylated state and can thus carry 1 *bit* of information. Theoretically, in the human genome, thousands to tens of thousands of *bits* could be passed on this way. But even then, this is much less than what is transmitted between human generations genetically. We estimated that genetically each gene caries about 800 *bits* of information, while the promoter methylation status for this gene carries just 1 or a few *bits*. As discussed in the previous sections, without combinatorial information-encoding mechanisms, no more than a few *bits* of

information can be encoded. Although epigenetic marks indeed are used combinatorically in somatic epigenetic inheritance, these are not used widely in transgenerational inheritance.

38. A single methylated nucleotide typically does not make a difference; what matters for gene regulation is regions of hundreds of nucleotides consistently methylated. The methylation 'status' of such a longer region is probably more stable than a single nucleotide methylation, but still the fidelity of long-term maintenance of such regions depends on the fidelity of individual methylated positions. Moreover, if information is coded by genomic regions rather than single nucleotides, the information density is smaller, and thus the channel capacity is smaller too. Information transmission via other mechanisms, such as chromatin marks, is even less stable (Richards, 2006; Reik, 2007).

39. Laland and Sterelny, 2006.

40. Laland et al., 1999; Post and Palkovacs, 2009.

41. Post and Palkovacs, 2009.

42. Williams, 1966, page 16.

43. Maynard Smith and Harper, 2003; Dawkins, 2004; Searcy and Nowicki, 2005; Laland and Janik, 2006; Danchin et al., 2011; Laland et al., 2014.

44. Fragaszy and Perry, 2003; Whiten, 2005.

45. Note that '*acquire in part through socially aided learning*' remains undefined. *Social* aspect and *culture* may be regarded as closely related, thus there is an element of circularity in this definition. One could define '*acquired through socially aided learning*' as what is NOT acquired through genetic or epigenetic inheritance or predetermined by the particular environment. For instance, some aspects of bird songs are determined genetically, some learned socially. There is still a weak spot in this definition: '*predetermined by the particular environment*' cannot always be distinguished from '*acquired socially*'. For instance, normally a newly fledged bird is in the environment where its parents sing. On the other hand, animals cannot crack nuts if there are no nuts in their habitat. Although this distinction is not clear cut, ingenious experiments have been carried out in animal behavioural studies, often proving one or the other beyond doubt.

46. Whiten et al., 1999.

47. Whiten, 2005.

48. Lonsdorf et al., 2004.

49. Biro et al., 2003.

50. Horner et al., 2006.

51. Hoppitt et al., 2008.

52. Beecher and Brenowitz, 2005.

53. Kroodsma et al., 1997; Leitner et al., 2002.

54. Marler and Tamura, 1964.

55. Wheatcroft and Qvarnström, 2017.

56. Gentner et al., 2006; Berwick et al., 2011; ten Cate and Okanoya, 2012.

57. Templeton et al., 2005.

58. Engesser et al., 2016.

59. Struhsaker, 1967; Maynard Smith and Harper, 2003.

60. Arnold and Zuberbuhler, 2006; Leroux et al., 2023.

61. Maynard Smith and Harper, 2003.

62. Azevedo et al., 2005.

63. Bonner, 1988.

64. McShea, 1996; Carroll, 2001.

65. Prochnik et al., 2010.

66. Lynch, 2007.

67. Sebe-Pedros et al., 2016.

68. McShea, 1996.

69. Rensch, 1948; Bonner, 1988; Bonner 2004.

70. Turner, 2011.

71. Malinsky et al., 2015.

72. Rainey and Travisano, 1998.

73. From Gould, 1996.

74. Williams, 1966, page 34.

75. Holland, 1973, 1992; Grefenstette, 1993; Mitchell, 1996.

76. For instance, if the first sequence is 00000 and the second is 11100, then using the first method and assuming the breakpoint at position 2, the crossover results in a new sequence 00100. Using the second method a recombination can result in any sequence ending in 00.

77. Note that if we accept Kolmogorov's definition of information, no algorithm can create new information—whatever sequence of *bits* the algorithm generates, this sequence contains no more information than what is needed to describe the algorithm. This is the essence of Kolmogorov complexity—there is only as much information in a sequence as in the shortest algorithm that can generate this description and thus an algorithm cannot generate new information. However, our algorithm is learning from the interactions with the environment, and thus taking the information from the environment and concentrating it into the genome.

78. MacKay, 2003, page 269.

79. Kimura, 1961.

80. Maynard Smith, 2000. As demonstrated more formally by David MacKay, Maynard Smith's conclusion is indeed true in the absence of crossover.

81. Maynard Smith, 2000.

82. Leff and Rex, 1990; Otsuka and Nozawa, 1998; Adami et al., 2000; Davies, 2019.

83. Norton, 2005.

84. Bier and Astumian, 1996; Bier, 1997; Ptacin et al., 2010.

85. Davies, 2019.

86. Such a simple compression would not distinguish between functional information and junk DNA, and thus would only provide the upper bound. To lower it, we would need to need to distinguish between two types of genetic variability: first, 'junk', which is not 'visible' to evolutionary selection and therefore is likely to be caused by neutral drift; and second, the variability in the genome

positions that have an impact on the individual's phenotype, and thus contribute to functional genome information. When estimating the amount of functional information in the genomes of a population, the first type of variability can be discarded, the second needs to be included.

87. Pennisi, 2017; Lewin et al., 2018.

Chapter VIII

1. Chimpanzees and bonobo lineages began diverging from each other about 2 million years ago, which is after their common ancestral lineage had already diverged from that of hominins, one branch of which evolved into humans.

2. Hauser et al., 2002. I regard Noam Chomsky's political views as insane but this does not change his contribution to science.

3. Bond et al., 2004.

4. Matsuzawa, 1985; Kawai and Matsuzawa, 2000.

5. Chomsky, 2015.

6. Berwick and Chomsky, 2016.

7. Yang et al., 2017.

8. Truppa et al., 2010.

9. In 1969, in an article published in the Russian journal *Mathematics in School,* Andrey Kolmogorov, wrote that 'creatures' in a finite world with many objects might be forced to invent infinity (Математика в Школе (1969) 3: 12–17).

10. Pinker and Jackendoff, 2005.

11. Fitch et al., 2005.

12. Jackendoff and Pinker, 2005.

13. Bickerton, 2007.

14. Bickerton, 2009.

15. Neurobiologists know a great deal about which human brain regions are associated with language, thus the shapes of skulls of early humans may give us some indication of the individuals' potential for language. However, this does not tell us if the skull 'owner' could indeed talk.

16. Fitch, 2000.

17. Podnieks, 2009.

18. Hauser et al., 2014.

19. Fitch, 2010.

20. Phonemes are distinct units of sound, distinguishable in a particular linguistic context, that is, by speakers of a particular language or language group.

21. Nowak and Krakauer, 1999; Nowak and Komarova, 2001.

22. Fitch, 2010.

23. Cognitive scientists talk about the *semiotic triangle*: (i) the objects and events in the real world; (ii) the concepts in the human mind related to these object or events; and (iii) the signals, such as words and signs, mapping on to these. In the *realist model,* there is a direct link between the objects/events in the world and the signals, while in the *cognitive model,* the link comes exclusively via the concepts in one's mind.
24. Miller, 1956.
25. Bickerton, 1990, page 224.
26. Laland, 2018.
27. Chomsky, 1965.
28. Fitch, 2010.
29. Let me note here that developing an algorithm for learning hierarchically nested *and so on* expressions from their finite examples was the topic of my PhD thesis.
30. Gold, 1967.
31. Nowak, 2006, page 263.
32. Fitch, 2010.
33. Langergraber et al., 2012.
34. Stringer, 2011.
35. Klein, 2017.
36. Langergraber et al., 2012.
37. Schlebusch et al., 2017.
38. Stringer, 2016.
39. The first sustained migration into the Middle East, possibly happened 70,000–80,000 years ago, and some migration even earlier, for instance, the first modern humans might in fact have reached Greece as early as 200,000 years ago (Harvati et al., 2019). Genome comparisons also show that the split between the ancestors of the contemporary African and non-African human populations might have happened already by 90,000 years ago as a split between *H. sapiens* Eastern and Western African populations, long before the expansion of the first into Eurasia (Scally and Durbin, 2012; Stewart and Stringer, 2012).
40. Higham et al., 2014.
41. Prufer et al., 2014.
42. Harmand et al., 2015.
43. These are the so-called Acheulean technologies.
44. Bickerton, 1990.
45. Stringer, 2011, page 241.
46. Klein, 2017.
47. Stringer, 2011.
48. Fitch, 2000, page 331; Dediu and Levinson, 2013.
49. Jackendoff, 1999.

50. Bolhuis et al., 2014; Klein, 2017; Tattersall, 2017.

51. In his book *Adam's Tongue*, Bickerton raises an interesting question: what could these first 'words' actually have been? We may never know, but we can study a less elusive question: what the minimal vocabulary that is useful in a particular context, for instance for scavenging, is?

52. Mallick et al., 2016.

53. Fagundes et al., 2007; Li and Durbin, 2011; Henn et al., 2012.

54. Atkinson, 2011; Perreault and Mathew, 2012.

55. It is likely that Neanderthals and anatomically modern humans or their direct ancestors lived side by side in different parts of the world at different times, thus their overlap 45,000–40,000 years ago possibly was only one in a series of recurrent interactions.

56. Chimpanzee and Analysis, 2005; Locke et al., 2011; Scally et al., 2012; Prüfer et al., 2012.

57. Several genes related to brain development have been amplified in humans (overall, there are over 30 genes that have been amplified specifically in the human lineage), while the genes related to the sense of smell have more members in the chimp genome (see Sudmant et al., 2013).

58. Having a large brain is costly—in humans, 20% of total energy consumption is taken up by brain and 40% of energy consumption at rest (Raichle and Gusnard, 2002). This does not help with fitness; therefore, either the information processing abilities enabled by a larger brain should compensate this spending of energy or a more efficient energy metabolism had to evolve in parallel to a larger brain. In the lineage leading to modern humans, most likely both happened.

59. Tattersall, 2008. Neanderthals had a slightly larger brain than modern humans, moreover for the last 100,000 years the average brain size of *H. sapiens* has slightly decreased.

60. Fiddes et al., 2018.

61. Mochida and Walsh, 2001.

62. Evans et al., 2005; Tang, 2006.

63. Dennis et al., 2012.

64. Enard et al., 2002.

65. Enard et al., 2009.

66. Paabo, 2014.

67. Noonan et al., 2006; Green et al., 2006; Prufer et al., 2014.

68. Bickerton, 2009, page 231.

69. Somel et al., 2013; Sousa et al., 2017.

70. Blum, 1963.

71. Schuster et al., 2010; Li and Durbin, 2011; Scally and Durbin, 2012 1000 Genomes Project et al., 2015.

72. Reich, 2018.

73. Williamson et al., 2007; Mathieson et al., 2015.

74. Jablonski and Chaplin, 2000.

75. Thompson et al., 2004.
76. Tishkoff et al., 2007; Bersaglieri et al., 2004.
77. Mekel-Bobrov et al., 2005.
78. Okbay et al., 2016.
79. Kong et al., 2017; Beauchamp, 2016.
80. Schmandt-Besserat, 1992.
81. Schmandt-Besserat, 1978, 1992; Zimansky 1993.
82. Cooper, 2004, page 71.
83. Englund, 2004.
84. Baines, 2004, page 151.
85. Trigger, 2004, page 61; Baines, 2004, page 175.
86. Houston, 2004.
87. Ifrah, 2001.
88. Trigger, 1998.
89. Gelb, 1963.
90. Houston, 2004.
91. Robertson, 2004.
92. Trigger, 2004, page 48.
93. Schumacher et al., 2007.
94. Although it is thought that the Chinese writing system emerged independently of other writing systems, it cannot be ruled out that it was influenced by the Mesopotamian/Egyptian writing. Given that Mesopotamian and Egyptian systems emerged only a couple of hundred years apart from each other in geographically proximal regions, it has been argued that there may be a link (Houston, 2004).
95. Jean, 1992.
96. Ifrah, 2001.
97. Moore first estimated the doubling time as a year but later revised it to two years.
98. Wu et al., 2022.
99. Goldman et al., 2013.

Chapter IX

1. Ifrah, 2001.
2. Ifrah, 2001.
3. The mass of the Sun is of about 2×10^{27} tonnes, thus if the Sun had been powered by burning coal, it would burn out in less than 10,000 years.

Bibliography

1000 Genomes Project Consortium (2015). A global reference for human genetic variation. *Nature* 526(7571): 68–74.

Aaronson, S (2008). The limits of quantum computers. *Sci Am* 298(3): 50–7.

Achaz, G, Rocha, EP, Netter, P, et al. (2002). Origin and fate of repeats in bacteria. *Nucleic Acids Res* 30(13): 2987–94.

Achtman, M and Wagner, M (2008). Microbial diversity and the genetic nature of microbial species. *Nat Rev Microbiol* 6(6): 431–40.

Adam, M, Murali, B, Glenn, NO, et al. (2008). Epigenetic inheritance based evolution of antibiotic resistance in bacteria. *BMC Evol Biol* 8: 52.

Adami, C, Ofria, C, and Collier, TC (2000). Evolution of biological complexity. *Proc Natl Acad Sci U S A* 97(9): 4463–8.

Adams, D (1979). The Hitch Hiker's Guide to the Galaxy. London: Pan Books.

Adleman, LM (1994). Molecular computation of solutions to combinatorial problems. *Science* 266(5187): 1021–4.

Ågren, JA (2021). *The Gene's-eye View of Evolution*. Oxford University Press.

Ahnert, SE, Marsh, JA, Hernandez, H, et al. (2015). Principles of assembly reveal a periodic table of protein complexes. *Science* 350(6266): aaa2245.

Aldridge, P and Hughes, KT (2002). Regulation of flagellar assembly. *Curr Opin Microbiol* 5(2): 160–5.

Amir, A and Nelson, DR (2012). Dislocation-mediated growth of bacterial cell walls. *Proc Natl Acad Sci U S A* 109(25): 9833–8.

Andam, CP and Gogarten, JP (2011). Biased gene transfer in microbial evolution. *Nat Rev Microbiol* 9(7): 543–55.

Anderson, PW (1972). More is different. *Science* 177(4047): 393–6.

Anfinsen, CB (1973). Principles that govern the folding of protein chains. *Science* 181(4096): 223–30.

Arendt, D (2003). Evolution of eyes and photoreceptor cell types. *Int J Dev Biol* 47(7–8): 563–71.

Arendt, D (2008). The evolution of cell types in animals: Emerging principles from molecular studies. *Nat Rev Genet* 9(11): 868–82.

Arendt, D, Benito-Gutierrez, E, Brunet, T, et al. (2015). Gastric pouches and the mucociliary sole: Setting the stage for nervous system evolution. *Philos Trans R Soc Lond B Biol Sci* 370(1684). doi: 10.1098/rstb.2015.0286. PMID: 26554050; PMCID: PMC4650134.

Arendt, D, Tosches, MA, and Marlow, H (2016). From nerve net to nerve ring, nerve cord and brain—Evolution of the nervous system. *Nat Rev Neurosci* 17(1): 61–72.

Arendt, D and Wittbrodt, J (2001). Reconstructing the eyes of Urbilateria. *Philos Trans R Soc Lond B Biol Sci* 356(1414): 1545–63.

Arnold, K and Zuberbuhler, K (2006). Language evolution: semantic combinations in primate calls. *Nature* 441(7091): 303.

Aron, J (2010). The life simulator. *New Scientist* (2765): 6–7.

Aspray, W (1990). *John von Neumann and the Origins of Modern Computing*. Cambridge, MA: MIT Press.

Atkinson, QD (2011). Phonemic diversity supports a serial founder effect model of language expansion from Africa. *Science* 332(6027): 346–9.

Azevedo, RB, Lohaus, R, Braun, V, et al. (2005). The simplicity of metazoan cell lineages. *Nature* 433(7022): 152–6.

Babu, M, Arnold, R, Bundalovic-Torma, C, et al. (2014). Quantitative genome-wide genetic interaction screens reveal global epistatic relationships of protein complexes in Escherichia coli. *PLoS Genet* 10(2): e1004120.

Babu, M and Teichmann, SA (2003). Evolution of transcription factors and the gene regulatory network in Escherichia coli. *Nucleic Acids Res* 31(4): 1234–44.

Baines, J (2004). The earliest Egyptian writing: Development, context, purpose. In SD Houston (ed.), *The First Writing*. Cambridge: Cambridge University Press, pp. 150–89.

Baker, D (2000). A surprising simplicity to protein folding. *Nature* 405(6782): 39–42.

Baker, TA and Bell, SP (1998). Polymerases and the replisome: Machines within machines. *Cell* 92(3): 295–305.

Barghoorn, ES and Tyler, SA (1965). Microorganisms from the Gunflint Chert: These structurally preserved Precambrian fossils from Ontario are the most ancient organisms known. *Science* 147(3658): 563–75.

Barlow, M (2009). What antimicrobial resistance has taught us about horizontal gene transfer. *Methods Mol Biol* 532: 397–411.

Barrett, JW, Garcke, H, and Nurnberg, R (2012). Numerical computations of faceted pattern formation in snow crystal growth. *Phys Rev E Stat Nonlin Soft Matter Phys* 86(1 Pt 1): 011604.

Barrick, JE, Yu, DS, Yoon, SH, et al. (2009). Genome evolution and adaptation in a long-term experiment with Escherichia coli. *Nature* 461(7268): 1243–7.

Batut, B, Knibbe, C, Marais, G, et al. (2014). Reductive genome evolution at both ends of the bacterial population size spectrum. *Nat Rev Microbiol* 12(12): 841–50.

Baumdicker, F, Hess, WR, and Pfaffelhuber, P (2012). The infinitely many genes model for the distributed genome of bacteria. *Genome Biol Evol* 4(4): 443–56.

Baxter, S (2003). *Revolutions in the Earth: James Hutton and the True Age of the World*. London: Weidenfeld & Nicolson.

Bayles, KW (2014). Bacterial programmed cell death: Making sense of a paradox. *Nat Rev Microbiol* 12(1): 63–9.

Beauchamp, JP (2016). Genetic evidence for natural selection in humans in the contemporary United States. *Proc Natl Acad Sci U S A* 113(28): 7774–9.

Beecher, MD and Brenowitz, EA (2005). Functional aspects of song learning in songbirds. *Trends Ecol Evol* 20(3): 143–9.

Bell, G and Mooers, AO (1997). Size and complexity among multicellular organisms. *Biol J Linn Soc Lond* 60(3): 345–63.

Benner, SA, Carrigan, MA, Ricardo, A, et al. (2006). Setting the stage: The history, chemistry, and geobiology behind RNA. In RF Gesteland, TR Cech, and JF Atkins (eds), Cold Spring Harbor, NY: Cold Spring Harbor Laboratory Press, pp. 1–22.

Berlekamp, ER, Conway, JH, and Guy, RK (1982). *Winning Ways: For Your Mathematical Plays*. London: Academic Press.

Berleman, JE and Kirby, JR (2009). Deciphering the hunting strategy of a bacterial wolfpack. *FEMS Microbiol Rev* 33(5): 942–57.

Berman, H, Henrick, K, and Nakamura, H (2003). Announcing the worldwide Protein Data Bank. *Nat Struct Biol* 10(12): 980.

Bersaglieri, T, Sabeti, PC, Patterson, N, et al. (2004). Genetic signatures of strong recent positive selection at the lactase gene. *Am J Hum Genet* 74(6): 1111–20.

Berwick, RC and Chomsky, N (2016). *Why Only Us: Language and Evolution*. Cambridge, MA: MIT Press.

Berwick, RC, Okanoya, K, Beckers, GJ, et al. (2011). Songs to syntax: The linguistics of birdsong. Trends Cogn Sci 15(3): 113–21.

Bi, EF and Lutkenhaus, J (1991). FtsZ ring structure associated with division in Escherichia coli. *Nature* 354(6349): 161–4.

Bickerton, D (1990). *Species and Language*. Chicago, IL: University of Chicago Press.

Bickerton, D (2007). Language evolution: A brief guide for linguists. *Lingua* 117(3): 510–26.

Bickerton, D (2009). *Adam's Tongue: How Humans Made Language, How Language Made Humans*. New York, NY: Hill and Wang.

Bier, M (1997). Brownian ratchets in physics and biology. *Contemp Phys* 38(6): 371–9.

Bier, M and Astumian, RD (1996). Biased Brownian motion as the operating principle for microscopic engines. *Bioelectrochem Bioenerg* 39(1): 67–75.

Biller, SJ, Berube, PM, Lindell, D, et al. (2015). Prochlorococcus: The structure and function of collective diversity. *Nat Rev Microbiol* 13(1): 13–27.

Biondi, EG, Reisinger, SJ, Skerker, JM, et al. (2006). Regulation of the bacterial cell cycle by an integrated genetic circuit. *Nature* 444(7121): 899–904.

Biro, D, Inoue-Nakamura, N, Tonooka, R, et al. (2003). Cultural innovation and transmission of tool use in wild chimpanzees: Evidence from field experiments. *Anim Cogn* 6(4): 213–23.

Blain, JC and Szostak, JW (2014). Progress toward synthetic cells. *Annu Rev Biochem* 83: 615–40.

Blattner, FR, Plunkett, G, 3rd, Bloch, CA, et al. (1997). The complete genome sequence of Escherichia coli K-12. *Science* 277(5331): 1453–62.

Blount, ZD, Barrick, JE, Davidson, CJ, et al. (2012). Genomic analysis of a key innovation in an experimental Escherichia coli population. *Nature* 489(7417): 513–18.

Blount, ZD, Borland, CZ, and Lenski, RE (2008). Historical contingency and the evolution of a key innovation in an experimental population of *Escherichia coli. Proc Natl Acad Sci U S A* 105(23): 7899–906.

Blum, HF (1963). On the origin and evolution of human culture. *American Scientist* 51(1): 32–47.

Bochner, BR (2009). Global phenotypic characterization of bacteria. *FEMS Microbiol Rev* 33(1): 191–205.

Bolhuis, JJ, Tattersall, I, Chomsky, N, et al. (2014). How could language have evolved? *PLoS Biol* 12(8): e1001934.

Bond, AB, Kamil, AC, and Balda, RP (2004). Pinyon jays use transitive inference to predict social dominance. *Nature* 430(7001): 778.

Bonner, JT (1988). *The Evolution of Complexity: By Means of Natural Selection*. Princeton, NJ: Princeton University Press.

Bonner, JT (2004). Perspective: the size-complexity rule. *Evolution* 58(9): 1883–90.

Borgdorff, MW and van Soolingen, D (2013). The re-emergence of tuberculosis: What have we learnt from molecular epidemiology? *Clin Microbiol Infect* 19(10): 889–901.

Boybat, I, Le Gallo, M, Nandakumar, SR, et al. (2018). Neuromorphic computing with multi-memristive synapses. *Nat Commun* 9(1): 2514.

Brasier, MD, Antcliffe, J, Saunders, M, et al. (2015). Changing the picture of Earth's earliest fossils (3.5-1.9 Ga). with new approaches and new discoveries. *Proc Natl Acad Sci U S A* 112(16): 4859–64.

Brenner, S (2012a). History of science. The revolution in the life sciences. *Science* 338(6113): 1427–8.

Brenner, S (2012b). Turing centenary: Life's code script. *Nature* 482(7386): 461.

Brocks, JJ, Logan, GA, Buick, R, et al. (1999). Archean molecular fossils and the early rise of eukaryotes. *Science* 285(5430): 1033–6.

Budin, I and Szostak, JW (2010). Expanding roles for diverse physical phenomena during the origin of life. *Annu Rev Biophys* 39: 245–63.

Burchfield, JD (1975). *Lord Kelvin and the Age of the Earth*. New York, NY: Science History Publications.

Burks, A (1970). Von Neumann's self-reproducing automata. In A Burks (ed.), *Essays on Cellular Automata*. Urbana, IL: University of Illinois Press, pp. 3–64.

Caiati, MD, Safiulina, VF, Fattorini, G, et al. (2013). PrPC controls via protein kinase A the direction of synaptic plasticity in the immature hippocampus. *J Neurosci* 33(7): 2973–83.

Cairns, J (1963). The bacterial chromosome and its manner of replication as seen by autoradiography. *J Mol Biol* 6: 208–13.

Cairns-Smith, AG (1966). The origin of life and the nature of the primitive gene. *J Theor Biol* 10(1): 53–88.

Callaway, E (2015). Mammoth genomes hold recipe for Arctic elephants. *Nature* 521(7550): 18–19.

Camazine, S (2003). *Self-organization in Biological Systems*. Princeton, NJ: Princeton University Press.

Carini, P, Steindler, L, Beszteri, S, et al. (2013). Nutrient requirements for growth of the extreme oligotroph 'Candidatus Pelagibacter ubique' HTCC1062 on a defined medium. *ISME J* 7(3): 592–602.

Carroll, SB (2001). Chance and necessity: The evolution of morphological complexity and diversity. *Nature* 409(6823): 1102–9.

Cavalier-Smith, T (2009). Predation and eukaryote cell origins: A coevolutionary perspective. *Int J Biochem Cell Biol* 41(2): 307–22.

Ceruzzi, PE (2003). *A History of Modern Computing*. Cambridge, MA: MIT Press.

Chaitin, GJ (1969). On the length of programs for computing finite binary sequences: Statistical considerations. *Journal of the ACM*. 16(1): 145–59.

Charlebois, RL and Doolittle, WF (2004). Computing prokaryotic gene ubiquity: Rescuing the core from extinction. *Genome Res* 14(12): 2469–77.

Chimpanzee, S and Analysis, C (2005). Initial sequence of the chimpanzee genome and comparison with the human genome. *Nature* 437(7055): 69–87.

Chisholm, SW, Olson, RJ, Zettler, ER, et al. (1988). A novel free-living prochlorophyte abundant in the oceanic euphotic zone. *Nature* 334(6180): 340–3.

Cho, H, McManus, HR, Dove, SL, et al. (2011). Nucleoid occlusion factor SlmA is a DNA-activated FtsZ polymerization antagonist. *Proc Natl Acad Sci U S A* 108(9): 3773–8.

Chomsky, N (1965). *Aspects of the Theory of Syntax*. Cambridge, MA: MIT Press.

Chomsky, N (2015). *The Minimalist Program*. Cambridge, MA: MIT Press.

Christen, B, Abeliuk, E, Collier, JM, et al. (2011). The essential genome of a bacterium. *Mol Syst Biol* 7: 528.

Christensen, CB, Christensen-Dalsgaard, J, and Madsen, PT (2015). Hearing of the African lungfish (Protopterus annectens) suggests underwater pressure detection and rudimentary aerial hearing in early tetrapods. *J Exp Biol* 218(Pt 3): 381–7.

Church, A (1932). A set of postulates for the foundation of logic. *Ann Math* 32(2): 346–66.

Church, A (1936). An unsolvable problem of elementary number theory. *Am J Math* 58(2): 345–63.

Church, GM, Gao, Y and Kosuri, S (2012). Next-generation digital information storage in DNA. *Science* 337(6102): 1628.

Ciccarelli, FD, Doerks, T, von Mering, C, et al. (2006). Toward automatic reconstruction of a highly resolved tree of life. *Science* 311(5765): 1283–7.

Cobb, M (2015). *Life's Greatest Secret: The Race to Crack the Genetic Code*. New York, NY: Basic Books.

Coleman, ML, Sullivan, MB, Martiny, AC, et al. (2006). Genomic islands and the ecology and evolution of Prochlorococcus. *Science* 311(5768): 1768–70.

Conant, GC and Wolfe, KH (2008). Turning a hobby into a job: How duplicated genes find new functions. *Nat Rev Genet* 9(12): 938–50.

Conrad, C, Wunsche, A, Tan, TH, et al. (2011). Micropilot: Automation of fluorescence microscopy-based imaging for systems biology. *Nat Methods* 8(3): 246–9.

Cooper, JS (2004). Babylonian beginnings: The origin of the cuneiform writing system in comparative perspecrive. In SD Houston (ed.), *The First Writing*. Cambridge: Cambridge University Press, 71–99.

Cooper, VS and Lenski, RE (2000). The population genetics of ecological specialization in evolving Escherichia coli populations. *Nature* 407(6805): 736–9.

Covert, AW, 3rd, Lenski, RE, Wilke, CO, et al. (2013). Experiments on the role of deleterious mutations as stepping stones in adaptive evolution. *Proc Natl Acad Sci U S A* 110(34): E3171–E3178.

Crick, F (1958). On protein synthesis. *Symp Soc Exp Biol* 12: 138–63.

Crick, F (1970). Central dogma of molecular biology. *Nature* 227(5258): 561–3.

Crick, F, Barnett, L, Brenner, S, et al. (1961). General nature of the genetic code for proteins. *Nature* 192: 1227–32.

Csuhaj-Varju, E, Freund, R, Kari, L, et al. (1995). DNA computing based on splicing: Universality results. *Pacific Symposium on Biocomputing*, Singapore, 1 December 1995. River Edge, NJ: World Scientific Publishing Company, pp. 179–90.

Cubas, P, Vincent, C, and Coen, E (1999). An epigenetic mutation responsible for natural variation in floral symmetry. *Nature* 401(6749): 157–61.

Dagan, T and Martin, W (2007). Ancestral genome sizes specify the minimum rate of lateral gene transfer during prokaryote evolution. *Proc Natl Acad Sci U S A* 104(3): 870–5.

Dalrymple, GB (2001). The age of the Earth in the twentieth century: A problem (mostly) solved. *Geological Society, London, Special Publications* 190(1): 205–21.

Danchin, E, Charmantier, A, Champagne, FA, et al. (2011). Beyond DNA: Integrating inclusive inheritance into an extended theory of evolution. *Nat Rev Genet* 12(7): 475–86.

Dandekar, T, Snel, B, Huynen, M, et al. (1998). Conservation of gene order: A fingerprint of proteins that physically interact. *Trends Biochem Sci* 23(9): 324–8.

Davies, PCW (2019). *The Demon in the Machine: How Hidden Webs of Information Are Solving the Mystery of Life*. London: Allen Lane.

Davis, M (1982). Why Gödel didn't have Church's thesis. *Inf Control* 54(1–2): 3–24.

Dawkins, R (1976). *The Selfish Gene*. Oxford: Oxford University Press.

Dawkins, R (1982). *The Extended Phenotype: the Long Reach of the Gene*. Oxford: Oxford University Press, 1982.

Dawkins, R (2004). Extended phenotype–but not too extended. A reply to Laland, Turner and Jablonka. *Biol Philosophinf Control* 19(3): 377–96.

Dayel, MJ, Alegado, RA, Fairclough, SR, et al. (2011). Cell differentiation and morphogenesis in the colony-forming choanoflagellate Salpingoeca rosetta. *Dev Biol* 357(1): 73–82.

de Silva, E, Kelley, LA, and Stumpf, MP (2004). The extent and importance of intragenic recombination. *Hum Genomics* 1(6): 410–20.

de Steenwinkel, JE, ten Kate, MT, de Knegt, GJ, et al. (2012). Drug susceptibility of Mycobacterium tuberculosis Beijing genotype and association with MDR TB. *Emerg Infect Dis* 18(4): 660–3.

Dediu, D and Levinson, SC (2013). On the antiquity of language: The reinterpretation of Neandertal linguistic capacities and its consequences. *Front Psychol* 4: 397.

Dennis, MY, Nuttle, X, Sudmant, PH, et al. (2012). Evolution of human-specific neural SRGAP2 genes by incomplete segmental duplication. *Cell* 149(4): 912–22.

Derr, J, Manapat, ML, Rajamani, S, et al. (2012). Prebiotically plausible mechanisms increase compositional diversity of nucleic acid sequences. *Nucleic Acids Res* 40(10): 4711–22.

Dobson, CM (2003). Protein folding and misfolding. *Nature* 426(6968): 884–90.

Dodd, MS, Papineau, D, Grenne, T, et al. (2017). Evidence for early life in Earth's oldest hydrothermal vent precipitates. *Nature* 543(7643): 60–4.

Dominguez-Escobar, J, Chastanet, A, Crevenna, AH, et al. (2011). Processive movement of MreB-associated cell wall biosynthetic complexes in bacteria. *Science* 333(6039): 225–8.

Donaldson-Matasci, MC, Bergstrom, CT, and Lachmann, M (2010). The fitness value of information. *Oikos* 119(2): 219–30.

Donoghue, PC and Antcliffe, JB (2010). Early life: Origins of multicellularity. *Nature* 466(7302): 41–2.

Draghi, J and Wagner, GP (2008). Evolution of evolvability in a developmental model. *Evolution* 62(2): 301–15.

Drake, JW (1991). A constant rate of spontaneous mutation in DNA-based microbes. *Proc Natl Acad Sci U S A* 88(16): 7160–4.

Dworkin, M (1996). Recent advances in the social and developmental biology of the myxobacteria. *Microbiol Rev* 60(1): 70–102.

Dyson, FJ (1999). *Origins of Life*. Cambridge: Cambridge University Press.

Eigen, M (2013). *From Strange Simplicity to Complex Familiarity: A Treatise on Matter, Information, Life and Thought*. Oxford: Oxford University Press.

Eigen, M and Schuster, P (1977). The hypercycle: A principle of self-organization. Part A: Emergence of the hypercycle. *Naturwissenschaftern* Nov;64(11): 541–65. Berlin.

El Albani, A, Bengtson, S, Canfield, DE, et al. (2010). Large colonial organisms with coordinated growth in oxygenated environments 2.1 Gyr ago. *Nature* 466(7302): 100–4.

Elena, SF, Cooper, VS and Lenski, RE (1996). Punctuated evolution caused by selection of rare beneficial mutations. *Science* 272(5269): 1802–4.

Ellis, RJ and Hartl, FU (1999). Principles of protein folding in the cellular environment. *Curr Opin Chem Biol* 9(1): 102–10.

Elowitz, MB, Levine, AJ, Siggia, ED, et al. (2002). Stochastic gene expression in a single cell. *Science* 297(5584): 1183–6.

Elowitz, MB, Surette, MG, Wolf, PE, et al. (1999). Protein mobility in the cytoplasm of Escherichia coli. *J Bacteriol* 181(1): 197–203.

Eme, L, Sharpe, SC, Brown, MW, et al. (2014). On the age of eukaryotes: Evaluating evidence from fossils and molecular clocks. *Cold Spring Harb Perspect Biol* 6(8) :a016139. doi: 10.1101/cshperspect.a016139. PMID: 25085908; PMCID: PMC4107988.

Enard, W, Gehre, S, Hammerschmidt, K, et al. (2009). A humanized version of Foxp2 affects cortico-basal ganglia circuits in mice. *Cell* 137(5): 961–71.

Enard, W, Przeworski, M, Fisher, SE, et al. (2002). Molecular evolution of FOXP2, a gene involved in speech and language. *Nature* 418(6900): 869.

Engesser, S, Ridley, AR and Townsend, SW (2016). Meaningful call combinations and compositional processing in the southern pied babbler. *Proc Natl Acad Sci U S A* 113(21): 5976–81.

Englund, RK (2004). The state of decipherment of proto-Elamite. In SD Houston (ed.), *The First Writing*. Cambridge University Press., 100–49.

Erickson, HP, Anderson, DE, and Osawa, M (2010). FtsZ in bacterial cytokinesis: Cytoskeleton and force generator all in one. *Microbiol Mol Biol Rev* 74(4): 504–28.

Evans, PD, Gilbert, SL, Mekel-Bobrov, N, et al. (2005). Microcephalin, a gene regulating brain size, continues to evolve adaptively in humans. *Science* 309(5741): 1717–20.

Fagundes, NJ, Ray, N, Beaumont, M, et al. (2007). Statistical evaluation of alternative models of human evolution. *Proc Natl Acad Sci U S A* 104(45): 17614–19.

Fang, F, Lin, YH, Pierce, BD, et al. (2015). A rhizobium radiobacter histidine kinase can employ both Boolean AND and OR logic gates to initiate pathogenesis. *Chembiochem* 16(15): 2183–90.

Feynman, RP, Leighton, RB, and Sands, ML (1963). *The Feynman Lectures on Physics*. Reading, MA: Addison-Wesley Pub. Co.

Fiddes, IT, Lodewijk, GA, Mooring, M, et al. (2018). Human-specific NOTCH2NL genes affect notch signaling and cortical neurogenesis. *Cell* 173(6): 1356–69 e1322.

Fiers, W, Contreras, R, Duerinck, F, et al. (1976). Complete nucleotide sequence of bacteriophage MS2 RNA: Primary and secondary structure of the replicase gene. *Nature* 260(5551): 500–7.

Fischer, B, Sandmann, T, Horn, T, et al. (2015). A map of directional genetic interactions in a metazoan cell. *eLife* 4:e05464.

Fitch, WT (2000). The evolution of speech: A comparative review. *Trends in Cognitive Sciences* 4(7): 258–67.

Fitch, WT (2010). *The Evolution of Language*. Cambridge: Cambridge University Press.

Fitch, WT, Hauser, MD, and Chomsky, N (2005). The evolution of the language faculty: Clarifications and implications. *Cognition* 97(2): 179–210; discussion 211-125.

Fleischmann, RD, Adams, MD, White, O, et al. (1995). Whole-genome random sequencing and assembly of Haemophilus influenzae Rd. *Science* 269(5223): 496–512.

Flores, E and Herrero, A (2010). Compartmentalized function through cell differentiation in filamentous cyanobacteria. *Nat Rev Microbiol* 8(1): 39–50.

Foo, YH, Spahn, C, Zhang, H, et al. (2015). Single cell super-resolution imaging of E. coli OmpR during environmental stress. *Integr Biol (Camb)* 7(10): 1297–308.

Forster, AC and Church, GM (2006). Towards synthesis of a minimal cell. *Mol Syst Biol* 2: 45.

Fowler, RG and Schaaper, RM (1997). The role of the mutT gene of Escherichia coli in maintaining replication fidelity. *FEMS Microbiol Rev* 21(1): 43–54.

Fox, D (2016). What sparked the Cambrian explosion? *Nature* 530(7590): 268–70.

Fraenkel, AS (1993). Complexity of protein folding. *Bull Math Biol* 55(6): 1199–210.

Fragaszy, DM and Perry, S (2003). *The Biology of Traditions: Models and Evidence*. Cambridge: Cambridge University Press.

Fraser, CM, Gocayne, JD, White, O, et al. (1995). The minimal gene complement of Mycoplasma genitalium. *Science* 270(5235): 397–403.

Frieden, TR, Sterling, T, Pablos-Mendez, A, et al. (1993). The emergence of drug-resistant tuberculosis in New York City. *N Engl J Med* 328(8): 521–6.

Frost, LS, Leplae, R, Summers, AO, et al. (2005). Mobile genetic elements: The agents of open source evolution. *Nat Rev Microbiol* 3(9): 722–32.

Furchtgott, L, Wingreen, NS, and Huang, KC (2011). Mechanisms for maintaining cell shape in rod-shaped Gram-negative bacteria. *Mol Microbiol* 81(2): 340–53.

Furnham, N, Sillitoe, I, Holliday, GL, et al. (2012). Exploring the evolution of novel enzyme functions within structurally defined protein superfamilies. *PLoS Comput Biol* 8(3): e1002403.

Gahlmann, A and Moerner, WE (2014). Exploring bacterial cell biology with single-molecule tracking and super-resolution imaging. *Nat Rev Microbiol* 12(1): 9–22.

Galardini, M, Koumoutsi, A, Herrera-Dominguez, L, et al. (2017). Phenotype inference in an Escherichia coli strain panel. *Elife* 6.

Galliot, B and Quiquand, M (2011). A two-step process in the emergence of neurogenesis. *Eur J Neurosci* 34(6): 847–62.

Gan, L, Chen, S, and Jensen, GJ (2008). Molecular organization of Gram-negative peptidoglycan. *Proc Natl Acad Sci U S A* 105(48): 18953–7.

Gardner, M (1970). The fantastic combinations of John Conway's new solitaire game 'life' by Martin Gardner. *Scientific American* 223: 120–3.

Garg, S, Lou, J, Jain, A, et al. (2022). Dynamic precision analog computing for neural networks. *IEEE Journal of Selected Topics in Quantum Electronics* 29(2): 1–12.

Garner, EC, Bernard, R, Wang, W, et al. (2011). Coupled, circumferential motions of the cell wall synthesis machinery and MreB filaments in B. subtilis. *Science* 333(6039): 222–5.

Gelb, IJ (1963). *A Study of Writing.* Chicago, IL: University of Chicago Press.

Gentner, TQ, Fenn, KM, Margoliash, D, et al. (2006). Recursive syntactic pattern learning by songbirds. *Nature* 440(7088): 1204–7.

Gerdes, SY, Scholle, MD, Campbell, JW, et al. (2003). Experimental determination and system level analysis of essential genes in Escherichia coli MG1655. *J Bacteriol* 185(19): 5673–84.

Gesteland, RF, Cech TR, and Atkins JF (eds) (2006). editors. *The RNA World.* Cold Spring Harbor, NY: Cold Spring Harbor Laboratory Press.

Geyer, F and Van der Zouwen, J (1994). Norbert Wiener and the social sciences. *Kybernetes* 23(6/7): 46–61.

Gibson, DG, Glass, JI, Lartigue, C, et al. (2010). Creation of a bacterial cell controlled by a chemically synthesized genome. *Science* 329(5987): 52–6.

Gilbert, P, Porter, SM, Sun, CY, et al. (2019). Biomineralization by particle attachment in early animals. *Proc Natl Acad Sci U S A* 116(36): 17659–65.

Gilbert, W (1986). Origin of life: The RNA world. *Nature* 319(6055).

Giovannoni, SJ, Tripp, HJ, Givan, S, et al. (2005). Genome streamlining in a cosmopolitan oceanic bacterium. *Science* 309(5738): 1242–5.

Goddard, MR (2016). Molecular evolution: Sex accelerates adaptation. *Nature* 531(7593): 176–7.

Gold, EM (1967). Language identification in the limit. *Inf Cont* 10(5): 447–4.

Goldman, N, Bertone, P, Chen, S, et al. (2013). Towards practical, high-capacity, low-maintenance information storage in synthesized DNA. *Nature* 494(7435): 77–80.

Goley, ED, Iniesta, AA, and Shapiro, L (2007). Cell cycle regulation in Caulobacter: Location, location, location. *J Cell Sci* 120(Pt 20): 3501–7.

Goley, ED, Yeh, YC, Hong, SH, et al. (2011). Assembly of the Caulobacter cell division machine. *Mol Microbiol* 80(6): 1680–98.

Gonzalez-Pastor, JE, San Millan, JL and Moreno, F (1994). The smallest known gene. *Nature* 369(6478): 281.

Good, BH, McDonald, MJ, Barrick, JE, et al. (2017). The dynamics of molecular evolution over 60,000 generations. *Nature* 551(7678): 45–50.

Gould, JL and Gould, CG (2007). *Animal Architects: Building and the Evolution of Intelligence.* New York, NY: Basic Books.

Gould, SJ (1989). *Wonderful Life: the Burgess Shale and the Nature of History.* New York, NY: W.W. Norton.

Gould, SJ (1996). *Full House: The Spread of Excellence from Plato to Darwin.* New York, NY: Harmony.

Gravner, JaG, D. (2008). Modeling snow crystal growth II: A mesoscopic lattice map with plausible dynamics. *Physica D* 237: 385–404.

Green, RE, Krause, J, Ptak, SE, et al. (2006). Analysis of one million base pairs of Neanderthal DNA. *Nature* 444(7117): 330–6.

Greer, EL, Maures, TJ, Ucar, D, et al. (2011). Transgenerational epigenetic inheritance of longevity in Caenorhabditis elegans. *Nature* 479(7373): 365–71.

Grefenstette, JJ (1993). Genetic algorithms and machine learning. *Proceedings of the Sixth Annual Conference on Computational Learning Theory.* New York, NY: ACM, 3–4.

Grefenstette, JJ, Ramsey, CL, and Schultz, AC (1990). Learning sequential decision rules using simulation models and competition. *Machine Learning* 5(4): 355–81.

Gremer, JR, Crone, EE and Lesica, P (2012). Are dormant plants hedging their bets? Demographic consequences of prolonged dormancy in variable environments. *The Am Nat* 179(3): 315–27.

Griffith, S, Goldwater, D, and Jacobson, JM (2005). Robotics: Self-replication from random parts. *Nature* 437(7059): 636.

Grosberg, RK and Strathmann, RR (2007). The evolution of multicellularity: A minor major transition? *Annu Rev Ecol Evol Syst* 38: 621–54.

Grossniklaus, U, Kelly, WG, Kelly, B, et al. (2013). Transgenerational epigenetic inheritance: How important is it? *Nat Rev Genet* 14(3): 228–35.

Guerrier-Takada, C, Gardiner, K, Marsh, T, et al. (1983). The RNA moiety of ribonuclease P is the catalytic subunit of the enzyme. *Cell* 35(3 Pt 2): 849–57.

Haeusser, DP and Levin, PA (2008). The great divide: Coordinating cell cycle events during bacterial growth and division. *Curr Opin Microbiol* 11(2): 94–9.

Han, TM and Runnegar, B (1992). Megascopic eukaryotic algae from the 2.1-billion-year-old negaunee iron-formation, Michigan. *Science* 257(5067): 232–5.

Harmand, S, Lewis, JE, Feibel, CS, et al. (2015). 3.3-million-year-old stone tools from Lomekwi 3, West Turkana, Kenya. *Nature* 521(7552): 310–15.

Hartley, RV (1928). Transmission of information. *Bell Labs Technical Journal* 7(3): 535–63.

Harvati, K, Roding, C, Bosman, AM, et al. (2019). Apidima Cave fossils provide earliest evidence of Homo sapiens in Eurasia. Nature 571(7766): 500–4.

Hauser, MD, Chomsky, N, and Fitch, WT (2002). The faculty of language: What is it, who has it, and how did it evolve? *Science* 298(5598): 1569–79.

Hauser, MD, Yang, C, Berwick, RC, et al. (2014). The mystery of language evolution. *Front Psychol* 5: 401.

Heard, E and Martienssen, RA (2014). Transgenerational epigenetic inheritance: Myths and mechanisms. *Cell* 157(1): 95–109.

Henderson, GP and Jensen, GJ (2006). Three-dimensional structure of Mycoplasma pneumoniae's attachment organelle and a model for its role in gliding motility. *Mol Microbiol* 60(2): 376–85.

Henn, BM, Cavalli-Sforza, LL, and Feldman, MW (2012). The great human expansion. *Proc Natl Acad Sci U S A* 109(44): 17758–64.

Hennies, J, Lleti, JMS, Schieber, NL, et al. (2020). AMST: Alignment to median smoothed template for focused ion beam scanning electron microscopy image stacks. *Sci Rep* 10(1): 2004.

Henz, SR, Huson, DH, Auch, AF, et al. (2005). Whole-genome prokaryotic phylogeny. *Bioinformatics* 21(10): 2329–35.

Herron, MD, Hackett, JD, Aylward, FO, et al. (2009). Triassic origin and early radiation of multicellular volvocine algae. *Proc Natl Acad Sci U S A* 106(9): 3254–8.

Herrou, J and Crosson, S (2011). Function, structure and mechanism of bacterial photosensory LOV proteins. *Nat Rev Microbiol* 9(10): 713–23.

Higham, T, Douka, K, Wood, R, et al. (2014). The timing and spatiotemporal patterning of Neanderthal disappearance. *Nature* 512(7514): 306–9.

Hoagland, MB, Stephenson, ML, Scott, JF, et al. (1958). A soluble ribonucleic acid intermediate in protein synthesis. *JBC* 231(1): 241–57.

Hofstadter, DR (1979). *Gödel, Escher, Bach: An Eternal Golden Braid*. New York, NY: Basic Books.

Holland, JH (1973). Genetic algorithms and the optimal allocation of trials. *SIAM J Comput* 2(2): 88–105.

Holland, JH (1992). Genetic algorithms. *Scientific American* 267(1): 66–73.

Holtzendorff, J, Hung, D, Brende, P, et al. (2004). Oscillating global regulators control the genetic circuit driving a bacterial cell cycle. *Science* 304(5673): 983–7.

Hoppitt, WJ, Brown, GR, Kendal, R, et al. (2008). Lessons from animal teaching. *Trends Ecol Evol* 23(9): 486–93.

Horner, V, Whiten, A, Flynn, E, et al. (2006). Faithful replication of foraging techniques along cultural transmission chains by chimpanzees and children. *Proc Natl Acad Sci U S A* 103(37): 13878–83.

Houston, SD (2004). *The First Writing: Script Invention as History and Process*. Cambridge: Cambridge University Press.

Hutchison, CA, 3rd, Chuang, RY, Noskov, VN, et al. (2016). Design and synthesis of a minimal bacterial genome. *Science* 351(6280): aad6253.

Hutchison, CA, Peterson, SN, Gill, SR, et al. (1999). Global transposon mutagenesis and a minimal Mycoplasma genome. *Science* 286(5447): 2165–9.

Ifrah, G (2001). *A Universal History of Computing: From the Abacus to the Quantum Computer.* New York, NY: John Wiley.

Ikeda, T, Asakura, S, and Kamiya, R (1989). Total reconstitution of Salmonella flagellar filaments from hook and purified flagellin and hook-associated proteins in vitro. *J Mol Biol* 209(1): 109–14.

Imhof, M and Schlotterer, C (2001). Fitness effects of advantageous mutations in evolving Escherichia coli populations. *Proc Natl Acad Sci U S A* 98(3): 1113–17.

International Telecommunication Union and Michaelis, AR (1965). *From Semaphore to Satellite.* Geneva: International Telecommunication Union.

Jablonka, E (2002). Information: Its interpretation, its inheritance, and its sharing. *Philos Sci* 69(4): 578–605.

Jablonka, E and Raz, G (2009). Transgenerational epigenetic inheritance: Prevalence, mechanisms, and implications for the study of heredity and evolution. *Q Rev Biol* 84(2): 131–76.

Jablonski, NG and Chaplin, G (2000). The evolution of human skin coloration. *J Hum Evol* 39(1): 57–106.

Jackendoff, R (1999). Possible stages in the evolution of the language capacity. *TiCS* 3(7): 272–9.

Jackendoff, R and Pinker, S (2005). The nature of the language faculty and its implications for evolution of language (Reply to Fitch, Hauser, and Chomsky). *Cognition* 97(2): 211–25.

Jaenicke, R (1998). Protein self-organization in vitro and in vivo: Partitioning between physical biochemistry and cell biology. *Biol Chem* 379(3): 237–43.

Jean, G (1992). *Writing: The Story of Alphabets and Scripts.* New York, NY: Abrams.

Jee, J, Rasouly, A, Shamovsky, I, et al. (2016). Rates and mechanisms of bacterial mutagenesis from maximum-depth sequencing. *Nature* 534(7609): 693–6.

Jeong, H, Barbe, V, Lee, CH, et al. (2009). Genome sequences of Escherichia coli B strains REL606 and BL21(DE3). *J Mol Biol* 394(4): 644–52.

Johannsen, W (1909). *Elemente der exakten Erblichkeitslehre.* Jena: Fischer.

Johnson, A and O'Donnell, M (2005). Cellular DNA replicases: Components and dynamics at the replication fork. *Annu Rev Biochem* 74: 283–315.

Johnston, WK, Unrau, PJ, Lawrence, MS, et al. (2001). RNA-catalyzed RNA polymerization: Accurate and general RNA-templated primer extension. *Science* 292(5520): 1319–25.

Jones, D, Metzger, HJ, Schatz, A, et al. (1944). Control of gram-negative bacteria in experimental animals by streptomycin. *Science* 100(2588): 103–5.

Jones, RAL (2004). *Soft Machines: Nanotechnology and Life.* Oxford: Oxford University Press.

Jorgensen, SL, Hannisdal, B, Lanzen, A, et al. (2012). Correlating microbial community profiles with geo-chemical data in highly stratified sediments from the Arctic Mid-Ocean Ridge. *Proc Natl Acad Sci U S A* 109(42): E2846–2855.

Joyce, GF and Orgel, LE (2006). Progress towards understanding the origin of the RNA world. In RF Gesteland, TR Cech, and JF Atkins (eds), *The RNA World.* Cold Spring Harbor, NY: Cold Spring Harbor Laboratory Press, p. 23.

Jumper, J, Evans, R, Pritzel, A, et al. (2021). Highly accurate protein structure prediction with AlphaFold. *Nature* 596(7873): 583–9.

Jun, S and Mulder, B (2006). Entropy-driven spatial organization of highly confined polymers: Lessons for the bacterial chromosome. *Proc Natl Acad Sci U S A* 103(33): 12388–93.

Jun, S and Wright, A (2010). Entropy as the driver of chromosome segregation. *Nat Rev Microbiol* 8(8): 600–7.

Kaern, M, Elston, TC, Blake, WJ, et al. (2005). Stochasticity in gene expression: From theories to phenotypes. *Nat Rev Genet* 6(6): 451–64.

Kamminga, H (1988). Historical perspective: The problem of the origin of life in the context of developments in biology. *Orig Life Evol Biosph* 18(1–2): 1–11.

Kashyap, PC, Marcobal, A, Ursell, LK, et al. (2013). Genetically dictated change in host mucus carbohydrate landscape exerts a diet-dependent effect on the gut microbiota. *Proc Natl Acad Sci U S A* 110(42): 17059–64.

Kauffman, SA (1993). *The Origins of Order: Self-Organization and Selection in Evolution.* New York, NY: Oxford University Press.

Kawai, N and Matsuzawa, T (2000). Numerical memory span in a chimpanzee. *Nature* 403(6765): 39–40.

Kawecki, TJ, Lenski, RE, Ebert, D, et al. (2012). Experimental evolution. *Trends Ecol Evol* 27(10): 547–60.

Keller, EF (1995). *Refiguring Life: Metaphors of Twentieth-century Biology.* New York, NY: Columbia University Press.

Kemeny (1955). Man viewed as a machine. *Scientific American* 192(4): 58–67

Kettler, GC, Martiny, AC, Huang, K, et al. (2007). Patterns and implications of gene gain and loss in the evolution of Prochlorococcus. *PLoS Genet* 3(12): e231.

Khan, AI, Dinh, DM, Schneider, D, et al. (2011). Negative epistasis between beneficial mutations in an evolving bacterial population. *Science* 332(6034): 1193–6.

Kibota, TT and Lynch, M (1996). Estimate of the genomic mutation rate deleterious to overall fitness in E. coli. *Nature* 381(6584): 694–6.

Kiekebusch, D, Michie, KA, Essen, LO, et al. (2012). Localized dimerization and nucleoid binding drive gradient formation by the bacterial cell division inhibitor MipZ. *Mol Cell* 46(3): 245–9.

Kimura, M (1961). Natural selection as the process of accumulating genetic information in adaptive evolution. *Genet Res* 2(01): 127–40.

Kimura, M (1968). Evolutionary rate at the molecular level. *Nature* 217(5129): 624–6.

Kimura, M (1983). *The Neutral Theory of Molecular Evolution.* New York, NY: Cambridge University Press.

King, N (2004). The unicellular ancestry of animal development. *Dev Cell* 7(3): 313–25.

King, N, Westbrook, MJ, Young, SL, et al. (2008). The genome of the choanoflagellate Monosiga brevicollis and the origin of metazoans. *Nature* 451(7180): 783–8.

Klein, RG (2017). Language and human evolution. *J Neurolinguistics* 43: 204–21.

Knoll, AH (2003). *Life on a Young Planet: The First Three Billion Years of Evolution on Earth.* Princeton, NJ: Princeton University Press.

Knoll, AH (2011). The multiple origins of complex multicellularity. *Ann Rev Earth Planet Sci* 39: 217–39.

Knoll, AH and Carroll, SB (1999). Early animal evolution: Emerging views from comparative biology and geology. *Science* 284(5423): 2129–37.

Kolmogorov, AN (1965). Three approaches to the quantitative definition of information. *Problemi Peredachi Informatsii* 1(1): 3–11.

Kolmogorov, AN (1968). Three approaches to the quantitative definition of information. *Int J Comput Math* 2(1–4): 157–68.

Kolter, R and Greenberg, EP (2006). Microbial sciences: The superficial life of microbes. *Nature* 441(7091): 300–2.

Kondrashov, AS (1988). Deleterious mutations and the evolution of sexual reproduction. *Nature* 336(6198): 435–40.

Kondrashov, FA (2012). Gene duplication as a mechanism of genomic adaptation to a changing environment. *Proc Biol Sci* 279(1749): 5048–57.

Kondrashov, FA, Rogozin, IB, Wolf, YI, et al. (2002). Selection in the evolution of gene duplications. *Genome Biol* 3(2): Research0008. doi: 10.1186/gb-2002-3-2-research0008. Epub 2002 Jan 14. PMID: 11864370; PMCID: PMC65685.

Kong, A, Frigge, ML, Thorleifsson, G, et al. (2017). Selection against variants in the genome associated with educational attainment. *Proc Natl Acad Sci U S A* 114(5): E727–E732.

Koonin, EV (2000). How many genes can make a cell: The minimal-gene-set concept. *Annu Rev Genomics Hum Genet* 1: 99–116.

Koonin, EV (2012). *The Logic of Chance: The Nature and Origin of Biological Evolution.* Upper Saddle River, NJ: Financial Times/Prentice Hall.

Koonin, EV, Fedorova, ND, Jackson, JD, et al. (2004). A comprehensive evolutionary classification of proteins encoded in complete eukaryotic genomes. *Genome Biol* 5(2): R7.

Koonin, EV and Wolf, YI (2009). The fundamental units, processes and patterns of evolution, and the tree of life conundrum. *Biol Direct* 4: 33.

Krause, DC and Balish, MF (2001). Structure, function, and assembly of the terminal organelle of Mycoplasma pneumoniae. *FEMS Microbiol Lett* 198(1): 1–7.

Kroodsma, DE, Houlihan, PW, Fallon, PA, et al. (1997). Song development by grey catbirds. *Anim Behav* 54(2): 457–64.

Kukulski, W, Schorb, M, Welsch, S, et al. (2011). Correlated fluorescence and 3D electron microscopy with high sensitivity and spatial precision. *J Cell Biol* 192(1): 111–19.

Kun, A, Santos, M, and Szathmary, E (2005). Real ribozymes suggest a relaxed error threshold. *Nat Genet* 37(9): 1008–11.

Kurner, J, Frangakis, AS, and Baumeister, W (2005). Cryo-electron tomography reveals the cytoskeletal structure of Spiroplasma melliferum. *Science* 307(5708): 436–8.

Kvitek, DJ and Sherlock, G (2011). Reciprocal sign epistasis between frequently experimentally evolved adaptive mutations causes a rugged fitness landscape. *PLoS Genet* 7(4): e1002056.

Laland, K (2018). How we became a different kind of animal: An evolved uniqueness. *Scientific American* 319(3): 33–9.

Laland, K, Uller, T, Feldman, M, et al. (2014). Does evolutionary theory need a rethink? *Nature* 514(7521): 161–4.

Laland, KN and Janik, VM (2006). The animal cultures debate. *Trends Ecol Evol* 21(10): 542–7.

Laland, KN, Odling-Smee, FJ, and Feldman, MW (1999). Evolutionary consequences of niche construction and their implications for ecology. *Proc Natl Acad Sci U S A* 96(18): 10242–7.

Laland, KN and Sterelny, K (2006). Perspective: Seven reasons (not) to neglect niche construction. *Evolution* 60(9): 1751–62.

Lamb, TD, Collin, SP, and Pugh, EN, Jr. (2007). Evolution of the vertebrate eye: Opsins, photoreceptors, retina and eye cup. *Nat Rev Neurosci* 8(12): 960–76.

Landauer, R (1961). Irreversibility and heat generation in the computing process. *IBM J Res Dev* 5(3): 183–91.

Lander, ES, Linton, LM, Birren, B, et al. (2001). Initial sequencing and analysis of the human genome. *Nature* 409(6822): 860–921.

Lane, N (2015). The Vital Question: Why Is Life the Way It Is? London: Profile Books.

Lane, N and Martin, W (2010). The energetics of genome complexity. *Nature* 467(7318): 929–34.

Lang, AS, Zhaxybayeva, O, and Beatty, JT (2012). Gene transfer agents: Phage-like elements of genetic exchange. *Nat Rev Microbiol* 10(7): 472–82.

Langergraber, KE, Prufer, K, Rowney, C, et al. (2012). Generation times in wild chimpanzees and gorillas suggest earlier divergence times in great ape and human evolution. *Proc Natl Acad Sci U S A* 109(39): 15716–21.

Langton, CG (1984). Self-reproduction in cellular automata. *Physica D* 10(1–2): 135–44.

Lartigue, C, Glass, JI, Alperovich, N, et al. (2007). Genome transplantation in bacteria: Changing one species to another. *Science* 317(5838): 632–8.

Laub, MT, McAdams, HH, Feldblyum, T, et al. (2000). Global analysis of the genetic network controlling a bacterial cell cycle. *Science* 290(5499): 2144–8.

Laub, MT, Shapiro, L, and McAdams, HH (2007). Systems biology of Caulobacter. *Annu Rev Genet* 41: 429–41.

Lawrence, JG and Ochman, H (1998). Molecular archaeology of the Escherichia coli genome. *Proc Natl Acad Sci U S A* 95(16): 9413–17.

Lee, H, Popodi, E, Tang, H, et al. (2012). Rate and molecular spectrum of spontaneous mutations in the bacterium Escherichia coli as determined by whole-genome sequencing. *Proc Natl Acad Sci U S A* 109(41): E2774–2783.

Leff, HS and Rex, AF (1990). Resource letter MD-1: Maxwell's demon. *Am J Phys* 58(3): 201–9.

Leigh, EG, Jr. (1971). *Adaptation and Diversity; Natural History and the Mathematics of Evolution [by] Egbert Giles Leigh, Jr.* San Francisco, CA: Freeman, Cooper.

Leitner, S, Nicholson, J, Leisler, B, et al. (2002). Song and the song control pathway in the brain can develop independently of exposure to song in the sedge warbler. *Proc Biol Sci* 269(1509): 2519–24.

Lem, S, (1973) *Solaris*. London: Arrow Books. (Translated by Kilmartin, J and Cox, S)

Lenski, RE, Rose, MR, Simpson, SC, et al. (1991). Long-term experimental evolution in Escherichia-coli 1. Adaptation and divergence during 2,000 generations. *Am Nat* 138(6): 1315–41.

Lenski, RE and Travisano, M (1994). Dynamics of adaptation and diversification: A 10,000-generation experiment with bacterial populations. *Proc Natl Acad Sci U S A* 91(15): 6808–14.

Leroux, M, Schel, AM, Wilke, C, et al. (2023) Call combinations and compositional processing in wild chimpanzees. *Nat Commun* 14(1): 2225.

Letunic, I and Bork, P (2011). Interactive Tree Of Life v2: Online annotation and display of phylogenetic trees made easy. *Nucleic Acids Res* 39 (Web Server issue): W475–W478.

Lewin, HA, Robinson, GE, Kress, WJ, et al. (2018). Earth BioGenome Project: Sequencing life for the future of life. *Proc Natl Acad Sci U S A* 115(17): 4325–33.

Li, H and Durbin, R (2011). Inference of human population history from individual whole-genome sequences. *Nature* 475(7357): 493–6.

Li, M and Vitányi, P (2008). *An Introduction to Kolmogorov Complexity and its Applications. Texts in Computer Science*. New York, NY: Springer.

Li, Y, Hsin, J, Zhao, L, et al. (2013). FtsZ protofilaments use a hinge-opening mechanism for constrictive force generation. *Science* 341(6144): 392–5.

Libbrecht, KG (2005). The physics of snow crystals. *Rep Prog Phys* 68: 855–95.

Libbrecht, KG (2006). *Field Guide to Snowflakes*. Minneapolis, MN: Voyageur Press.

Libbrecht, KG (2007). The formation of snow crystals. *American Scientist* 95: 52–9.

Libbrecht, KG (2013). Quantitative modeling of faceted ice crystal growth from water vapor using cellular automata. *Journal of Computational Methods in Physics*. DOI: 10.1155/2013/174806.

Lipowsky, R (1991). The conformation of membranes. *Nature* 349(6309): 475–81.

Lluch-Senar, M, Querol, E, and Pinol, J (2010). Cell division in a minimal bacterium in the absence of ftsZ. *Mol Microbiol* 78(2): 278–89.

Locke, DP, Hillier, LW, Warren, WC, et al. (2011). Comparative and demographic analysis of orang-utan genomes. *Nature* 469(7331): 529.

Lonsdorf, EV, Eberly, LE and Pusey, AE (2004). Sex differences in learning in chimpanzees. *Nature* 428(6984): 715–16.

Loose, M, Fischer-Friedrich, E, Ries, J, et al. (2008). Spatial regulators for bacterial cell division self-organize into surface waves in vitro. *Science* 320(5877): 789–92.

Luisi, PL (2002). Toward the engineering of minimal living cells. *Anat Rec* 268(3): 208–14.

Luisi, PL (2006). *The Emergence of Life: From Chemical Origins to Synthetic Biology*. Cambridge: Cambridge University Press.

Lynch, M (2002). Genomics. Gene duplication and evolution. *Science* 297(5583): 945–7.

Lynch, M (2007). *The Origins of Genome Architecture*. Sunderland, MA: Sinauer Associates.

Lynch, M (2010). Evolution of the mutation rate. *Trends Genet* 26(8): 345–52.

Lynch, M and Conery, JS (2000). The evolutionary fate and consequences of duplicate genes. *Science* 290(5494): 1151–5.

Lynch, M and Marinov, GK (2015). The bioenergetic costs of a gene. *Proc Natl Acad Sci U S A* 112(51): 15690–5.

Lynch, M and Marinov, GK (2016). Reply to Lane and Martin: Mitochondria do not boost the bioenergetic capacity of eukaryotic cells. *Proc Natl Acad Sci U S A* 113(6): E667–E668.

Macia, J and Sole, R (2014). How to make a synthetic multicellular computer. *PLoS One* 9(2): e81248.

MacKay, DJC (2003). *Information Theory, Inference and Learning Algorithms*. Cambridge: Cambridge University Press.

MacLean, RC, Hall, AR, Perron, GG, et al. (2010). The population genetics of antibiotic resistance: Integrating molecular mechanisms and treatment contexts. *Nat Rev Genet* 11(6): 405–14.

Macnab, RM (2003). How bacteria assemble flagella. *Annu Rev Microbiol* 57: 77–100.

Malinsky, M, Challis, RJ, Tyers, AM, et al. (2015). Genomic islands of speciation separate cichlid ecomorphs in an East African crater lake. *Science* 350(6267): 1493–8.

Mallick, S, Li, H, Lipson, M, et al. (2016). The Simons Genome Diversity Project: 300 genomes from 142 diverse populations. *Nature* 538(7624): 201–6.

Manning, K, Tor, M, Poole, M, et al. (2006). A naturally occurring epigenetic mutation in a gene encoding an SBP-box transcription factor inhibits tomato fruit ripening. *Nat Genet* 38(8): 948–52.

Marler, P and Tamura, M (1964). Culturally transmitted patterns of vocal behavior in sparrows. *Science* 146(3650): 1483–6.

Marraffini, LA and Sontheimer, EJ (2010). CRISPR interference: RNA-directed adaptive immunity in bacteria and archaea. *Nat Rev Genet* 11(3): 181–90.

Marsh, JA and Teichmann, SA (2015). Structure, dynamics, assembly, and evolution of protein complexes. *Annu Rev Biochem* 84: 551–75.

Marshall, WF, Young, KD, Swaffer, M, et al. (2012). What determines cell size? *BMC Biol* 10: 101.

Martin, W, Baross, J, Kelley, D, et al. (2008). Hydrothermal vents and the origin of life. *Nat Rev Microbiol* 6(11): 805–14.

Martinez Cuesta, S, Furnham, N, Rahman, SA, et al. (2014). The evolution of enzyme function in the isomerases. *Curr Opin Struct Biol* 26: 121–30.

Mathieson, I, Lazaridis, I, Rohland, N, et al. (2015). Genome-wide patterns of selection in 230 ancient Eurasians. *Nature* 528(7583): 499–503.

Matroule, JY, Lam, H, Burnette, DT, et al. (2004). Cytokinesis monitoring during development; rapid pole-to-pole shuttling of a signaling protein by localized kinase and phosphatase in Caulobacter. *Cell* 118(5): 579–90.

Matsuzawa, T (1985). Use of numbers by a chimpanzee. *Nature* 315(6014): 57–9.

Maynard Smith, J (1970). Natural selection and the concept of a protein space. *Nature* 225(5232): 563–4.

Maynard Smith, J (1971). The origin and maintenance of sex. *Group Selection*. GC Williams (ed.), London: Routledge, pp. 163–75.

Maynard Smith, J (1983). Models of evolution. *Proc Royal Soc B* 219: 315–25.

Maynard Smith, J (2000). The concept of information in biology. *Philosophy of Science* 67(2), 177–94.

Maynard Smith, J and Harper, DD (2003). *Animal SWignals*. Oxford: Oxford University Press.

Maynard Smith, J and Szathmáry, ER (1995). *The Major Transitions in Evolution*. Oxford: W.H. Freeman/Spektrum.

Maynard-Smith, J (1978). *The Evolution of Sex*. Cambridge: Cambridge University Press.

McClelland, M, Florea, L, Sanderson, K, et al. (2000). Comparison of the Escherichia coli K-12 genome with sampled genomes of a Klebsiella pneumoniae and three salmonella enterica serovars, Typhimurium, Typhi and Paratyphi. *Nucleic Acids Res* 28(24): 4974–86.

McDonald, MJ, Rice, DP and Desai, MM (2016). Sex speeds adaptation by altering the dynamics of molecular evolution. *Nature* 531(7593): 233–6.

McShea, DW (1996). Perspective: Metazoan complexity and evolution: Is there a trend? *Evolution* 50(2): 477–92.

Medini, D, Donati, C, Tettelin, H, et al. (2005). The microbial pan-genome. *Curr Opin Genet Dev* 15(6): 589–94.

Mekel-Bobrov, N, Gilbert, SL, Evans, PD, et al. (2005). Ongoing adaptive evolution of ASPM, a brain size determinant in Homo sapiens. *Science* 309(5741): 1720–2.

Mercier, R, Kawai, Y, and Errington, J (2013). Excess membrane synthesis drives a primitive mode of cell proliferation. *Cell* 152(5): 997–1007.

Miller, GA (1956). The magical number seven, plus or minus two: Some limits on our capacity for processing information. *Psychol Rev* 63(2): 81.

Miller, W, Drautz, DI, Ratan, A, et al. (2008). Sequencing the nuclear genome of the extinct woolly mammoth. *Nature* 456(7220): 387–90.

Minamino, T, Imada, K, and Namba, K (2008). Mechanisms of type III protein export for bacterial flagellar assembly. *Mol Biosyst* 4(11): 1105–15.

Miska, EA and Ferguson-Smith, AC (2016). Transgenerational inheritance: Models and mechanisms of non-DNA sequence-based inheritance. *Science* 354(6308): 59–63.

Mitchell, M (1996). *An Introduction to Genetic Algorithms*. Cambridge, MA: MIT Press.

Mochida, GH and Walsh, CA (2001). Molecular genetics of human microcephaly. *Curr Opin Neurol* 14(2): 151–6.

Monahan, LG and Harry, EJ (2013). Identifying how bacterial cells find their middle: A new perspective. *Mol Microbiol* 87(2): 231–4.

Monod, J and Wainhouse, A (1972). *Chance and Necessity: An Essay on the Natural Philosophy of Modern Biology*. London: Collins.

Moorbath, S (2005). Palaeobiology: Dating earliest life. *Nature* 434(7030): 155.

Moore, EF (1970). Machine models of self-reproduction. In A Burks (ed.), *Essays on Cellular Automata*. Urbana, IL: University of Illinois Press, pp. 187–203.

Moran, PA (1958). Random processes in genetics. *Mathematical Proceedings of the Cambridge Philosophical Society* 54(01): 60–71.

Morgan, HD, Santos, F, Green, K, et al. (2005). Epigenetic reprogramming in mammals. *Hum Mol Genet* 14 Spec No 1: R47–R58.

Morgan, HD, Sutherland, HG, Martin, DI, et al. (1999). Epigenetic inheritance at the agouti locus in the mouse. *Nat Genet* 23(3): 314–18.

Morris, JJ, Johnson, ZI, Szul, MJ, et al. (2011). Dependence of the cyanobacterium Prochlorococcus on hydrogen peroxide scavenging microbes for growth at the ocean's surface. *PLoS One* 6(2): e16805.

Morris, JJ, Lenski, RE, and Zinser, ER (2012). The Black Queen hypothesis: Evolution of dependencies through adaptive gene loss. *MBio* 3(2).

Morris, RM, Rappe, MS, Connon, SA, et al. (2002). SAR11 clade dominates ocean surface bacterioplankton communities. *Nature* 420(6917): 806–810.

Morris, SC (2000). The Cambrian 'explosion': Slow-fuse or megatonnage? *Proc Natl Acad Sci U S A* 97(9): 4426–9.

Mott, ML and Berger, JM (2007). DNA replication initiation: Mechanisms and regulation in bacteria. *Nat Rev Microbiol* 5(5): 343–54.

Muller, HJ (1932). Some genetic aspects of sex. *Am Nat* 66(703): 118–38.

Muller, HJ (1964). The relation of recombination to mutational advance. *Mutation Research/Fundamental and Molecular Mechanisms of Mutagenesis* 1(1): 2–9.

Nakaya, UD (1954). *Snow Crystals Natural and Artificial*. Cambridge, MA: Harvard University Press.

Nichols, RJ, Sen, S, Choo, YJ, et al. (2011). Phenotypic landscape of a bacterial cell. *Cell* 144(1): 143–56.

Nielsen, C (2008). Six major steps in animal evolution: Are we derived sponge larvae? *Evol Dev* 10(2): 241–57.

Nierhaus, KH (1991). The assembly of prokaryotic ribosomes. *Biochimie* 73(6): 739–55.

Nierhaus, KH and Dohme, F (1974). Total reconstitution of functionally active 50S ribosomal subunits from Escherichia coli. *Proc Natl Acad Sci U S A* 71(12): 4713–17.

Nierman, WC, Feldblyum, TV, Laub, MT, et al. (2001). Complete genome sequence of Caulobacter crescentus. *Proc Natl Acad Sci U S A* 98(7): 4136–41.

Noonan, JP, Coop, G, Kudaravalli, S, et al. (2006). Sequencing and analysis of Neanderthal genomic DNA. *Science* 314(5802): 1113–18.

Norton, JD (2005). Eaters of the lotus: Landauer's principle and the return of the Maxwell's demon. *Stud Hist Philos Sco B* 36(2): 375–441.

Nowak, MA (2006). *Evolutionary Dynamics: Exploring the Equations of Life*. Cambridge, MA: Belknap Press of Harvard University Press.

Nowak, MA and Komarova, NL (2001). Towards an evolutionary theory of language. *Trends Cogn Sci* 5(7): 288–95.

Nowak, MA and Krakauer, DC (1999). The evolution of language. *Proc Natl Acad Sci U S A* 96(14): 8028–33.

Nowak, MA and Ohtsuki, H (2008). Prevolutionary dynamics and the origin of evolution. *Proc Natl Acad Sci U S A* 105(39): 14924–7.

Nurse, P (2020). *What is Life?* Oxford: David Flickling Books.

Nyquist, H (1928). Thermal agitation of electric charge in conductors. *Physical Review* 32(1): 110.

O'Malley, MA (2010). The first eukaryote cell: An unfinished history of contestation. *Stud Hist Philos Biol Biomed Sci* 41(3): 212–24.

Obermayer, B, Krammer, H, Braun, D, et al. (2011). Emergence of information transmission in a prebiotic RNA reactor. *Phys Rev Lett* 107(1): 018101.

Ochman, H, Elwyn, S, and Moran, NA (1999). Calibrating bacterial evolution. *Proc Natl Acad Sci U S A* 96(22): 12638–43.

Ochman, H, Lawrence, JG, and Groisman, EA (2000). Lateral gene transfer and the nature of bacterial innovation. *Nature* 405(6784): 299–304.

Ochman, H and Wilson, AC (1987). Evolution in bacteria: Evidence for a universal substitution rate in cellular genomes. *J Mol Evol* 26(1–2): 74–86.

Oehler, JH and Schopf, JW (1971). Artificial microfossils: Experimental studies of permineralization of blue-green algae in silica. *Science* 174(4015): 1229–31.

Ohno, S (1970). *Evolution by Gene Duplication* New York, NY: Springer Science+Business Media.

Ohtsuki, H, Hauert, C, Lieberman, E, et al. (2006). A simple rule for the evolution of cooperation on graphs and social networks. *Nature* 441(7092): 502–5.

Okbay, A, Beauchamp, JP, Fontana, MA, et al. (2016). Genome-wide association study identifies 74 loci associated with educational attainment. *Nature* 533(7604): 539–42.

Orengo, CA, Michie, AD, Jones, S, et al. (1997). CATH—a hierarchic classification of protein domain structures. *Structure* 5(8): 1093–108.

Osawa, M, Anderson, DE and Erickson, HP (2008). Reconstitution of contractile FtsZ rings in liposomes. *Science* 320(5877): 792–4.

Otsuka, J and Nozawa, Y (1998). Self-reproducing system can behave as Maxwell's demon: Theoretical illustration under prebiotic conditions. *J Theor Biol* 194(2): 205–21.

Ouzounis, CA and Karp, PD (2000). Global properties of the metabolic map of Escherichia coli. *Genome Res* 10(4): 568–76.

Paabo, S (2014). The human condition—A molecular approach. *Cell* 157(1): 216–26.

Pennisi, E (2013). The man who bottled evolution. *Science* 342(6160): 790–3.

Pennisi, E (2017). Biologists propose to sequence the DNA of all life on Earth. *Science*, 355 (6328): 894–5

Penrose, LS (1959). Self-reproducing machines. *Scientific American* 200(6): 105–14.

Penrose, L and Penrose, R (1957). A self-reproducing analogue. *Nature* 179, 1183. https://doi.org/10.1038/1791183a0

Perfeito, L, Fernandes, L, Mota, C, et al. (2007). Adaptive mutations in bacteria: High rate and small effects. *Science* 317(5839): 813–15.

Perreault, C and Mathew, S (2012). Dating the origin of language using phonemic diversity. *PLoS One* 7(4): e35289.

Perutz, MF (1989). *Is Science Necessary?: Essays on Science and Scientists*. Barrie & Jenkins.

Pesavento, U (1995). An implementation of von Neumann's self-reproducing machine. *Artificial Life* 2(4): 337–54.

Peters, SE and Gaines, RR (2012). Formation of the 'Great Unconformity' as a trigger for the Cambrian explosion. *Nature* 484(7394): 363–6.

Phillips, PC (2008). Epistasis—the essential role of gene interactions in the structure and evolution of genetic systems. *Nat Rev Genet* 9(11): 855–67.

Pigliucci, M (2008). Is evolvability evolvable? *Nat Rev Genet* 9(1): 75–82.

Pinker, S and Jackendoff, R (2005). The faculty of language: What's special about it? *Cognition* 95(2): 201–36.

Pittis, AA and Gabaldon, T (2016). Late acquisition of mitochondria by a host with chimaeric prokaryotic ancestry. *Nature* 531(7592): 101–4.

Plaga, W and Ulrich, SH (1999). Intercellular signalling in Stigmatella aurantiaca. *Curr Opin Microbiol* 2(6): 593–7.

Podnieks, K (2009). Towards model-based model of cognition. *The Reasoner* 3(6): 5–6.

Poole, AM and Gribaldo, S (2014). Eukaryotic origins: How and when was the mitochondrion acquired? *Cold Spring Harb Perspect Biol* 6(12): a015990.

Porter, KR, Claude, A, and Fullam, EF (1945). A study of tissue culture cells by electron microscopy. *J Exp Med* 81(3): 233–46.

Post, DM and Palkovacs, EP (2009). Eco-evolutionary feedbacks in community and ecosystem ecology: Interactions between the ecological theatre and the evolutionary play. *Philos Trans R Soc Lond B Biol Sci* 364(1523): 1629–40.

Post, EL (1943). Formal reductions of the general combinatorial decision problem. *Am J Math* 65(2): 197–215.

Poundstone, W and Wainwright, RT (1985). *The Recursive Universe: Cosmic Complexity and the Limits of Scientific Knowledge.* Oxford: Oxford University Press.

Pradzynski, CC, Forck, RM, Zeuch, T, et al. (2012). A fully size-resolved perspective on the crystallization of water clusters. *Science* 337(6101): 1529–32.

Prochnik, SE, Umen, J, Nedelcu, AM, et al. (2010). Genomic analysis of organismal complexity in the multicellular green alga Volvox carteri. *Science* 329(5988): 223–6.

Proskurowski, G, Lilley, MD, Seewald, JS, et al. (2008). Abiogenic hydrocarbon production at lost city hydrothermal field. *Science* 319(5863): 604–7.

Prüfer, K, Munch, K, Hellmann, I, et al. (2012). The bonobo genome compared with the chimpanzee and human genomes. *Nature* 486(7404): 527.

Prufer, K, Racimo, F, Patterson, N, et al. (2014). The complete genome sequence of a Neanderthal from the Altai Mountains. *Nature* 505(7481): 43–9.

Przybilski, R and Hammann, C (2006). The hammerhead ribozyme structure brought in line. *Chembiochem* 7(11): 1641–4.

Ptacin, JL, Gahlmann, A, Bowman, GR, et al. (2014). Bacterial scaffold directs pole-specific centromere segregation. *Proc Natl Acad Sci U S A* 111(19): E2046–E2055.

Ptacin, JL, Lee, SF, Garner, EC, et al. (2010). A spindle-like apparatus guides bacterial chromosome segregation. *Nat Cell Biol* 12(8): 791–8.

Rabin, MO (1980). Probabilistic algorithm for testing primality. *J Number Theory* 12(1): 128–8.

Raichle, ME and Gusnard, DA (2002). Appraising the brain's energy budget. *Proc Natl Acad Sci U S A* 99(16): 10237–10239.

Rainey, PB and Travisano, M (1998). Adaptive radiation in a heterogeneous environment. *Nature* 394(6688): 69–72.

Raj, A and van Oudenaarden, A (2008). Nature, nurture, or chance: Stochastic gene expression and its consequences. *Cell* 135(2): 216–26.

Rappe, MS, Connon, SA, Vergin, KL, et al. (2002). Cultivation of the ubiquitous SAR11 marine bacterio-plankton clade. *Nature* 418(6898): 630–3.

Rees, DC, Johnson, E, and Lewinson, O (2009). ABC transporters: The power to change. *Nat Rev Mol Cell Biol* 10(3): 218–27.

Reich, DA (2018). *Who We Are and How We Got Here: Ancient DNA and the New Science of the Human Past.* Oxford: Oxford University Press

Reik, W (2007). Stability and flexibility of epigenetic gene regulation in mammalian development. *Nature* 447(7143): 425–32.

Reik, W, Dean, W and Walter, J (2001). Epigenetic reprogramming in mammalian development. *Science* 293(5532): 1089–93.

Remy, JJ (2010). Stable inheritance of an acquired behavior in Caenorhabditis elegans. *Curr Biol* 20(20): R877–R878.

Rensch, B (1948). Histological changes correlated with evolutionary changes of body size. *Evolution* 2(3): 218–30.

Richards, EJ (2006). Inherited epigenetic variation—revisiting soft inheritance. *Nat Rev Genet* 7(5): 395–401.

Rivoire, O and Leibler, S (2011). The value of information for populations in varying environments. *J Stat Phys* 142(6): 1124–66.

Robertson, JS (2004). The possibility and actuality of writing. In SD Houston (ed.), *The First Writing.* Cambridge: Cambridge University Press, 16–38.

Rocap, G, Larimer, FW, Lamerdin, J, et al. (2003). Genome divergence in two Prochlorococcus ecotypes reflects oceanic niche differentiation. *Nature* 424(6952): 1042–7.

Rohland, N, Malaspinas, AS, Pollack, JL, et al. (2007). Proboscidean mitogenomics: Chronology and mode of elephant evolution using mastodon as outgroup. *PLoS Biol* 5(8): e207.

Roller, A (1974). *Discovering the Basis of Life: An Introduction to Molecular Biology.* New York, NY: McGraw-Hill.

Romero, D and Palacios, R (1997). Gene amplification and genomic plasticity in prokaryotes. *Annu Rev Genet* 31: 91–111.

Romero, P, Wagg, J, Green, ML, et al. (2005). Computational prediction of human metabolic pathways from the complete human genome. *Genome Biol* 6(1): R2.

Rong, J, Niu, Z, Lee, LA, et al. (2011). Self-assembly of viral particles. *Curr Opin Colloid Interface Sci* 16(6): 441–50.

Rosato, A, Strandburg, KJ, Prinz, F, et al. (1987). Why the Brazil nuts are on top: Size segregation of particulate matter by shaking. *Phys Rev Lett* 58(10): 1038–40.

Rosenfeld, N, Young, JW, Alon, U, et al. (2005). Gene regulation at the single-cell level. *Science* 307(5717): 1962–5.

Rothfield, L, Taghbalout, A, and Shih, YL (2005). Spatial control of bacterial division-site placement. *Nat Rev Microbiol* 3(12): 959–68.

Rudwick, MJSA. (2014). *Earth's Deep History: How It Was Discovered and Why It Matters.* Chicago, IL: University of Chicago Press.

Russell, B (1914) *Our knowledge of the external world : as a field for scientific method in philosophy.* [S.l.]: Open Court Publishing.

Ryan, CJ, Roguev, A, Patrick, K, et al. (2012). Hierarchical modularity and the evolution of genetic interactomes across species. *Mol Cell* 46(5): 691–704.

Samson, RY and Bell, SD (2011). Cell cycles and cell division in the archaea. *Curr Opin Microbiol* 14(3): 350–6.

Sandegren, L and Andersson, DI (2009). Bacterial gene amplification: Implications for the evolution of antibiotic resistance. *Nat Rev Microbiol* 7(8): 578–88.

Sanger, F, Air, GM, Barrell, BG, et al. (1977). Nucleotide sequence of bacteriophage phi X174 DNA. *Nature* 265(5596): 687–95.

Sanz, C, Morgan, D and Gulick, S (2004). New insights into chimpanzees, tools, and termites from the Congo Basin. *Am Nat* 164(5): 567–81.

Sarkisyan, KS, Bolotin, DA, Meer, MV, et al. (2016). Local fitness landscape of the green fluorescent protein. *Nature* 533(7603): 397–401.

Sawasaki, T, Ogasawara, T, Morishita, R, et al. (2002). A cell-free protein synthesis system for high-throughput proteomics. *Proc Natl Acad Sci U S A* 99(23): 14652–7.

Scally, A and Durbin, R (2012). Revising the human mutation rate: Implications for understanding human evolution. *Nat Rev Genet* 13(10): 745–53.

Scally, A, Dutheil, JY, Hillier, LW, et al. (2012). Insights into hominid evolution from the gorilla genome sequence. *Nature* 483(7388): 169.

Scheffer, LK, Xu, CS, Januszewski, M, et al. (2020). A connectome and analysis of the adult Drosophila central brain. eLife 9:e57443.

Schlebusch, CM, Malmstrom, H, Gunther, T, et al. (2017). Southern African ancient genomes estimate modern human divergence to 350,000 to 260,000 years ago. *Science* 358(6363): 652–655.

Schlitt, T and Brazma, A (2007). Current approaches to gene regulatory network modelling. *BMC Bioinformatics* 8(Suppl 6): S9.

Schmandt-Besserat, D (1978). The earliest precursor of writing. *Scientific American* 238(6): 50–9.

Schmandt-Besserat, D (1992). *How writing came about.* Austin, TX: University of Texas Press.

Schmidhuber, J (2015). Deep learning in neural networks: An overview. *Neural Networks* 61: 85–117.

Schneiker, S, Perlova, O, Kaiser, O, et al. (2007). Complete genome sequence of the myxobacterium Sorangium cellulosum. *Nat Biotechnol* 25(11): 1281–9.

Schopf, JW (1993). Microfossils of the Early Archean Apex chert: New evidence of the antiquity of life. *Science* 260: 640–6.

Schopf, JW and Packer, BM (1987). Early Archean (3.3-billion to 3.5-billion-year-old) microfossils from Warrawoona Group, Australia. *Science* 237: 70–3.

Schrödinger, E (1944). *What is Life?* Cambridge: Cambridge University Press.

Schrum, JP, Zhu, TF, and Szostak, JW (2010). The origins of cellular life. *Cold Spring Harb Perspect Biol* 2(9): a002212.

Schulz, HN and Schulz, HD (2005). Large sulfur bacteria and the formation of phosphorite. *Science* 307(5708): 416–18.

Schumacher, J, Hoffmann, P, Schmal, C, et al. (2007). Genetics of dyslexia: The evolving landscape. *J Med Genet* 44(5): 289–97.

Schuster, SC, Miller, W, Ratan, A, et al. (2010). Complete Khoisan and Bantu genomes from southern Africa. *Nature* 463(7283): 943–7.

Searcy, WA and Nowicki, S (2005). *The Evolution of Animal Communication: Reliability and Deception in Signaling Systems*. Princeton, NJ: Princeton University Press.

Sebe-Pedros, A, Ballare, C, Parra-Acero, H, et al. (2016). The dynamic regulatory genome of Capsaspora and the origin of animal multicellularity. *Cell* 165(5): 1224–37.

Sender, R, Fuchs, S, and Milo, R (2016). Revised Estimates for the Number of Human and Bacteria Cells in the Body. *PLoS Biol* 14(8): e1002533.

Serres, MH, Kerr, AR, McCormack, TJ, et al. (2009). Evolution by leaps: Gene duplication in bacteria. *Biol Direct* 4: 46.

Seyfarth, RM, Cheney, DL, Bergman, T, et al. (2010). The central importance of information in studies of animal communication. *Anim Behav* 80(1): 3–8.

Shannon, CE (1948). A mathematical theory of communication. *The Bell System Technical Journal* 27(3): 379–423.

Shannon, CE (1949). Communication in the presence of noise. *Proceedings of the IRE* 37(1): 10–21.

Sharpe, SC, Eme, L, Brown, MW, et al. (2015). Timing the origins of multicellular eukaryotes through phylogenomics and relaxed molecular clock analyses. I Ruiz-Trillo and AM Nedelcu (eds), *Evolutionary Transitions to Multicellular Life*. Berlin: Springer, pp. 3–29.

Shen, X, Collier, J, Dill, D, et al. (2008). Architecture and inherent robustness of a bacterial cell-cycle control system. *Proc Natl Acad Sci U S A* 105(32): 11340–5.

Shik, JZ, Hou, C, Kay, A, et al. (2012). Towards a general life-history model of the superorganism: Predicting the survival, growth and reproduction of ant societies. *Biol Lett* 8(6): 1059–62.

Sigal, A, Milo, R, Cohen, A, et al. (2006). Variability and memory of protein levels in human cells. *Nature* 444(7119): 643–6.

Sipper, M (1998). Fifty years of research on self-replication: An overview. *Artificial Life* 4(3): 237–57.

Sniegowski, PD and Gerrish, PJ (2010). Beneficial mutations and the dynamics of adaptation in asexual populations. *Philos Trans R Soc Lond B Biol Sci* 365(1544): 1255–63.

Somel, M, Liu, X, and Khaitovich, P (2013). Human brain evolution: Transcripts, metabolites and their regulators. *Nat Rev Neurosci* 14(2): 112–27.

Soucy, SM, Huang, J, and Gogarten, JP (2015). Horizontal gene transfer: Building the web of life. *Nat Rev Genet* 16(8): 472–82.

Sousa, AMM, Meyer, KA, Santpere, G, et al. (2017). Evolution of the human nervous system function, structure, and development. *Cell* 170(2): 226–47.

Spang, A, Saw, JH, Jorgensen, SL, et al. (2015). Complex archaea that bridge the gap between prokaryotes and eukaryotes. *Nature* 521(7551): 173–9.

Srivastava, M, Simakov, O, Chapman, J, et al. (2010). The Amphimedon queenslandica genome and the evolution of animal complexity. *Nature* 466(7307): 720–6.

Stajich, JE, Berbee, ML, Blackwell, M, et al. (2009). The fungi. *Curr Biol* 19(18): R840–R845.

Stevens, KE and Sebert, ME (2011). Frequent beneficial mutations during single-colony serial transfer of Streptococcus pneumoniae. *PLoS Genet* 7(8): e1002232.

Stewart, JR and Stringer, CB (2012). Human evolution out of Africa: The role of refugia and climate change. *Science* 335(6074): 1317–21.

Stringer, C (2011). *The Origin of Our Species*. London: Allen Lane.

Stringer, C (2016). The origin and evolution of Homo sapiens. *Philos Trans R Soc Lond B Biol Sci* 371(1698).

Struhsaker, TT (1967). Auditory communication among vervet monkeys (Cercopithecus aethiops). In SA Altmann (ed.), *Social Communication among Primates*. Chicago, IL: University of Chicago Press, pp. 281–324.

Su, PT, Liao, CT, Roan, JR, et al. (2012). Bacterial colony from two-dimensional division to three-dimensional development. *PLoS One* 7(11): e48098.

Sudmant, PH, Huddleston, J, Catacchio, CR, et al. (2013). Evolution and diversity of copy number variation in the great ape lineage. *Genome Res* 23(9): 1373–82.

Swade, D (2000). *The Cogwheel Brain: Charles Babbage and the Quest to Build the First Computer*. London: Little, Brown.

Szostak, JW, Bartel, DP, and Luisi, PL (2001). Synthesizing life. *Nature* 409(6818): 387–90.

Tang, BL (2006). Molecular genetic determinants of human brain size. *Biochem Biophys Res Commun* 345(3): 911–16.

Tattersall, I (2008). An evolutionary framework for the acquisition of symbolic cognition by Homo sapiens. *Comp Cogn Behav* 3(1): 99–114.

Tattersall, I (2017). The material record and the antiquity of language. *Neurosci Biobehav Rev* 81(Pt B): 247–54.

Templeton, CN, Greene, E, and Davis, K (2005). Allometry of alarm calls: Black-capped chickadees encode information about predator size. *Science* 308(5730): 1934–7.

ten Cate, C and Okanoya, K (2012). Revisiting the syntactic abilities of non-human animals: Natural vocalizations and artificial grammar learning. *Philos Trans R Soc Lond B Biol Sci* 367(1598): 1984–94.

Tenaillon, O, Barrick, JE, Ribeck, N, et al. (2016). Tempo and mode of genome evolution in a 50,000-generation experiment. *Nature* 536(7615): 165–70.

Tenaillon, O, Skurnik, D, Picard, B, et al. (2010). The population genetics of commensal Escherichia coli. *Nat Rev Microbiol* 8(3): 207–17.

Thompson, EE, Kuttab-Boulos, H, Witonsky, D, et al. (2004). CYP3A variation and the evolution of salt-sensitivity variants. *Am J Hum Genet* 75(6): 1059–69.

Tishkoff, SA, Reed, FA, Ranciaro, A, et al. (2007). Convergent adaptation of human lactase persistence in Africa and Europe. *Nat Genet* 39(1): 31–40.

Tong, AH, Lesage, G, Bader, GD, et al. (2004). Global mapping of the yeast genetic interaction network. *Science* 303(5659): 808–13.

Toprak, E, Veres, A, Michel, JB, et al. (2012). Evolutionary paths to antibiotic resistance under dynamically sustained drug selection. *Nat Genet* 44(1): 101–5.

Traub, P and Nomura, M (1968). Structure and function of E. coli ribosomes. V. Reconstitution of functionally active 30S ribosomal particles from RNA and proteins. *Proc Natl Acad Sci U S A* 59(3): 777–84.

Traulsen, A and Nowak, MA (2006). Evolution of cooperation by multilevel selection. *Proc Natl Acad Sci U S A* 103(29): 10952–5.

Travisano, M, Mongold, JA, Bennett, AF, et al. (1995). Experimental tests of the roles of adaptation, chance, and history in evolution. *Science* 267(5194): 87–90.

Trigger, BG (1998). Writing systems: A case study in cultural evolution. *Nor Archaeol Rev* 31(1): 39–62.

Trigger, BG (2004). Writing systems: A case study in cultural evolution. In SD Houston (ed.), *The First Writing*. Cambridge: Cambridge University Press, pp. 39–68.

Truppa, V, Sovrano, VA, Spinozzi, G, et al. (2010) Processing of visual hierarchical stimuli by fish (Xenotoca eiseni). *Behav Brain Res* 207(1): 51–60.

Turing, AM (1936). On computable numbers, with an application to the Entscheidungsproblem. *J. of Math* 58(345–363): 5.

Turing, AM (1952). The chemical basis of morphogenesis. *Philos Trans R Soc Lond B Biol Sci* 237(641): 37–72.

Turing, AM (1950). Computing machinery and intelligence. *Mind* 49: 433–60.

Turner, DD (2011). *Paleontology: A Philosophical Introduction*. Cambridge: Cambridge University Press.

Typas, A, Banzhaf, M, Gross, CA, et al. (2012). From the regulation of peptidoglycan synthesis to bacterial growth and morphology. *Nat Rev Microbiol* 10(2): 123–36.

Ulam, SM (1970). On recursively defined geometrical objects and patterns of growth In A Burks (ed.), *Essays on Cellular Automata*. Chicago, IL: University of Illinois Press, pp. 232–43.

Unger, R and Moult, J (1993). Finding the lowest free energy conformation of a protein is an NP-hard problem: Proof and implications. *Bull Math Biol* 55(6): 1183–98.

UniProt, C (2015). UniProt: A hub for protein information. *Nucleic Acids Res* 43(Database issue): D204–D212.

Van Leeuwenhoek, A (1684). Microscopical observations about animals in the scurf of the teeth. *Philos Trans R Soc Lond B Biol Sci* 14: 568–74.

van Soolingen, D, Qian, L, de Haas, PE, et al. (1995). Predominance of a single genotype of Mycobacterium tuberculosis in countries of east Asia. *J Clin Microbiol* 33(12): 3234–8.

VanRullen, R and Koch, C (2003). Is perception discrete or continuous? Trends Cogn Sci 7(5): 207–13.

Vasas, V, Fernando, C, Santos, M, et al. (2012). Evolution before genes. *Biol Direct* 7: 1; discussion 1.

Velicer, GJ (2003). Social strife in the microbial world. *Trends Microbiol* 11(7): 330–7.

Viklund, J, Ettema, TJ, and Andersson, SG (2012). Independent genome reduction and phylogenetic reclassification of the oceanic SAR11 clade. *Mol Biol Evol* 29(2): 599–615.

Vollmer, W and Bertsche, U (2008). Murein (peptidoglycan) structure, architecture and biosynthesis in Escherichia coli. *Biochim Biophys Acta* 1778(9): 1714–34.

Von Neumann, J (1951). *The General and Logical Theory of Automata, Cerebral Mechanisms in Behavior. The Hixon Symposium.* New York, NY: John Wiley & Sons. 1–41.

Von Neumann, J (1956). Probabilistic logics and the synthesis of reliable organisms from unreliable components. *Automata Studies* 34: 43–98.

Von Neumann, J (1993). First draft of a report on the EVDAC. *IEEE Ann Hist Comput* 15(4): 27–75.

Von Neumann, J and Burks, AW (1966). *Theory of Self-reproducing Automata.* Urbana, IL: University of Illinois Press.

Vos, M and Didelot, X (2009). A comparison of homologous recombination rates in bacteria and archaea. *ISME J* 3(2): 199–208.

Walsh, C (2000). Molecular mechanisms that confer antibacterial drug resistance. *Nature* 406(6797): 775–81.

Wang, H, Bier, R, Zgleszewski, L, et al. (2020). Distinct distribution of archaea from soil to freshwater to estuary: Implications of archaeal composition and function in different environments. *Front Microbiol* 11: 576661.

Wang, X, Montero Llopis, P, and Rudner, DZ (2013). Organization and segregation of bacterial chromosomes. *Nat Rev Genet* 14(3): 191–203.

Watson, JD (1968). *The Double Helix: A Personal Account of the Discovery of the Structure of DNA.* London: Weidenfeld & Nicolson.

Watson, JD and Crick, FH (1953). Genetical implications of the structure of deoxyribonucleic acid. *Nature* 171, 964–7. https://doi.org/10.1038/171964b0.

Watson, JD and Crick, FH (1953). Molecular structure of nucleic acids; a structure for deoxyribose nucleic acid. *Nature* 171(4356): 737–8.

Watson, RA and Szathmary, E (2016). How can evolution learn? *Trends Ecol Evol* 31(2): 147–57.

West, SA, Fisher, RM, Gardner, A, et al. (2015). Major evolutionary transitions in individuality. *Proc Natl Acad Sci U S A* 112(33): 10112–19.

Westergard, L, Christensen, HM, and Harris, DA (2007). The cellular prion protein (PrP(C)): Its physiological function and role in disease. *Biochim Biophys Acta* 1772(6): 629–44.

Wheatcroft, D and Qvarnström, A (2017). Genetic divergence of early song discrimination between two young songbird species. *Nature Ecol Evol* 1(7): 0192.

Whiten, A (2005). The second inheritance system of chimpanzees and humans. *Nature* 437(7055): 52–5.

Whiten, A, Goodall, J, McGrew, WC, et al. (1999). Cultures in chimpanzees. *Nature* 399(6737): 682–5.

Whitman, WB, Coleman, DC, and Wiebe, WJ (1998). Prokaryotes: The unseen majority. *Proc Natl Acad Sci U S A* 95(12): 6578–83.

Wielgoss, S, Barrick, JE, Tenaillon, O, et al. (2011). Mutation rate inferred from synonymous substitutions in a long-term evolution experiment with Escherichia coli. *G3 (Bethesda)* 1(3): 183–86.

Wielgoss, S, Barrick, JE, Tenaillon, O, et al. (2013). Mutation rate dynamics in a bacterial population reflect tension between adaptation and genetic load. *Proc Natl Acad Sci U S A* 110(1): 222–7.

Wilde, SA, Valley, JW, Peck, WH, et al. (2001). Evidence from detrital zircons for the existence of continental crust and oceans on the Earth 4.4 Gyr ago. *Nature* 409(6817): 175–8.

Williams, D, Gogarten, JP, and Papke, RT (2012). Quantifying homologous replacement of loci between haloarchaeal species. *Genome Biol Evol* 4(12): 1223–44.

Williams, GC (1966). *Adaptation and Natural Selection: A Critique of Some Current Evolutionary Thought.* Princeton, NJ: Princeton University Press.

Williams, GC (1992). *Natural Selection: Domains, Levels, and Challenges*. Oxford: Oxford University Press.

Williams, TA, Foster, PG, Cox, CJ, et al. (2013). An archaeal origin of eukaryotes supports only two primary domains of life. *Nature* 504(7479): 231–6.

Williamson, SH, Hubisz, MJ, Clark, AG, et al. (2007). Localizing recent adaptive evolution in the human genome. *PLoS Genet* 3(6): e90.

Wiser, MJ, Ribeck, N, and Lenski, RE (2013). Long-term dynamics of adaptation in asexual populations. *Science* 342(6164): 1364–7.

Wochner, A, Attwater, J, Coulson, A, et al. (2011). Ribozyme-catalyzed transcription of an active ribozyme. *Science* 332(6026): 209–12.

Woese, CR (1987). Bacterial evolution. *Microbiol Rev* 51(2): 221–271.

Woese, CR and Fox, GE (1977). Phylogenetic structure of the prokaryotic domain: The primary kingdoms. *Proc Natl Acad Sci U S A* 74(11): 5088–90.

Woods, D and Neary, T (2009). The complexity of small universal Turing machines: A survey. *Theoretical Computer Science* 410(4): 443–50.

Woods, RJ, Barrick, JE, Cooper, TF, et al. (2011). Second-order selection for evolvability in a large Escherichia coli population. *Science* 331(6023): 1433–6.

Wootton, D (2015). *The Invention of Science: A New History of the Scientific Revolution*. London: Allen Lane.

Wu, F, Tian, H, Shen, Y, et al. (2022). Vertical MoS2 transistors with sub-1-nm gate lengths. *Nature* 603(7900): 259–64.

Yamanaka, K, Fang, L, and Inouye, M (1998). The CspA family in Escherichia coli: Multiple gene duplication for stress adaptation. *Mol Microbiol* 27(2): 247–55.

Yang, C, Crain, S, Berwick, RC, et al. (2017). The growth of language: Universal grammar, experience, and principles of computation. *Neurosci Biobehav Rev* 81(Pt B): 103–19.

Yao, X, Jericho, M, Pink, D, et al. (1999). Thickness and elasticity of gram-negative murein sacculi measured by atomic force microscopy. *J Bacteriol* 181(22): 6865–75.

Zappa, F (1971). 'Frank Zappa', directed by Roelef Kiers. Broadcast by the Dutch public broadcaster VPRO Television in 1971 in the Netherlands. https://www.pinterest.co.uk/pin/803188914784598238/ 40:50. The transcript is on https://www.donlope.net/fz/videos/Frank_Zappa_VPRO.html

Zaug, AJ, Been, MD, and Cech, TR (1986). The Tetrahymena ribozyme acts like an RNA restriction endonuclease. *Nature* 324(6096): 429–33.

Zhaxybayeva, O, Doolittle, WF, Papke, RT, et al. (2009). Intertwined evolutionary histories of marine Synechococcus and Prochlorococcus marinus. *Genome Biol Evol* 1: 325–39.

Zhirnov, V, Zadegan, RM, Sandhu, GS, et al. (2016). Nucleic acid memory. *Nature Materials* 15(4): 366–70.

Zimansky, P (1993). Before writing. Volume I: From counting to cuneiform. *J Field Archaeol* 20(4): 513–17.

Zliobaite, I and Stenseth, NC (2016). Improving adaptation through evolution and learning: A response to Watson and Szathmary. *Trends Ecol Evol* 31(12): 892–3.

Index

For the benefit of digital users, indexed terms that span two pages (e.g., 52–53) may, on occasion, appear on only one of those pages.